Springer Series in Materials Science

Volume 301

The Springer Series in Materials Science covers the complete spectrum of materials research and technology, including fundamental principles, physical properties, materials theory and design. Recognizing the increasing importance of materials science in future device technologies, the book titles in this series reflect the state-of-the-art in understanding and controlling the structure and properties of all important classes of materials.

More information about this series at http://www.springer.com/series/856

Arvind Shah
Editor

Solar Cells and Modules

Springer

Editor
Arvind Shah
EPFL (PV-Lab)
Neuchâtel, Switzerland

ISSN 0933-033X ISSN 2196-2812 (electronic)
Springer Series in Materials Science
ISBN 978-3-030-46485-1 ISBN 978-3-030-46487-5 (eBook)
https://doi.org/10.1007/978-3-030-46487-5

This Springer imprint is published by the registered company Springer Nature Switzerland AG
The registered company address is: Gewerbestrasse 11, 6330 Cham, Switzerland

Preface

The coming years will—without any doubt—see a tremendous change in the world—climate change, globalization, political upheavals, unprecedented migration of populations and economic instability are some keywords that come to my mind in this context. Many of us are concerned about these changes and think, in particular, of climate change as a huge menace looming right in front of us.

The message of this book is, however, that there is absolutely no reason for fear; on the contrary, these changes will open up unprecedented opportunities, and new perspectives, if only we are prepared to rapidly act: collectively, collaboratively and with courage. One of the major opportunities, which is already clearly visible, lies in the field of photovoltaics (PV). This book will tell you how and why photovoltaics will constitute a decisive factor in contributing towards a very positive and favourable development of the whole world.

Let us look forward to the year 2050, just far enough that we have the time to modify by then the whole world's supply of energy; just near enough, so that fairly accurate predictions can be made. By 2050, we should be able to supply 50% of the world's total electricity with PV. Why focus on photovoltaics, rather than on other forms of renewable energy? This is because PV is truly unique. It is the only form of renewable energy that can be rapidly deployed at the necessary scale in every region of the world—on rural buildings, within cities, on mountains, in deserts and even on the surface of lakes and of the sea.

Of course, one can put PV to use in giant installations, in the same way, as one builds large gas power stations or huge hydroelectric dams. But photovoltaic systems have tremendous flexibility, in size, form and shape. Therefore, they should preferably be used in a decentralized manner—as near as possible to the current consumer, they can indeed be installed in small units, on the roof of a house, or in the midst of a pond. But will such small dispersed units really have any effect on the world's supply of electricity? Yes, they will—because there will be many millions of them.

As my mentor and guide, Shree Chamanlal Gupta of the Aurobindo Ashram in Pondicherry—the man who convinced me to start the photovoltaic laboratory in Neuchâtel—was always repeating to me:

Illustration sponsored by Ernst Schweizer AG, Hedingen, Switzerland; illustration drawn by Michael Meier

"Remember, Arvind, the key to solve global problems is to push ahead with local solutions".

Let us now look in more detail at the "Energy Challenge". 50% of the world's total electricity by 2050 is a tremendous amount of energy; this would be equivalent to about 12,000 TWh per year at today's (2020) levels of electricity consumption. This compares with 800 TWh per year that is generated by PV at present. So, this would be an increase by a factor of 15, within a period of just 30 years. As electricity consumption is steadily rising, mainly due to economic growth in China, India, Indonesia, Africa and South America, the actual increase we need, for PV, will probably be around a factor of 30.

Can this be done? From a purely technical point of view, the answer is certainly: «YES, WE CAN». But we have six other points to be considered: (a) raw materials and electricity needed for the production of PV modules; (b) durability and reliability of PV modules—if PV is to become a pillar of humanity's energy supply, it should be productive for 50 years or more—as all other technologies for the production of electricity do and not just for 25 to 30 years; (c) storage—solar electricity comes in, in an irregular and often unpredictable fashion, depending on climate and weather, and it has to be stored in some way before it can be used; (d) areas needed for the deployment of that huge amount of PV; (e) integration of PV into the energy system; (f) ecological production and recycling of PV modules.

In view of these six points, we will need the collaboration of many actors: certainly, the contribution of physicists, chemists, materials scientists and engineers. But we will also require the help of politicians, businessmen, bankers, home owners, architects, industry leaders and regional planners.

For this reason, this book is intended for a wide spectrum of readers—far wider than the narrow crowd of PV and solar energy specialists. As we are here looking into the near future—into the coming 30 years—this book is specially intended for use by the younger generation—by students and even youngsters from high school. We have therefore tried to keep this book as accessible as possible—accessible also to those who do not have a specialized scientific or technical education in photovoltaics or in any related fields. At the same time, we strived to be up-to-date and include in this book the latest developments in PV. This was a difficult task, and I do not know whether we have succeeded or not.

But as a reader of this book, do tell us whether we have really succeeded. Do tell us where we have gone wrong. Have we written a passage that you simply could not understand? Have we made a statement that you consider being wrong? Did we leave out some information that you consider important? Are there oversights or mistakes in this book?

You can very easily write to us: you have the email addresses and also the postal addresses of everyone who has contributed to this book. We promise to reply to all messages, which come to us. We are very much looking forward to hear from you.

Neuchâtel, Switzerland

Arvind Shah
Editor of the book
on behalf of all its authors

Acknowledgements

The editor, Arvind Shah, is indebted to the following people who have significantly contributed to the present book:

- Ambigapathy, R.: for his corrections of Chaps. 2 and 4 and for help with EXCEL
- Antognini, L. M.: for help with EXCEL
- Bacha, S.: for assistance with Chap. 12
- Bailat, J.: for reviewing Chap. 6
- Beck, B.: for expert assistance with basic physical concepts
- Boccard, M.: for his corrections of Chap. 7 and for help with Sect. 3.6
- Bourée, J.-E.: for corrections of Chap. 3
- Curtins, H.: for helpful and encouraging comments on Chap. 3
- Etienne, B.: for encouragement and helpful inputs for Chap. 11
- F Haug, F.-J.: for an expert review of Chap. 4
- Fiala, P.: for his help in drawing and editing figures
- Fischer, D.: for having detected contradictions between Chaps. 1 and 13
- Frischknecht, R.: for contributing to the discussion on sustainability of PV
- Gordon, I.: for reviewing Chap. 5
- Guekos, G.: for advice and help with Chap. 3
- Hofstetter, D.: for advice on basic physical concepts
- Keppner, H.: for advice and help with Chap. 3
- Korte, L.: for his corrections of Chap. 7
- Lux Steiner, M.: for having spotted an important mistake in Chap. 3
- Meier, H.: for assistance with Chap. 6
- Monokroussos, C.: for helpful comments on LETID in Chap. 10
- Narasimhan, K. L.: for advice on Chaps. 3 and 6
- Schock, H.-W.: for his corrections of Chap. 8
- Schweizer, H. R.: for general encouragement with the book project

- Slikker, T.: for collaboration on the question of temperature coefficients
- Smestad, G.: for a multitude of actions in improving the Book
- Topič, M.: for help with Chap. 2
- Verlinden, P.: for his corrections of Chap. 5
- Wei, L.: for comments on PID in Chap. 10

Contents

Editor and Contributors

About the Editor

Arvind Shah is the founder of the Photovoltaics Research Laboratory (PV Laboratory), at the Institute of Microtechnology (IMT), in Neuchâtel, Switzerland.

PV laboratory, Neuchâtel, has, since 1985, done pioneering work in the establishment of low-cost production methods for solar cells based on silicon. In this context, PV laboratory introduced, in 1987, a novel plasma-assisted deposition method called "VHF deposition" permitting a significant increase in the deposition rate for thin-film silicon layers. In 1994, PV laboratory, Neuchâtel, introduced microcrystalline silicon, deposited by VHF plasma, and with very low oxygen content, as novel absorber layer, within thin-film solar cells. From 2003 onwards, VHF deposition has been adopted by many industries, in Europe, USA and Japan.

PV laboratory, Neuchâtel, does also significant work on the development of transparent conductive oxides, as contact layers for solar cells, with enhanced light-trapping properties.

PV laboratory, Neuchâtel, is, furthermore, active in demonstrating novel methods for the design and fabrication of lightweight, low-cost flexible solar cells.

From 1979 to 2005, Arvind was a professor at the University of Neuchâtel. From 1987 to 2005, he was additionally a part-time professor at the EPFL Lausanne.

In 1975, he founded and co-directed the Centre for Electronics Design and Technology (CEDT) at the Indian Institute of Science in Bangalore. CEDT is now one of India's leading University Centres in the field of electronics. It has a strong industrial orientation.

From 1968 to 1975, he was lecturer and R&D group leader at the Department of Industrial Research of the ETH Zürich.

Since 2006, he has been active as a scientific consultant to the PV laboratory and to various Industries, in Europe, India and the USA.

Since 2008, he has been an active member of the Green Party of Switzerland; from 2009 to 2013, he was a member of the legislative assembly of the Canton of Neuchâtel.

He received the Swiss Solar Prize, together with Johannes Meier in 2005. He received the Becquerel Award in 2007.

Contributors

Christophe Ballif EPFL and CSEM, Neuchâtel, Switzerland

Adinath Funde School of Energy Studies, Savitribai Phule Pune University, Pune, India

Sylvère Leu Meyer Burger Technology A.G., Gwatt, Switzerland

Urs Muntwyler University of Applied Sciences, Burgdorf, Switzerland

Stefan Nowak NET Nowak Energy & Technology, St. Ursen, Switzerland

Mauro Pravettoni SERIS, National University of Singapore, Singapore, Singapore

Alessandro Romeo University of Verona, Verona, Italy

Arvind Shah EPFL (PV-Lab), Neuchâtel, Switzerland

Detlef Sontag Meyer Burger Technology A.G., Gwatt, Switzerland

Alessandro Virtuani EPFL (PV-Lab), Neuchâtel, Switzerland

Chapter 1
Introduction

Christophe Ballif

Abstract A short review on the potential of solar energy and on the dramatic improvements of photovoltaics (PV), which took place since the 50's, is given. From pioneering work on the first 6% efficient silicon solar cell in 1954 to today's mainstream modules with close to 20% efficiency, technology development and market have been intimately linked. Strong market growth has brought global PV installations to well over 600 Gigawatts (GW_p) cumulative capacity by 2020. After a short description of the various PV technologies, we show how industrialisation has driven a rapid decrease of manufacturing costs. Continuous technological improvements to processes and production equipment have led to a continuous increase in module efficiency, and to a reduction of embedded energy. The low electricity price achieved today by photovoltaic systems makes it the potential major source of electricity for decarbonising the planet—through a mix of large centralised plants and of smaller distributed systems. Integration of PV into Buildings, being particularly attractive for the future, is described in detail.

1.1 Photovoltaics: Potential and Orders of Magnitude

Photovoltaics (PV) is now seen as the major source of electricity for the second half of the century. It will become the strongest contributor to the decarbonisation of the world's energy system for the following reasons:

- Sunlight is abundant and available to everybody.
- PV is already now, in many regions of the World, the source of electricity with the absolutely lowest cost; and it is rapidly becoming even cheaper.
- PV is a "clean" technology with low CO_2 emissions. Electricity from PV can in future conveniently power electric vehicles and be used for heating/cooling needs. It can already do so, in many cases, more economically than fossil fuels.

C. Ballif (✉)
EPFL and CSEM, Neuchâtel, Switzerland
e-mail: christophe.ballif@epfl.ch

A. Shah (ed.), *Solar Cells and Modules*, Springer Series in Materials Science 301,
https://doi.org/10.1007/978-3-030-46487-5_1

- PV electricity is manageable: it can be dispatched and used on demand, thanks to the possibility of storage, e.g. in batteries or via pumped hydroelectric systems.
- With further decreasing costs, it will become convenient and cost-effective to transform solar electricity into chemical fuels (power-to-gas), enabling versatile long-term storage.

Hence, photovoltaics not only has the potential to become, in the foreseeable future, the World's major source of electricity—it could also become, by the second half of the century, the major source of energy, in general, via the transformation and storage of electricity.

This book is primarily concerned with the core component of a photovoltaic (PV) system—with solar cells and modules; it will describe the different types of solar cells and their assembly into entire modules, as well as various aspects of their application.

1.1.1 Is There Enough Energy from the Sun?

When the sun shines, it typically brings a power of 1000 W per m^2 on the ground. Thus, an area covered by a PV module with an efficiency of 20% (a typical value for a high-quality module today) will provide a peak power of approximately 200 W (referred to as Watt Peak or W_p). Depending on the location, every year the sun brings on each square metre of the ground 800–2700 kWh, as illustrated in Fig. 1.1. A well-oriented module at 40° latitude (Rome, New York, Beijing) receives around 1500–1600 kWh/m^2 every year, i.e. the energy equivalent of one barrel of oil (159 litres).

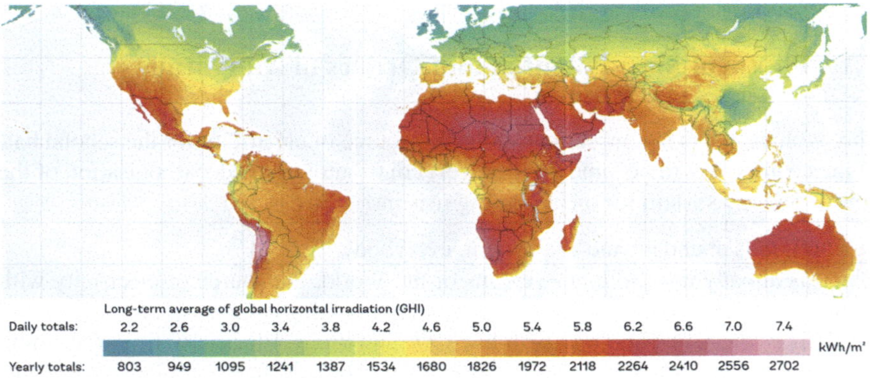

Fig. 1.1 Daily and yearly total global horizontal irradiation (GHI) values—given per m^2 and representing the energy received by a flat, horizontal surface. (Map obtained from the "Global Solar Atlas 2.0, a free, web-based application developed and operated by the company Solargis s.r.o. on behalf of the World Bank Group, utilizing Solargis data, with funding provided by the Energy Sector Management Assistance Program (ESMAP). *For additional information:* https://globalsolaratlas.info)

This is a high amount of energy! Equipping a roof with 1 m^2 of 20% efficient panels allows you to drive a car, with an electric engine, 1800 km annually, as far as 159 litres of petrol does for a typical combustion-engine car.

Indeed, a simple back-of-the-envelope calculation shows that for most industrialized countries in the world, the available roof surfaces could already provide for a significant fraction, in the range of 30–100%, of the electricity needs, if covered with PV[1].

As a case study, we can consider the small (40,000 km^2) and densely populated country of Switzerland (8.5 million people). A recent study [1] shows that the current energy system, which is based on 75% of energy from fossil fuels, could be almost entirely decarbonised by installing 50 GW_p of PV panels. This is less than the potential surfaces available on roofs and façades (see an example of a façade in Fig. 1.2b)—they have an estimated potential of ~70 GW_p. This massive penetration of PV should be accompanied by curtailment of PV production at certain times in the year; it would take place simultaneously with the following steps: (a) a shift to electro-mobility; (b) an improved thermal isolation of all buildings: (c) a widespread employment of heat pumps for the heating of most buildings. Thereby, a large part of the fossil energy presently used could be totally suppressed. Hydroelectric power and electric cars could, in such a scenario, provide most of the required system flexibility.

The World's global electricity consumption was, in 2018, totally ~22,000 TWh (1 TWh = 1 Terawatt hour = 1000 GWh). The current total world energy demand of ~160,000 TWh, largely based on fossil energy, could be reduced, thanks to the gain in efficiency if one switches to energy systems based on electricity. Hence, going to a zero-carbon society could be technically done by installing 50–60 TW_p of PV, assuming a gain by a factor 3 in energy efficiency—by switching e.g., from fossil fuel cars to electric cars, from direct fossil fuel heating to heat pumps. If photovoltaics is to cover around 2/3 of the renewable energy required by 2050 with around 34 TW_p, over 1000 GW_p (or 1 TW_p) of new PV panels should be installed in average every year until 2050 [2]. This means increasing—by at least a factor 10—the current

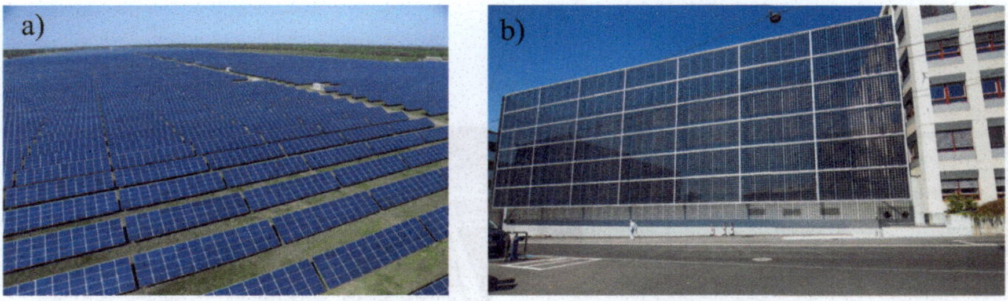

Fig. 1.2 **a** Large 43 MW_p solar farm at Starokazache (Ukrainia); **b** Solar screen with bifacial silicon heterojunction solar cells on the façade of CSEM (Switzerland)

[1]Another alternative to harness the sun's abundant energy is biomass, but the area required to grow these crops is much larger than for PV, because it is a factor 20–100 times less efficient in terms of final energy and would require surfaces which are simply not available.

production. The required area, with 20% efficient modules covering half the surface of the solar power plants (Fig. 1.2a), would be in the range of 340,000 km^2. This corresponds to a modest 3% of the area of the Sahara desert, or, alternatively, to 3.4% of the territory of the USA or China.

1.2 Photovoltaics: A Choice of Technology

There are many possible semiconductors with which to make solar cells, and there are different ways to process these materials into solar cells. However, the three main commercialized categories can be summarised as follows:

- Crystalline silicon solar cells (c-Si): these are based on silicon wafers cut from ingots, which are either mono- or multicrystalline. The wafers, which are typically 120–180 micrometres (μm) thick, are processed into solar cells; the latter are then interconnected by soldering before they are packaged in a module. c-Si constitutes at present for more than 90% of all solar cells.
- Thin-film solar cells: thin layers of semiconductors (typically 0.1–5 μm thick), which are deposited directly onto glass substrates, or on foils. Examples of materials used are CdTe, Cu(In,Ga)Se_2 (CIGS), amorphous silicon (a-Si) and perovskites. Between the processing steps, the solar cells are usually patterned and interconnected by a conductive layer, in a so-called monolithic integration.
- III-V multi-junction solar cells: originally developed for space applications, these solar cells are grown epitaxially on crystalline wafers and can reach efficiencies over 35%. They are too costly to be directly used for power generation on earth—But light can be focused on them, with a concentration factor from 200 to 1000, leading to concentrated photovoltaics (CPV). Despite high cell efficiencies, the delicate system aspects (need for light focussing and for highly accurate sun tracking), have not allowed CPV to gain sizeable market shares.

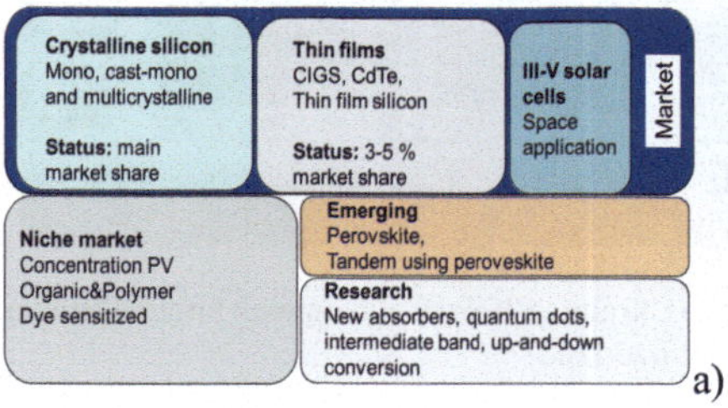

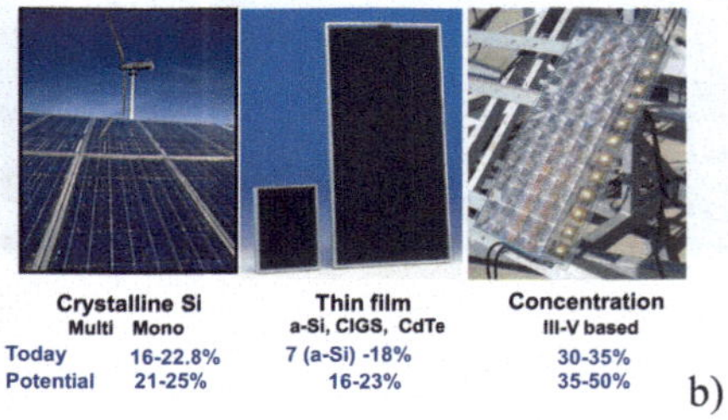

Fig. 1.3 **a** Classification of various PV technologies; **b** three major commercial photovoltaic technologies, with today's best module commercial efficiency and their estimated practical long-term potential

Figure 1.3 gives a brief overview of the various solar cell technologies, including those in a pre-commercial phase. It also indicates the typical efficiency of commercially available PV modules, and their long-term efficiency potential. It should be noted that individual record solar cell efficiencies are typically 15–30% higher than the efficiencies of commercial modules[2].

The focus of this book will be on those solar technologies, which are most widely applied, in particular on crystalline silicon (c-Si), because of its high share (95%) of the PV market: This branch of PV technology will be discussed in **Chaps**. 5 **and** 7. **Chapter** 8 will treat existing thin-film technologies, which make up most of the remaining 5% of the PV module market.

Chapters 2, 3 and 4, next up, will discuss basic concepts used for all PV Technologies:

- **Chapter** 2 will describe the solar spectrum under various atmospheric conditions
- **Chapter** 3 deals with the Basic Theory of Solar Cells
- **Chapter** 4 describes losses within solar cells—optical losses and electrical (recombination) losses.

Chapter 6 will discuss amorphous silicon layers and solar cells; the latter are today mostly used for indoor applications or for the internet of things (IOT). The full fabrication of modules will be treated in **Chap**. 9, whereas the system aspects of PV will be presented in **Chaps**. 10–12. Finally, **Chap**. 13 will focus on the role of PV in the global energy system.

Considering the continuous PV market growth, and a future annual volume production in the range of TW_p, research is still ongoing, both for improving existing commercialized technologies, and for figuring out processes allowing for higher efficiency. In parallel, one is preparing for future technologies, in particular those which will be able to surpass the efficiency of crystalline silicon, as will be discussed later.

1.3 Photovoltaics: Technology Evolution

By switching from selenium-based solar cells, which had an efficiency of 0.5% in 1952, to silicon, Chapin et al. were able to demonstrate 6% efficient solar cells in 1954 [3]. These results triggered research and commercialisation, initially mostly for space applications. For several decades, the terrestrial PV market was limited to off-grid applications. The low manufacturing volumes translated into high prices per Watt-peak ($7–8 per W_p in 1990), for both c-Si and thin film (a-Si): this prevented massive deployment of PV for power generation.

---

[2]They are several reasons for the efficiency difference between record solar cells and commercial modules; those include: the space between the solar cells, the non-active area close to the module edges, the electrical losses in interconnection ribbons, the simplified processes used in mass production, the reduced homogeneity for large devices, etc.

From the 1980's to the early 1990's, the most important technological bricks for the realisation of high performance industrial solar cells were developed. Those were inspired by microelectronics research in the case of silicon, and related to pure PV research for amorphous silicon, CIGS, and CdTe. The challenge, was to find a way to reduce manufacturing costs down by a factor 20–30 to make PV a more competitive source of electricity.

In the years 2000–2010, a stronger market development was triggered by incentives in several countries. This created a high demand for PV modules, with healthy margins for module makers. This is illustrated in Fig. 1.4a by the high average module price between 2005 and 2010, which was then well above the production costs. However, the strong market growth (+30% per year from 2000 to 2010), mostly based on c-Si, led to a lack of refined polysilicon from 2007 to 2010, with solar grade Si material reaching up to $400/kg (compared to $30–50/kg earlier). The silicon shortage had two major effects. First, it led to large investment into polysilicon production plants. Second, it also led to increased investments in thin-film technologies, such as CIGS, CdTe, and thin-film silicon, which typically utilize 100–1000 times less semiconductor material than c-Si solar cells. After several decades of research, thin-film companies started mass production in the early years of the twenty-first Century, with some companies reaching multi-gigawatt production capacity.

With a market still growing in size, a large part of the new production capacity for solar cells and solar modules took place in Asian countries, and in particular in China. The bottleneck in silicon feedstock, which was quickly overcome, because it was not inherent to the technology, eventually led to plummeting silicon feed-stock prices. With many companies looking just at the long-term high-volume potential, there was, however, a massive over-investment in the production capacity for solar modules. This led to PV module oversupply from 2012 to 2015. The resulting decrease in selling prices, in particular for silicon-based modules, often to levels below production

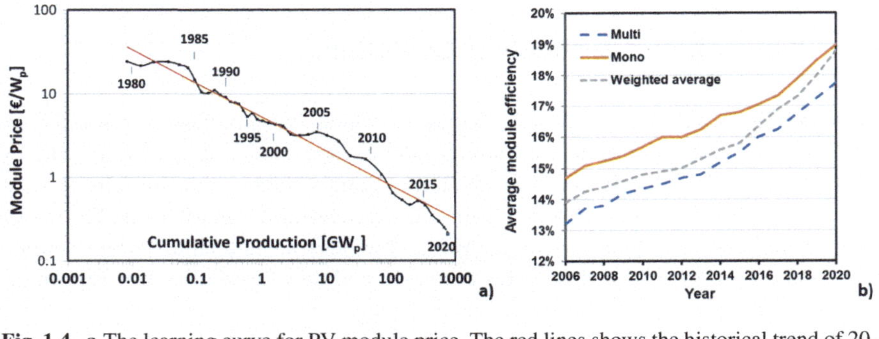

Fig. 1.4 **a** The learning curve for PV module price. The red lines shows the historical trend of 20–24% reduction in module manufacturing cost with each doubling of cumulative production. This rate of reduction might even have accelerated in the last 5 years; **b** Average efficiencies for mono- and multicrystalline PV modules over time. The weighted average considers the market share of mono and multicrystalline silicon. *Source* Fraunhofer ISE: Photovoltaics Report, updated: March 2019, and in-house estimation for 2019 and 2020

costs, forced many companies to stop activities, in particular companies with low production volumes.

This did not prevent mass industrialisation to continue with a further volume growth of around 25% per year in the decade starting in 2010. Indeed, the low price of modules led to lower and lower electricity prices, triggering further market development. The mark of 100 GW_p annual PV module production was reached in 2018, with the strongest market share (95%) for crystalline silicon, and the remaining 5% for thin-films led by CdTe, followed by CIGS and thin-film silicon.

The most striking feature of PV market development is the evolution of the PV module prices, illustrated in Fig. 1.4a. The price per Watt-peak has gone down a factor of 35 since the 1990's, reaching today prices the range of 20–30 Eurocents per W_p for standard PV modules. From the 1980's to now, the average reduction rate is between 20 and 24%. This means 20–24% reduction in module price for each doubling of cumulative production. The price decrease, dictated by offer and demand, reflects directly the manufacturing costs: standard PV modules have now become commodities with a low profit margin and a strongly competitive environment. Hence, there is an excellent correlation between selling price and manufacturing costs.

The low modules prices, a similar reduction in the costs of inverters, and partially on the engineering and mounting costs, have brought PV electricity to an amazingly low price. With total investments in the range of 50–70 €cts/W_p, large, ground-mounted, solar parks can now produce electricity at 3.6–4.5 €cts/kWh in Central Europe and as low as 1.8–2.7 €cts/kWh in sunnier regions.

1.4 Photovoltaics: Manufacturing Chain and Efficiency Increases

In addition to sheer volume effects, each solar technology benefits from continuous improvements linked to R&D, and can gain from developments made for other sectors. For instance, silicon technologies benefited originally from the immense amount of work done in microelectronics. One reason for the success of c-Si technologies can be found in the ease with which the manufacturing chain for c-Si from sand to module, can be split into individual production facilities, as illustrated in Fig. 1.5. Each step can indeed be optimised independently, with improvements almost on a daily basis, at the levels of polysilicon purification, ingot manufacturing, wafer casting or pulling, wafer sawing, solar cell processing, and module lamination.

Another key feature of the c-Si industry is the continuous increase in module efficiency. Over the last decade, an absolute efficiency improvement of 0.3–0.4% per year has taken place both for mono- and multicrystalline Si, as illustrated in Fig. 1.4b. This progress was first obtained using the so-called Aluminium Al-Back surface field (Al-BSF) process illustrated in Fig. 1.5, and continued by a shift to the PERC (passivated emitter and rear contact) technology. In 2020, the following typical average efficiencies were obtained for commercial PERC mono-crystalline products:

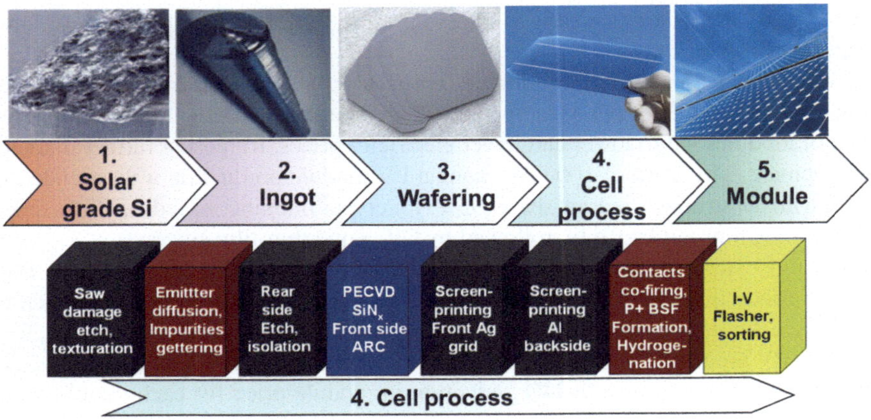

Fig. 1.5 Top: The main manufacturing steps of c-Si modules; Bottom: The six process steps, followed by power measurement on an I-V flasher, of the classical Al-BSF silicon solar cell

Cells 21.5–22.5%, modules 19–20%. In the coming decade, the cell efficiency will further increase and the difference in efficiency between solar cells and modules efficiencies will be further reduced, e.g. by using half-cells, or by shingling narrower solar cells on top of each other to avoid the presence of busbars at the front of the solar cells. From Fig. 1.3b, one can extrapolate an average efficiency for modules based on monocrystalline Si of 22–23% before 2030. Standard modules by then will hence reach today's best commercial module efficiency, up to 22.8% [4]. The latter modules are based on a more complex manufacturing process and sell at a higher price. Concurrently, PV module manufacturing costs will continue to decrease [5, 6]. Assuming a market growth of 16% annually until 2030, a learning rate of 20–24%, could lead to module manufacturing costs down to 10–12.6 €cts/W_p, for low and high efficiency "standard" c-Si modules, respectively, i.e. 21–31 €/m^2.

1.5 Photovoltaics: Impact of Technology on Energy Pay-Back Time

While PV systems generate electricity from a renewable source, their production has an environmental impact. Thin-film technologies have low energy payback times because of the small amount of semiconductor used in their fabrication. Crystalline silicon technology had, initially, a more critical starting position, because of the large quantity of silicon required, but some major technical modifications have made c-Si technology "greener"[3]. These are:

[3]"greener" meaning "more ecologically compatible".

- *Reduced energy in the preparation of pure polysilicon*: the most energy-intensive step is crystallization of polysilicon from a purified gas containing silicon. This is done in what is known as a "Siemens reactor". Today's reactors use multiple filaments or tubular filaments, to speed up polysilicon deposition, and highly-reflective coated jars to keep the wall colder. They can produce up to 10 tons of polysilicon per run, ensuring energy usage in the range of 40–50 kWh per kg of silicon [7], against 130–250 kWh per kg a decade ago.
- *Reduction in the amount of silicon per wafer thanks to advanced multi-wire sawing*: Between 2016 and 2019, the entire industry switched from SiC-slurry based multi-wire sawing to diamond wire sawing. In the latter process, steel wires incorporating small diamonds are used to cut the ingot into wafers. The typical kerf loss (material losses) between two wafers of 150–200 microns has been quickly reduced down to 60 microns, allowing an increase of 30–40% in the number of wafers sawn from the same ingot. This is accompanied by a regular decrease in wafer thickness (Fig. 1.6).
- *Reduction of the amount of silicon for a given module power, through efficiency increase*: the 3–4% gain in efficiency during the last decade allows a direct reduction per W_p of all material volumes (silicon, encapsulation polymers, metallization materials, glass).

Rough calculations show that silicon usage has been reduced from 10 g/W_p down to 3–4 g/W_p over the last decade, as illustrated in Fig. 1.6. With state-of-the-art processes, the typical energy consumption is estimated in the range of 0.8–1 kWh/W_p to produce a module (from sand to the finished product), meaning that the module energy payback time is in the range of six months up to one year depending on the location of the Solar system. The CO_2 equivalent emission depends on the source of energy used to make electricity. For instance, values of 300 g of CO_2 equivalent per

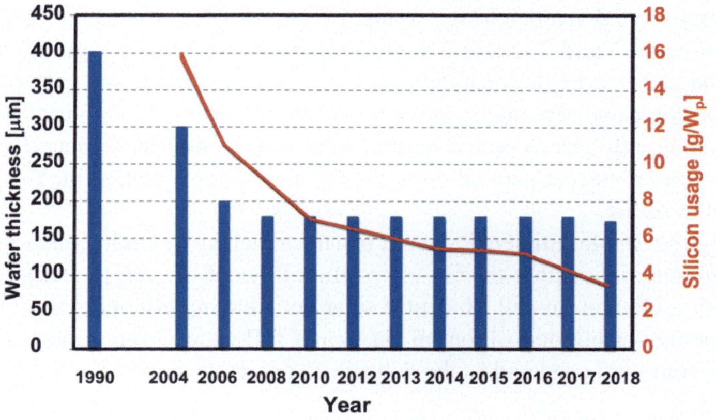

Fig. 1.6 Silicon wafer thickness [µm] and silicon usage [g/W_p] as a function of the years. *Data source* Fraunhofer ISE: Photovoltaics Report, updated: March 2019

W_p at the module level are now reported for mono-crystalline Si modules[4]. Assuming a module lifetime of 25 years, the corresponding module emissions depends on the location, but would be in the typical range of 10–13 g of CO_2 per kWh, to be compared to 400 g for a gas power plant and 900 g for a modern coal power plant. Efficiency improvements, which lead to a direct reduction per Watt-peak of all other material contributions, such as glass and encapsulates, and the ongoing reduction in wafer thickness (Fig. 1.6), will continue to improve the ecological impact of photovoltaics, even more if the sources of electricity used for the production of PV modules are also decarbonized. Finally, thin-film modules can have even less embodied energy thanks to the reduced usage of semiconductors.

1.6 Beyond Silicon Single-Junction Solar Cells

Current record efficiencies for solar cells of size larger than 1 cm^2: 22% for CdTe, 23.3% for CIGS, 26.7% for c-Si [8]. What could come next, in terms of efficiency? A possibility is given by single-junction GaAs solar cells, where record cells reach up to 29.1% [8]. The 2.4% difference between the record values for GaAs and c-Si can be attributed in a large part to intrinsic limitation of silicon, namely its indirect bandgap and Auger recombination. These material properties limit the efficiency of Silicon solar cells to 29.4–29.6% [9, 10].

The only proven concept to increase efficiency significantly is the combination of solar cells in a *multi-junction configuration*, i.e. where solar cells are stacked on top of each other. This allows for a better utilization of the light spectrum, thanks to the fact that each partial cell within a multi-junction configuration can be optimized for a part of the solar spectrum. The top cell absorbs the short-wavelength light (blue, green) and delivers a high voltage. The bottom cell absorbs the long-wavelength light (red, infrared) and delivers a lower voltage. Figure 1.7a illustrates the two classical configurations of 4 and 2 terminal devices. In the 4-terminal configuration the two partial solar cells are made separately and work independently, and each partial cell needs to be contacted separately. In the monolithic 2-terminal configuration, the top solar cell is directly grown on the bottom solar cell. It is easier to manufacture, but requires a similar current generation in the top and bottom cells, as the two cells are connected in series.

The highest stable efficiencies were usually reached by multi-junction devices made from materials within the GaAs system (alloys of Ga, Al, In etc.,), in combination with a Ge bottom cell. Recently solar cells having efficiencies up to 38.8% with 5 junctions using deposition on GaAs and InP wafers were reported [12]. As costly substrates are used and as the deposition process is expensive, such cells are,

[4]These carbon footprint values are certified and required for PV tenders in France. See for instance https://www.pv-magazine.com/press-releases/q-cells-modules-earn-further-low-carbon-certification-for-french-tenders/.

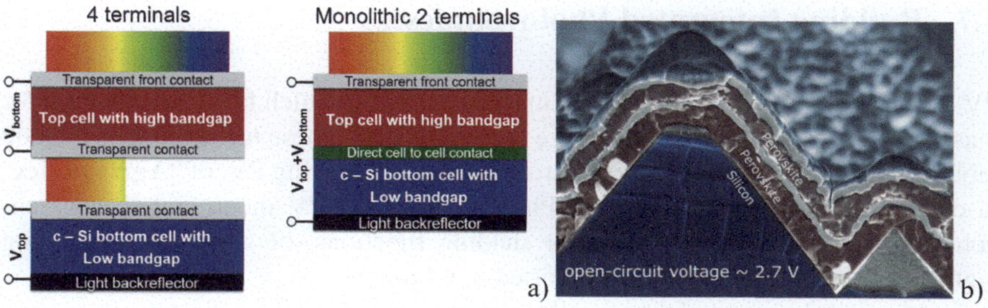

Fig. 1.7 **a** Example of two basic configurations for multi-junction solar cells. Left: 4-terminal solar cell, right: monolithic 2-terminal solar cell. More junctions can be added; **b** Scanning electron microscopy cross-section image of the first proof of concept of a triple junction 2 terminal monolithic tandem perovskite/perovskite/silicon reaching an open 2.7 V open circuit voltage, with a potential to reach over 35% efficiency (from [11])

however, about 200–1000 times costlier than c-Si cells per W_p, and they are therefore only used for niche applications, such as PV in space. A potential application domain for these cells is in the field of concentrated photovoltaics (CPV), where the area of solar cells is reduced by a factor of 200–1000 by focusing the light. This allows for system efficiencies over 30%. CPV requires a complex system to track the sun accurately and a high cleanliness to focus the light efficiently. It will be briefly mentioned in **Chap**. 10.

Let us therefore consider here only flat plate-modules without concentration. Using c-Si as "low cost" bottom cells, record efficiencies could be obtained combining GaAs on silicon and GaAs/InP on silicon at 32.8% and 35.9% respectively, in 4-terminal configuration [13]. Even though the cost problem linked to GaAs or GaInP persists, these results shows that silicon can form an ideal bottom cell for multi-junction cells.

In this context, tandem cells with potentially low manufacturing costs could be based on the combination of a Perovskite (PK) top cell with a silicon bottom cell. In 2018, the first tandem devices in 2-terminal configuration with efficiencies over 25% were reported [14], with a record now at 29.1% [8]. The efficiency potential for such devices is over 30%—and even higher if triple-junction PK/PK/Si configurations are considered (see an example in Fig. 1.7b). The major challenge here is the demonstration of reliable products, as PK devices are more sensitive to extrinsic and intrinsic degradation phenomena. As such tandems are not yet commercially available[5] they will not be treated in this Book.

Considering the growing importance of photovoltaics, pushing efficiencies to their limit in the laboratory and in mass production, investigating new material systems to break efficiency barriers will continue for the decade to come to be a topic of high interest, both for academia and industry.

[5]The stability of Perovskite cells has increased dramatically over the last years, but a full control the reliability of such tandems will still require a large research effort.

1.7 Building Integrated Photovoltaics

PV is increasingly deployed in niche markets, some of which have the potential for massive increase—because of the large areas available. These niche markets include mobility and space applications, the growing areas of floating PV, of "**Agrivoltaics**" (a combination of **Agri**culture and Photo**voltaics**), of PV incorporated in greenhouses, of in PV carports with their shading functions, of sound barriers along highways. These applications are reported in **Chap**. 11.

In the general context of the so-called "***energy transition***,", i.e. the transition from an energy supply based on fossil fuels and on nuclear reactors to an energy supply based on renewable energy sources (wind, solar, biomass, geothermal, etc.), the topic of building-integrated photovoltaics (BIPV) is gaining interest [15]. In BIPV, the PV panel substitutes building components, such as roof tiles, or façade elements.

BIPV has now the potential to extend its market share for the following reasons [15]:

First, the base cost of PV panels per m^2 is now below the cost of many standard façade or high-quality roofing materials. It can, hence, save the costs of traditional building materials, both in new buildings and in renovation projects

Second, more and more countries are in the process of passing legislation, which prescribes that a significant fraction of the energy consumed by buildings has to be produced on-site. This should create a larger market, especially for PV façades.

Finally, with more automation in manufacturing, the fabrication of BIPV elements should become possible at a lower cost, making a more favourable economic case for BIPV.

Third, the last years have seen the emergence of transformative techniques, which allow a complete change of the visual appearance of PV modules: This opens new opportunities for architects and for aesthetic sensitive home owners. Figure 1.8a, b illustrate the case of crystalline silicon modules, which have been modified to appear white or terra-cotta coloured using special filters. There are indeed almost no technical restrictions any more to what can be done visually with PV panels: thus, a long-time hurdle for many designers and architects is being removed [15].

Finally, with more automation in manufacturing, the fabrication of BIPV elements will also be possible at a lower cost, making the case more economically favourable (Fig. 1.8).

Noticeably though, bringing a full integration of PV in buildings requires coordinating the work of the many entities involved in the building process: property owners, financers, architects, builders, tenants, etc. This task is sometimes more difficult than the solving of pure technological problems. Still, by creating awareness, by communicating through demonstration projects, and by stimulating demands through incentives or regulations, they are good prospects for the growth of this sector. It will come as a complement to the more traditional building-added PV (BAPV), where PV elements are added onto existing roof or façade infrastructures.

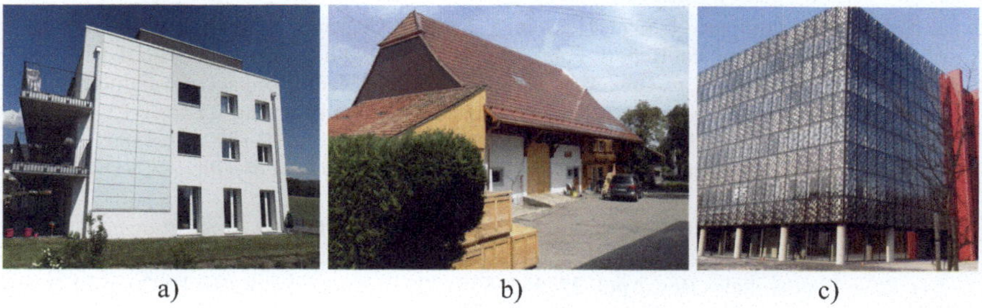

Fig. 1.8 BIPV examples; **a** Façade with white crystalline silicon modules (source Solaxess/CSEM). **b** Old roof on farmhouse renovated with terra-cotta crystalline silicon modules (source CSEM). **c** Translucent panels with c-Si cells (Tornay, Belgique, picture by Issol)

1.8 PV in Future Energy Systems

The deployment of PV will continue further and should even accelerate, through large solar parks, as addition on top or as integral part of small private houses or large industrial buildings, by realising micro-grids in remote areas, by installing floating power plants, or, even, as a direct electricity source in mobility applications. PV combined with hydro, wind, biomass, geothermal, can create the backbone of a fully decarbonised energy system. To arrive there, PV has to be used to provide the primary energy for mobility as well as for heating/cooling tasks within buildings and industry. With progress in electrochemical storage and with the possibility to transform electricity into fuel, there is the potential for PV to become the main provider of Energy, in all its forms. This would be a major contribution in reducing global warming. As will be discussed in **Chap**. 13, a positive regulatory framework, a technical adaptation of the electrical grid and novel forms of energy management are all needed, so as to enable PV to become worldwide the major source of electricity and energy.

References

1. R. Nordmann, "Sonne für den Klimaschutz—Ein Solarplan für die Schweiz" (Zytglogge Verlag, 2019)
2. N. M. Haegel et al., Terawatt-scale photovoltaics: Transform global energy. Science **364**, 836–838 (2019)
3. M. Chapin, C. S. Fuller, and G. L. Pearson, A New silicon p-n junction photocell for converting solar radiation into electricalpower. J. Appl. Phys. **25**, 676 (1954). https://doi.org/10.1063/1.172171125
4. SunPower® X-Series Residential DC|X22-370|SunPower, (n.d.)

5. J. N. Mayer, S. Philipps, N. Saad, Hussein, T. Schlegl, C. Senkpiel, Current and Future costs of photovoltiacs, Technical Report. February 2015. https://doi.org/10.13140/rg.2.1.1371.7206
6. E. Vartiainen, G. Masson, C. Breyer, D. Moser, E. Román, Medina impact of weighted average cost of capital, capital expenditure, and other parameters on future utility-scale PV levelised cost of electricity, 2019 Progress in Photovoltaics, https://doi.org/10.1002/pip.3189
7. S. K. Chunduri, M. Schmela, Taiyang News "Market Survey on Polysilicon CVD Reactors, 2017", online version
8. M. A. Green, Y. Hishikawa, W. Warta, E. D. Dunlop, D. H. Levi, J. Hohl-Ebinger, A. W. Y. W. H. Ho-Baillie, M. Yoshita, A. W. Y. W. H. Ho-Baillie, solar cell efficiency tables (version 54). Prog. Photovoltaics Res. Appl. **27**, 565–575 (2019). https://doi.org/10.1002/pip.3171 and follow-up editions. For the 29.1%, check NREL Best Research-Cell Efficiency Chart, http://www.nrel.gov/pv/cell-efficiency.html (status April 2020)
9. A. Richter, M. Hermle, S. W. Glunz, Reassessment of the limiting efficiency for crystalline silicon solar cells. IEEE J. Photovoltaics. **3**, 1184–1191 (2013). https://doi.org/10.1109/jphotov.2013.2270351
10. S. Schäfer, R. Brendel, Accurate calculation of the absorptance enhances efficiency limit of crystalline silicon solar cells with lambertian light trapping. IEEE J Photovoltaics (2018). https://doi.org/10.1109/JPHOTOV.2018.2824024
11. J. Werner, F. Sahli, D. Ren, J. Luo, F. Fu, J. J. Diaz Leon, A. Walter, B. A. Kamino, B. Niesen, M. Graetzel, S. Nicolay, Q. Jeangros, C. Ballif, Textured perovskite/perovskite/silicon monolithic triple junction solar cells with open-circuit voltage up to 2.69 V, ACS Energy Lett. **3**, 9, 2052-2058
12. P. T. Chiu, D. L. Law, R. L. Woo, S. Singer, D. Bhusari, W. D. Hong, A. Zakaria, J. C. Boisvert, S. Mesropian, R. R. King, N. H. Karam, 35.8% space and 38.8% terrestrial 5 J direct bonded cells. Proceedings of the 40th IEEE photovoltaic specialist conference, Denver, June 2014; 11–13
13. S. Essig, C. Allebé, T. Remo, J. F. Geisz, M. A. Steiner, K. Horowitz, L. Barraud, J. S. Ward, M. Schnabel, A. Descoeudres, D. L. Young, M. Woodhouse, M. Despeisse, C. Ballif, A. Tamboli, Raising the one-sun conversion efficiency of III–V/Si solar cells to 32.8% for two junctions and 35.9% for three junctions. Nat. Energy. **2**, 17144 (2017). https://doi.org/10.1038/nenergy.2017.144
14. F. Sahli, J. Werner, B. A. Kamino, M. Bräuninger, R. Monnard, B. Paviet-Salomon, L. Barraud, L. Ding, J. J. Diaz Leon, D. Sacchetto, G. Cattaneo, M. Despeisse, M. Boccard, S. Nicolay, Q. Jeangros, B. Niesen, and C. Ballif, Fully textured monolithic perovskite/silicon tandem solar cells with 25,2% power conversion efficiency. Nat. Mater. **17**, 820 (2018)
15. C. Ballif, L.E. Perret-Aebi, S. Lufkin, E. Rey, Integrated thinking for photovoltaics in buildings. Nat. Energy **3**, 438 (2018)

Christophe Ballif is Head of the Photovoltaics-Laboratory (PV-Lab) of IMT Neuchâtel, now an integral part of the Swiss Federal Institute of Technology, Lausanne (EPFL). He is also Director of the PV Centre within CSEM, a research and technology organization (RTO) and a public–private partnership located in Neuchâtel, dedicated to technology transfer. His research interests include materials for PV, high-efficiency crystalline silicon solar cells and multijunction solar cells, module technology, Building-Integrated Photovoltaics (BIPV), and Energy systems. Christophe contributed to numerous innovations and products. He received the Becquerel award in 2016. Prior to his assignment in Neuchâtel (2004) he was Researcher at the EMPA in Thun, after having spent several years at the Fraunhofer ISE (Germany), working on crystalline silicon photovoltaics. Christophe accomplished his postdoctoral research at NREL (Golden, US) on compound semiconductor solar cells (CIGS and CdTe). He received his Ph.D. in physics from EPFL, Switzerland, in 1998. He is a newly elected member of the Swiss Academy of Technical Science and a corresponding member of the Slovenian Academy of Science and Engineering.

Chapter 2
Solar Spectra

Adinath Funde and Arvind Shah

Abstract Every living being on Earth is using either a direct form of energy from the sun i.e. sunlight, as in the case of plants, or a derived form as in the case of animals. Solar energy is, thus, the basis of all forms of life on Earth. In the present era for human beings, there is an ever increasing demand of energy for use in every walk of life e.g. industry, transport, technology driven gadgets, healthcare etc. The future livelihood of human beings demands an extensive use of solar energy by photovoltaics technology for sustainable development. The adoption of solar photovoltaic technology is ultimately dependent on the nature of sunlight that reaches the Earth. The sunlight reaching a location on the surface of the Earth is affected by (a) sun-earth geometry and (b) atmospheric effects. This chapter addresses the effects of Earth's atmosphere on sunlight before it reaches the Earth at a given location. It gives the various spectra of sunlight, as received, for different times of the day and for different atmospheric and environmental conditions. It also discusses the reflection of sunlight by the ground surface, i.e. the so-called "albedo". Finally, it gives some representative examples for the spectra of indoor lighting and other low-light sources.

[1] The words "spectrum" and "spectral irradiance" are synonyms and will be interchangeably used throughout the book.

A. Funde (✉)
School of Energy Studies, Savitribai Phule Pune University, Ganeshkhind, Pune, India
e-mail: adinathf@gmail.com

A. Shah
EPFL (PV-Lab), Neuchâtel, Switzerland
e-mail: arvind.shah@epfl.ch

A. Shah (ed.), *Solar Cells and Modules*, Springer Series in Materials Science 301,
https://doi.org/10.1007/978-3-030-46487-5_2

2.1 Interaction of Sunlight and the Earth's Atmosphere

2.1.1 The Solar Spectrum: Nature of Solar Energy Reaching the Earth

According to the principles of Physics, every hot body emits radiation to its surroundings. Planck's Law describes the spectrum[1] of the radiation emitted by a hot surface. The temperature of the Sun's outer surface is 5778 K, which leads to the idealized Black Body Spectrum shown in Fig. 2.1 (red curve). Moreover, the actual spectrum measured outside the Earth's atmosphere (also referred to as "AM0", the blue curve of Fig. 2.1) closely resembles this idealized Black Body Spectrum. The term AM0 stands for "Air Mass 0": It means that the solar radiation has not encountered any interaction with the atmosphere of the earth. The term air mass ratio is elaborated in Sect. 2.1.2 of this chapter. It should be noted at this point that the solar irradiance spectrum spans between wavelength range 250–4000 nm. However, 99% of the solar irradiance is covered in the wavelength range of 250–3000 nm.

The spectral irradiance at individual wavelengths added together (integrated) over the entire spectrum from 250 to 3000 nm of AM0 curve, amounts to 1367 W/m^2 for AM0 case.[2] The third (black) curve in Fig. 2.1 is the actual nature of solar radiation

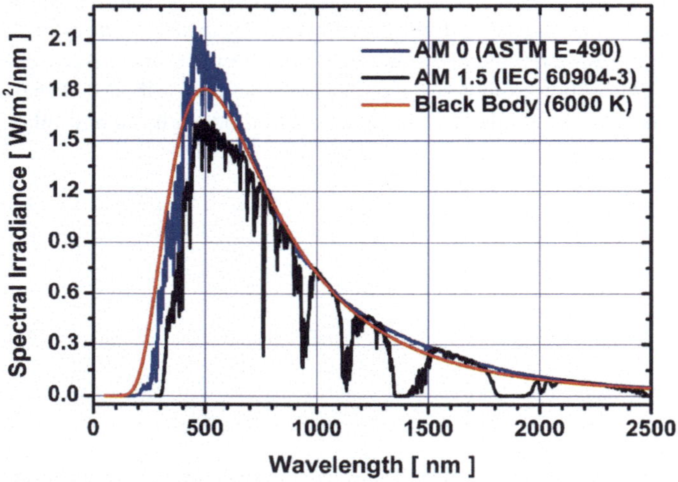

Fig. 2.1 Spectral distribution of the solar spectrum received on the earth's surface (AM1.5) and outside the atmosphere (AM0); given here are the standardized spectra according to international standards IEC 60904-3 [1] (for AM1.5, black curve) and ASTM E-490 [2] (for AM0, blue curve); the AM0 spectrum is compared with the radiation of a black body at 6000 K (red curve)

[2]The quantity 1367 W/m^2 is also termed as "solar constant".

spectrum reaching the earth surface at an Air mass ratio of 1.5 (AM1.5) after having interactions with the atmosphere.

The atmosphere interacts both with incoming sunlight as well as outgoing light from earth. Two main processes take place: light scattering and light absorption. Whereas light scattering redistributes any light energy in the atmosphere without changing the wavelength characteristics, light absorption converts the light energy to internal energy of the absorbing molecules and eventually transfers it to the surrounding atmosphere as heat. Now, the solar spectrum changes significantly in terms of its spectral nature, as the sunlight passes through the atmosphere. This is due to the fact that solar radiation is absorbed in the atmosphere by different interactions with atmospheric constituents as shown by the black cure in the Fig. 2.1. The individual wavelengths of the spectrum have different extents of interaction with the atmospheric constituents. Some of the effects and interactions of incoming sunlight with the atmosphere are described hereunder:

(a) **Reflection of light**

Solar radiation is reflected in the atmosphere and this reduces the radiation reaching the Earth. Major reflection of sunlight takes place from the clouds by particles of liquid state or frozen state of water. The reflection in such a situation can vary from 40 to 90% depending on density of clouds and composition of clouds. The other constituents of atmosphere also contribute to reflection, but their percentage of reflection is small.

(b) **Absorption of light**

When the solar radiation interacts with the gases and particles in the atmosphere, some of the latter have the ability to absorb incoming solar radiation. Absorption in the atmosphere can be defined as a process in which solar radiation is captured by the atmospheric constituents. This captured energy heats up these gases and particles. The captured energy is then re-radiated in the form of heat or long-wavelength radiation. A large variety of molecules in the atmosphere, e.g. water vapour, O_2, O_3, CO_2 molecules, etc. exhibit absorption in the region of the solar spectrum: Thus, partial absorption of sunlight alters the nature of sunlight reaching the Earth's surface. Due to this process, some of the wavelengths are practically absent in the incident sunlight, especially in the infrared region (above 800 nm wavelength) of the solar spectrum as noted for the black curve in Fig. 2.1.

(c) **Rayleigh scattering**

When sunlight falls on particles that are smaller than its wavelength, then Raleigh scattering takes place. Rayleigh scattering is strongly dependent on the wavelength λ; it is proportional to $1/\lambda^4$. Therefore, the amount of Rayleigh scattering increases as the wavelength decreases. This is the reason why the shorter wavelengths of sunlight i.e. blue light are strongly scattered, which is why the sky is blue and the sea is blue. The Rayleigh scattering phenomenon also explains the fact that the solar spectrum has a higher content of red and infrared in the morning and in the evening, in comparison with the solar spectrum at midday: consequently, the sun itself looks reddish at sunrise and sunset.

(d) **Scattering by aerosols and dust particles**

This type of scattering is also termed as "Mie Scattering" and originates from scattering particles in the atmosphere that are larger than the wavelength of sunlight. Mie scattering is caused by one or many of the pollen grains, dust, smoke, water droplets, and other particles in the lower portion of the atmosphere. The extent of Mie scattering at a given location on the Earth depends on the local environmental conditions. Thus, the attenuation of sunlight in the atmosphere is stronger in industrial regions and in densely populated areas.

2.1.2 Nature of Solar Irradiance Received on Earth's Surface

There are a variety of factors affecting the total amount of solar irradiance received (W m^{-2}) and the intensity of solar irradiance at each wavelength (W m^{-2} nm^{-1}), which is the spectral irradiance (or simply the spectrum), at a given location on earth. These factors include latitude, season, time of day, cloud cover and altitude. The nature of solar spectrum received outside the atmosphere is different from the solar spectrum received at sea level on the Earth's surface, at over-head sun position. This is because of the atmospheric effects (mentioned above in Sect. 2.1.1) encountered by the solar radiation when travelling from outside the atmosphere to a given location on Earth—absorption, reflection or scattering in the atmosphere. Figure 2.2 depicts the different atmospheric phenomena encountered by solar radiation.

The Air-Mass Ratio (AM)

Here, "air-mass" literally means the thickness of atmosphere crossed by the solar radiation before it reaches a given location on earth. Thus, outside the atmosphere, the air mass traversed by solar radiation is zero. The "air-mass ratio" is, as shown in Fig. 2.3 is the ratio of the atmospheric path traversed, in a given case, by the solar radiation to the atmospheric path traversed by the solar radiation if it had come in perpendicular to the Earth's surface (in overhead sun position). Thus, the air-mass ratio outside the atmosphere is referred to as air-mass ratio = 0 or simply AM0. The total length of atmosphere at sea level in overhead sun position is normalized to be 1. Thus, the air mass ratio at this situation is air-mass ratio = 1 or simply AM1.

At any other zenith angle θ_z, the air-mass ratio can be estimated as: Air-mass ratio $= \frac{Y}{X}$, where Y is the actual path travelled by the solar radiation and X is the unit length of air mass when the sun's position is overhead at that location. With simple trigonometry this quantity turns out to be $\sec \theta_z = 1/\cos(\theta_z)$.

Solar PV modules are characterized by their performance specifications IEC 61215 [3] under standard test conditions (**STC**), which are:

(1) Solar irradiance of 1000 W/m^2

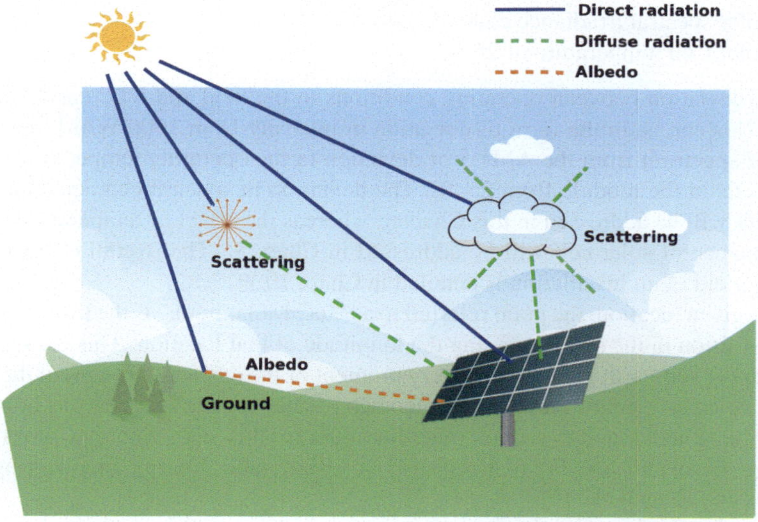

Fig. 2.2 Atmospheric scattering and reflections of solar radiation received on a receiver surface. The radiation received on the receiver surface (solar panel) comprises both: direct radiation and diffuse radiation. It also comprises the albedo (see Sect. 2.2)

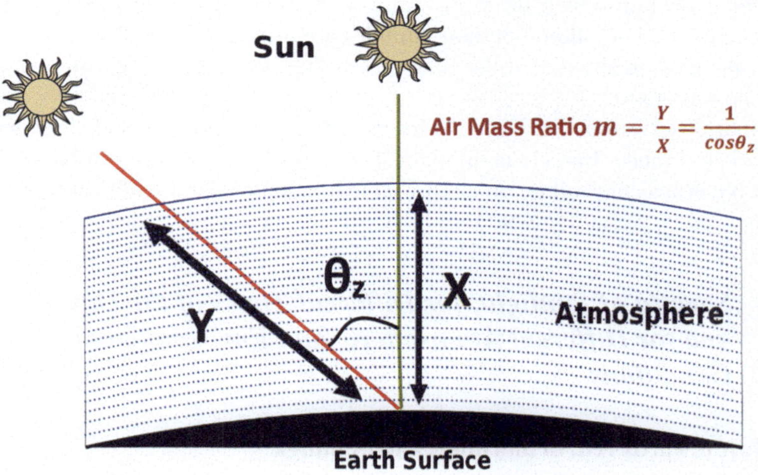

Fig. 2.3 The air mass ratio is the **ratio** of the atmospheric column at a given time, **to** the unit air mass at mean sea level at over-head sun position

(2) Solar spectral irradiance equivalent to AM1.5
(3) Solar Cell temperature of 25 °C.

The deviation between operating conditions in the field and STC can be significant. This can be in the form of deviation in intensity from 1000 W/m^2, deviation in solar spectrum from the AM1.5 or deviation in the operating temperature of the solar cells in the module from 25 °C. The deviation in spectral characteristics and intensity will be addressed in this Chapter, whereas the effect of temperature on the performance of solar cell will be addressed in Chap. 3. (The overall effect on the energy yield of an installation is detailed in Chap. 10.)

At a given location, the noon referred to as "local solar noon" is the time when the sun's position in the sky is crossing the longitude of that location. This comes from the varying sun-earth geometry at every moment of the day and throughout the year. The detailed account of the sun-earth geometry is beyond the scope of this book and readers are encouraged to consult other literature for this. e.g. "Energy Conversion" by Yogi Goswami and Frank Kreith [4] or "Renewable Energy Engineering" by Jenkins and Ekanayake [5].

At the local solar noon, the air-mass ratio experienced at different latitudes will be different as the path traversed by solar radiation through the atmosphere (and, thus, the air mass ratio) will be different for different latitudes: e.g. the cities of Astana in Kazakhstan, Peshawar in Pakistan, Ahmedabad, Mumbai and Kochi in India and Colombo in Sri Lanka have the same "local solar noon" because their longitudes are the same, near 75 °E, although their latitudes are different. Therefore, at any given time of the day, on any day of the year, these cities will experience different values of the air mass ratio.

Similar observations can be made for the cities of Cambridge and Greenwich in UK, Paris in France, Barcelona in Spain, Tiaret in Algeria, Niamey in Niger, Benin City in Nigeria as they all lie on almost same longitude position of near 0° E/W.

2.1.3 Spectra of Sunlight, for Different Times of the Day and for Different Atmospheric and Environmental Conditions

(a) **Shift towards red, in mornings and evenings**

There is a significant alteration in the nature of the solar spectrum at different values of the air-mass ratio—as the composition of the atmosphere is different for different air mass ratios. Even between mornings and evenings the solar spectrum on the surface of the Earth is different, because of changes in the atmospheric constituents: in the morning the air has a different relative humidity than in the evening. However, in both cases, there is a relative enhancement of the red part of the solar spectrum. Figure 2.4 shows the variation in the solar spectrum, as measured by King et al. [6], for three different times of the day at Sacramento (California, USA).

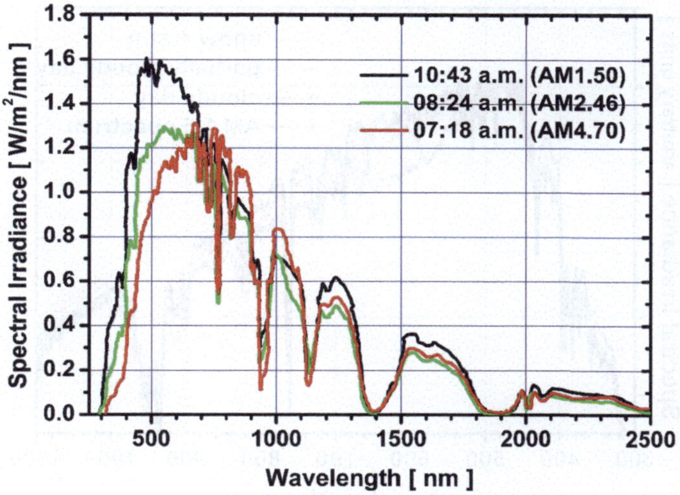

Fig. 2.4 Variation in the solar spectrum measured by King et al. [6], for three different times of the day (different air mass ratio values). The spectra are normalised to the intensity of AM1.5 spectrum, see text

Thereby three different curves result: the red curve for 7.18 a.m. (AM4.70); the blue curve for 8.24 a.m. (AM2.46) and the black curve for 10.43 a.m. (AM1.5). The curves in Fig. 2.4 are normalized, so as to obtain the same area beneath each curve. This area corresponds to a solar radiation of 856 W/m^2 (which is the solar radiation received at 10:43 a.m., whereas earlier in the morning, less solar radiation is received). Thanks to the normalization, the relative changes in the spectral irradiance at different wavelengths, for different times of the day, become clearly visible. However, the information on the total solar radiation, for the different times of the day, is lost.

In Fig. 2.4, we can clearly see that in the early morning (at 7:18 a.m. and 8:24 a.m.), the solar spectrum contains relatively more red light components than later in the day (at 10:43 a.m.).

(b) **Shift of spectrum towards blue on cloudy days, and towards red in the presence of snow in the nearby environment**

The atmospheric conditions of clouds also change the solar spectrum as shown in Fig. 2.5. Similarly, the presence of snow and ice, or of large water surfaces will also modify the solar spectrum, as received by a PV module.

In Fig. 2.5, we can clearly see that under cloudy conditions, the solar spectrum contains relatively more blue light or light with short wavelengths. With partially cloudy conditions, the spectrum shifts towards the AM1.5 solar spectrum. However, in presence of snow, and clear sky condition,[3] the solar spectrum contains more

[3]Clear-sky condition is generally defined as the absence of visible clouds in the sky dome.

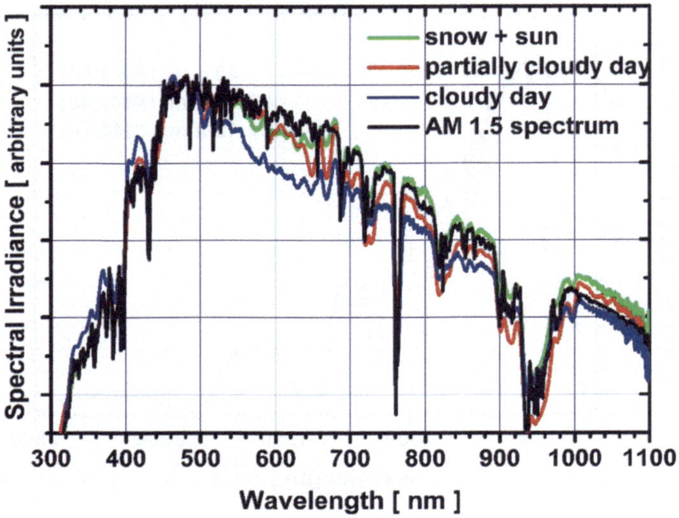

Fig. 2.5 Plot of irradiance spectrum with snow and cloudy conditions, as measured by Kristijan Brecl and Marko Topic, University of Ljubljana, Slovenia [7]. Normalization has been done in a similar way, as for Fig. 2.4

components of the red and infrared i.e. of long wavelengths, in comparison with the AM1.5 spectrum.

2.2 Albedo

The general meaning of albedo is "the proportion of the incident light or radiation that is reflected by a surface, typically that of a planet or moon". It is commonly used in astronomy to describe the reflective properties of planets, satellites, and asteroids. In our context, albedo is the fraction of sunlight reflected by the Earth's surface (or "ground"). Depending on the nature and composition of the ground, the albedo changes in terms of wavelength and intensity. An albedo value of 0 indicates complete absorption of sunlight by the ground surface and 1 indicates full reflection. The values in between the two indicate the fraction of sunlight reflected by the ground surface depending on its nature. Table 2.1 shows albedo values for different ground surfaces. The highest value for albedo is for ground covered by snow.

The albedo radiation on a PV module can be received on both the front side and the back side of a PV module. The contribution of albedo on the front side is significant only if there are mountains or high rise terrains in the vicinity of the PV installation. Albedo values are important in the case of "Bifacial solar PV modules" (see Chaps. 5, 7, 9, and 10). The higher the albedo value is, for the terrain where the modules are mounted, the more effective bifacial modules will be (Fig. 2.6).

Table 2.1 Albedo value of different types of grounds, adopted from reference [8]

Material	Albedo value
Grass	0.25
Lawn	0.18–0.23
Woods	0.05–0.18
Asphalt	0.15
Concrete, clean	0.30
Concrete, weathered	0.20
Snow, new	0.80–0.90
Snow, old	0.45–0.70

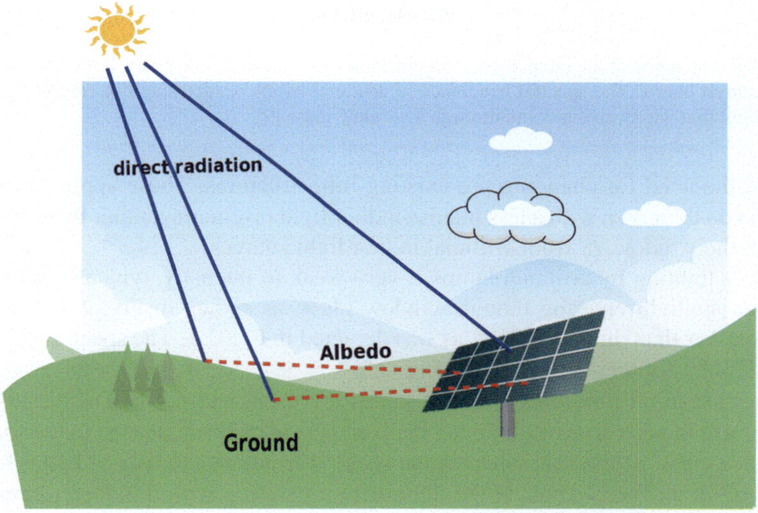

Fig. 2.6 The albedo here is ground reflection, governed by the nature of the ground surrounding the PV panel

2.3 Indoor Lighting

The development of specific solar cell technology like amorphous silicon solar cells (see Chap. 6)[4] to deliver significant electrical output in indoor and low-level light conditions is likely to open up, in future, several new applications. Examples are self-powered devices, such as intelligent sensors that communicate wirelessly, cutting down maintenance costs and ensuring flexibility to building management systems

[4] Amorphous silicon solar cells are already used to power calculator and so-called “solar watches”. Calculators are the ideal application, because they do not need any storage. Almost all hand-held calculators therefore use amorphous silicon solar cells. In “solar watches”, the storage problem is a major one, preventing this application from becoming generally used.

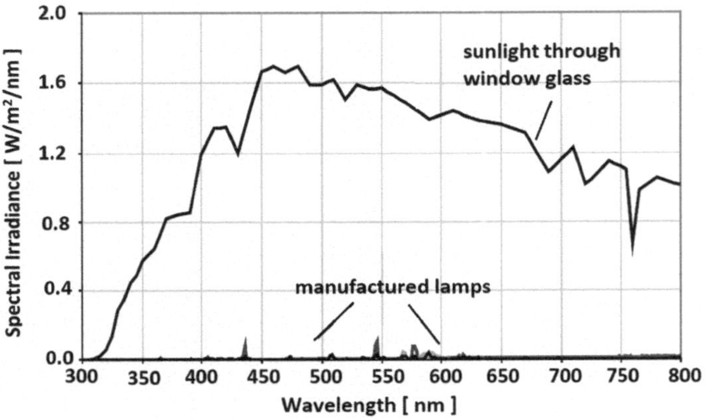

Fig. 2.7 Comparison of the intensity of sunlight through window glass with the light intensity from manufactured lamps. The spectral irradiance of artificial lamps is very small in comparison with the sunlight that we receive indoors through a window glass [9]

without the need for changing pre-existing infrastructures. These applications can function on their own with a less intense indoor light originating either from sunlight through the window, or from artificial indoor light sources.

Indoor lighting by artificial lamps is very weak in intensity, typically 200 times less than sunlight entering though window glass, as shown in Fig. 2.7. It is very much weaker than direct sunlight, as was depicted in Fig. 2.1. The spectral nature of artificial light sources is depicted in Fig. 2.8.

Note that in all museums, halogen lamps are used, and not LEDs, because the spectral irradiance curve of a halogen lamp comes, in its form, nearest to the spectral irradiance curve of the sun, whereas the spectral irradiance curves of LEDs have a totally different form. Thus, if we illuminate a painting with a halogen lamp, the colours of the painting are rendered correctly—whereas this is not at all the case if the painting would be illuminated by LEDs. One could even argue that halogen lamps are best suited for all situations, like reading, where good lighting and comfort are important, as the human eye is genetically adapted to sunlight, and halogen lamps imitate sunlight better than LEDs.[5]

Solar Cell Applications of Indoor Light

There are a large variety of applications, where solar cells are used under indoor conditions. This is the case of most pocket calculators, which are generally fitted with amorphous silicon solar cells, and are destined to work even at very low light intensities. In fact, we would expect our calculator to work if the illumination level

[5] There have been investigations (albeit controversial ones), which claim that very intense LED lights are harmful to the eye and lead to macular degeneration. Given that lighting generally accounts for less than 5% of Electricity Consumption, the authors hold that it is basically unjustified to prohibit, in the EU and in Switzerland, in the near future, the sales of all incandescent lamps, including halogen lamps, in the name of Energy Saving.

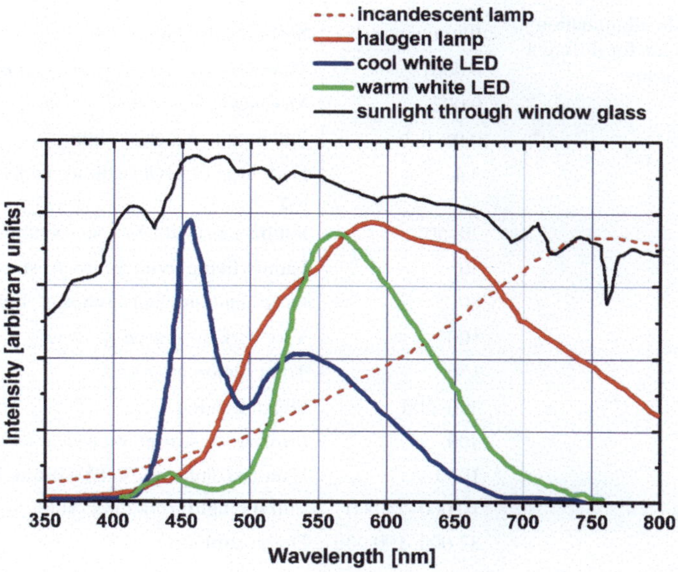

Fig. 2.8 The spectra of different artificial indoor light sources and sunlight through window glass. The spectral intensity of the different indoor lamps are individually adjusted (normalized), so that their maxima are approximately at the same height. Thus, this figure allows for a spectral comparison of the different lamps, but gives no information on their intensity. Information on the intensity of indoor lighting sources is given in the preceding Fig. 2.7. Reproduced from [9, 10]

is just about 100 lx (see Sect. 2.4, for the definition of "lux"). Other applications, where solar cells have to provide electric energy from indoor lighting are:

1. Remote controls (for TV sets, audio amplifiers, electrically driven window blinds etc.), in order to eliminate the need for batteries
2. Temperature sensors
3. Smoke sensors for fire alarms
4. Internet of things (IOT): smart sensors and controls.

2.4 "Lux" as a Unit of Light Measurement

The lux (symbol: lx) is the unit of illuminance derived from the SI system,[6] measuring luminous flux per unit area. It is equal to one lumen per square metre. It is used as measure of the light intensity as perceived by the human eye. It is important

[6]The **International System of Units** (**SI**, abbreviated from the **French** *Système international (d'unités)*) is the modern form of the **metric system** and is the most widely used **system of measurement**.

Table 2.2 The illumination intensities in lux for different lighting situations

Illuminance (lx)	Surfaces illuminated by
0.0001	Moonless, overcast night sky (**starlight**)
0.002	Moonless clear night sky with airglow
0.05–0.3	Full moon on a clear night
3.4	Dark limit of **civil twilight** under a clear sky
20–50	Public areas with dark surroundings
50	Family living room lights (Australia, 1998)
80	Office building hallway/**toilet** lighting
100	Very dark overcast day
150	Train station platforms
320–500	Office lighting
400	**Sunrise** or **sunset** on a clear day.
1000	Overcast day; typical **TV studio** lighting
10,000–25,000	Full **daylight** (not direct sun)
32,000–100,000	Direct **sunlight**

to distinguish between illuminance (measured in lux) and irradiance (measured, as we have already seen, in $W \times m^{-2}$): while the irradiance is the flux of power at all wavelengths (practically from 250 to 3000 nm, as we saw, for the Solar spectrum), the illuminance is the flux of power of visible light (hence the term "luminous flux" used above). The illuminance is thereby weighted according to the *luminosity function*, a standardized model of human visual brightness perception. In English, "lux" is used in both the singular and plural form (Table 2.2).

2.5 Moonlight

Although we generally do not realize that: Moonlight is ever so much weaker than Indoor Lighting. Even the full Moon provides typically only about 0.05–0.1 lx illuminance, i.e. about 20,000 times less than Indoor Lighting and some ten million times less than Direct Sunlight. It is therefore of absolute no interest to try to make "lunar panels". The only way to generate electricity in the Night is to use batteries, thermoelectric generators or other types of energy conversion and storage device. It is therefore surprising that we can see so much during a Full Moon Night. This is a tribute to the adaptability of the Human Eye. The illuminance provided by moonlight varies greatly depending on its phase; in the phase of "New Moon", the general level of illuminance will go down to around 0.003 lx. Nevertheless, modern measurement equipment allows us to measure precisely both intensity and spectrum of moonlight, as shown schematically in Fig. 2.9.

Fig. 2.9 Moonlight has an irradiance level that is about ten million times lower than direct sunlight. Courtesy: Dji-Illustration, Neuchâtel

2.6 Irradiance and Irradiation

'Irradiance' is an instantaneous quantity describing the flux of solar radiation incident on a surface with the unit $W \times m^{-2}$. Thus, irradiance is solar 'power density'. The irradiance value changes very much, in a given location, depending on the moment of the day. The typical variation of solar irradiance in tropical regions on a clear sunny day is a bell curve, as shown in Fig. 2.10.

By referring to 'spectral irradiance' one refers to the 'spectral power density'. Thus, the unit of spectral irradiance or spectral power density is $W \times m^{-2} \times nm^{-1}$.

Irradiation[7] is the integration of irradiance over a specified period of time ($MJ \times m^{-2}$ per hour, day, week, month, year as the case may be). Thus, it is simply the 'time integral of irradiance' or 'energy' received at a given location on earth. Irradiation measures the solar radiation energy density usually in $Wh \times m^{-2}$.

---

[7] Alternative terms used for irradiation are a) solar insolation b) radiant exposure. Although the International Electrotechnical Vocabulary (IEV [11]) defines "irradiation" as "exposure to ionizing radiation", the term is routinely used in the PV community instead of "radiant exposure" (i.e. the radiant energy incident on an element of the surface). An alternative term commonly used is "insolation".

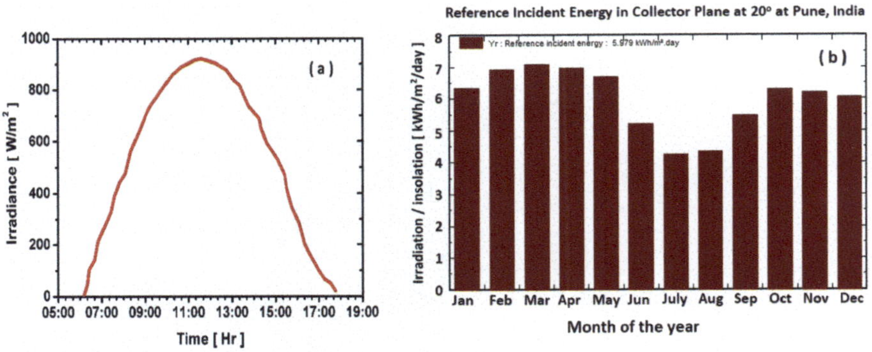

Fig. 2.10 **a** Irradiance variation through a typical clear-sky day of the month of April at Pune, India (measured data) and **b** The monthly-average of daily global solar radiant exposure throughout a year at Pune, India. The data of (**b**) is based on the Pvsyst® software simulated data for the location of Pune, India, derived from the MeteoNorm® synthetic data

The daily and yearly global irradiance give us, thus, a precise indication of the energy of incident sunlight to a specified area and during a specified period of time. One needs to know these quantities, at a given location to accurately predict the future performance of a solar power plant (see also Chap. 10 especially Sect. 10.2.5).

For better clarity of the terms used, following terms are drawn from the IEC Glossary. Web address: http://std.iec.ch/glossary.

[Following elaboration is referring to Publication: IEC TS 61836, ed. 2.0 (2007–12): Solar photovoltaic energy systems—Terms, definitions and symbols].

Irradiance:

(Symbol: G) (Unit: W × m^{-2})

electromagnetic radiated power per unit of area.

(a) direct irradiance: irradiance from the sun's disk and from the circumsolar region of the sky within a subtended angle of 8.7×10^{-2}rad (5°)
(b) diffuse irradiance: irradiance excluding that portion which contributes to direct irradiance
(c) global irradiance: sum of the direct and diffuse irradiance (or simply 'irradiance')
(d) global horizontal irradiance (GHI) global irradiance on a horizontal surface
    NOTE Global horizontal irradiance, GHI usually takes into consideration the albedo, although the albedo is often negligible in GHI.
(e) direct normal irradiance (DNI): irradiance received from a small solid angle centered on the sun's disc on a plane perpendicular to the sun's rays
(f) in-plane irradiance, or plane of array irradiance (POAI): irradiance received from the sun as a combination of direct normal irradiance and all forms of diffuse light in the plane of the PV array
(g) integrated irradiance: continuously integrated spectral irradiance over a specified range of wavelengths

NOTE: If the spectral range is limited, the range is to be stated. If not, then the irradiance is integrated over the total or almost total range of wavelengths. Integrated irradiance is measured by a pyranometer (see Chap. 10).

(h) spectral irradiance: (Symbol: E_l) (Unit: W $\times$ m^{-2} $\times$ nm^{-1}): irradiance per unit wavelength

Irradiation:
(Symbol: H) (Unit: J $\times$ m^{-2})
irradiance integrated over a specified time interval

(a) diffuse irradiation: diffuse irradiance integrated over a specified time interval
(b) direct irradiation: direct irradiance integrated over a specified time interval
(c) global irradiation: global irradiance integrated over a specified time interval
(d) total irradiation: (Symbol: H_T): total irradiance integrated over a specified time interval.

References

1. Standard Solar Constant and Zero Air Mass Solar Spectral Irradiance Tables, **ASTM E490-00a**. West Conshohocken, PA: ASTM International; (2014)
2. Measurement principles for terrestrial photovoltaic (PV) solar devices with reference spectral irradiance data. Int. Electrotechm. Comm. **IEC 60904-3**: Edition 4.0-Part 3 (2019)
3. Terrestrial photovoltaic (PV) modules—design qualification and type approval—Part 1–3: special requirements for testing of thin-film amorphous silicon based photovoltaic (PV) modules. Int. Electrotechm. Comm. **IEC 61215-1-3** (2016)
4. D. Yogi Goswami, Solar Energy Resources, in *Energy Conversion*, ed. by D. Goswami, F. Kreith (CRC Press, Boca Raton, 2017). https://doi.org/10.1201/9781315374192
5. N. Jenkins, J. Ekanayake, The Solar Energy Resource. *Renewable Energy Engineering* (Cambridge University Press, Cambridge, 2017), pp. 120–137. https://doi.org/10.1017/9781139236256.005
6. D.L. King, J.A. Kratochvil, W.E. Boyson, Measuring Solar Spectral and Angle-of-Incidence Effects on PV Modules and Solar Irradiance Sensors. *26th IEEE PV Specialists Conference*, pp. 1113–1116 (1997). https://doi.org/10.1109/pvsc.1997.654283
7. Data provided by Kristijan Brecl and Marko Topic, University of Ljubljana, Slovenia in personal communication (October 2019)
8. K. Mertens, *Photovoltaics—Fundamentals, Technology and Practice* (pp. 21–42), 1st Edn. (Wiley Ltd., 2014)
9. J. Quill, G. Fedor, P. Brennan, E. Everett, Quantifying the Indoor Light Environment, *Q-Lab Corporation*, (2007). https://www.q-lab.com/resources/technical-bulletins.aspx LX-5026—Quantifying Indoor Light. Accessed 25 Nov 2019
10. S. Kim, M. Jahandar, J.H. Jeong, D.C. Lim, Recent Progress in Solar Cell Technology for Low-Light Indoor Applications. Curr. Altern. Energy **3**, 3 (2019). https://doi.org/10.2174/1570180816666190112141857
11. International Electrotechnical Vocabulary. www.electropedia.org. Accessed on 25 Dec 2019

Adinath Funde is Assistant Professor at School of Energy Studies, Savitribai Phule Pune University (formerly University of Pune), Pune, India from the year 2011. He is coordinator for Solar Photovoltaic and Energy Storage initiatives of the department, School of Energy Studies of his University. His broader research interests are Renewable Energy Conversion and Energy Storage. His present specific research interests include low cost alternative materials for solar photovoltaics, battery electrode materials for sodium ion chemistry, and iron flow batteries. He has worked as Visiting Post-doctoral Fellow at Aalto University, Finland, where he worked on hybrid structure solar cells. He is member of several professional organizations. Adinath obtained his Ph.D. in Physics from University of Pune in 2011 on research in thin-film solar cells.

Arvind Shah is the Founder of the Photovoltaics Research Laboratory (PV-Lab), at the Institute of Microtechnology (IMT), in Neuchâtel, Switzerland.

PV laboratory, Neuchâtel, has done pioneering work in the establishment of low-cost production methods for solar cells based on silicon: It introduced a novel plasma-assisted deposition method called "VHF deposition" permitting a significant increase in the deposition rate for thin-film silicon layers. It also introduced microcrystalline silicon, deposited by VHF plasma, and with very low oxygen content, as novel absorber layer, within thin-film solar cells.

From 1979 to 2005, Arvind was a professor at the University of Neuchâtel. From 1987 to 2005, he was additionally a part-time professor at the EPFL Lausanne.

In 1975, he founded and co-directed the Centre for Electronics Design and Technology (CEDT) at the Indian Institute of Science in Bangalore. CEDT is now one of India's leading University Centres in the field of electronics.

Since 2006, he has been active as a scientific consultant to the PV laboratory and to various Industries, in Europe, India and the USA.

He received the Swiss Solar Prize, together with Johannes Meier in 2005. He received the Becquerel Award in 2007.

Chapter 3
Solar Cells: Basics

Arvind Shah

Abstract The present chapter is a central chapter of this book. In this chapter, we will attempt to explain and illustrate the functioning of a solar cell. It is divided into six sections:

- Section 3.1 explains the interaction between Light and a Semiconductor, like silicon—which is the main material used in solar cells. This section involves physical concepts, which may possibly be new for the Reader, but should not be too difficult to understand. As a result of light being absorbed in the solar cell, electrons and "holes" (= "missing" electrons) are generated within the semiconductor.
- Section 3.2 is the "key section" of the present chapter. As most solar cells are constituted of semiconductor diodes, this section will introduce the mathematical description of a diode. The characteristics of such a device can, in general, be described by an exponential function—a fact that is well established, both theoretically and experimentally. The exponential function used to describe diodes (and solar cells), has two main parameters:
  - The diode ideality factor *n*. Basically, *n* should be equal to 1, but for practical reasons *n* is usually higher than 1. The interpretation of this fact goes beyond the scope of the present book.
  - The diode reverse saturation current density J_0. This parameter should be as low as possible, to ensure a high performance of the diode and, thus, of the entire solar cell.

  Afterwards, we will introduce a light source. The diode becomes a photodiode or a solar cell. This is simply done by adding a photo-generation term to the diode equation, a procedure that is well established, but not easy to justify theoretically.
- Section 3.3 describes how the internal electric field present in all diodes is instrumental in separating electrons and "holes" and, thus, providing a current to the contacts of the solar cell.
- Section 3.4 deals with the electrical characteristics of the solar cell: Equivalent circuits and key parameters.

A. Shah (✉)
EPFL (PV-Lab), Neuchâtel, Switzerland
e-mail: arvind.shah@epfl.ch

A. Shah (ed.), *Solar Cells and Modules*, Springer Series in Materials Science 301,
https://doi.org/10.1007/978-3-030-46487-5_3

- Section 3.5 describes the limits for solar cell conversion efficiency, and, also, how these limits are affected by operating conditions: by temperature, and by the intensity of the incoming light.
- Section 3.6 introduces two important "tools", which are used to evaluate solar cells: the measurement of the cell's "Spectral Response" and of its "Quantum Efficiency". Section 3.6 will explain the difference between these two methods, as well as their practical use.

This chapter is based on concepts from Solid-State Physics and from Quantum Theory. Unfortunately it is not possible to explain, in a precise manner, the functioning of a solar cell without using these concepts. We have therefore thought it useful to summarize these concepts in the following "Preamble".

Preamble: Some basic terms from semiconductor physics

1. **Materials. There are basically three kinds of solid materials**
   (a) ***Insulators***—they do not conduct electric current (Example: most textiles, glass, porcelain and "plastics").
   (b) ***Conductors***—they can easily conduct electric current (Example: Metals, like gold, silver, copper, iron, mercury etc.). (Here, we are considering the whole range of conductors, from those with very high conductivity, like silver, copper, gold and aluminium, over those with medium conductivity, like nickel, iron and tin, to those with low conductivity, like lead, stainless steel and mercury.)
   (c) ***Semi-Conductors***—they can conduct an electric current only under certain circumstances (Example: photoconductors—conducting current only if illuminated).
2. **Band diagrams of semiconductors (and also of other solid materials)**
   (a) The ***band diagrams*** characterize the behaviour of electrons within the semiconductor.
   (b) The band diagram (as we use it in this book) is a plot with the Energy E on the vertical axis and the location x on the horizontal axis.
   (c) There exists a "***valence band***". When the valence band is fully occupied, the electrons are not free to move around—they are tied to an atom. However, if an electron "goes missing", it leaves a "hole" behind. The "hole" behaves just like an electron but it has a positive charge, whereas the electron has a negative charge. To understand a little better what a hole is like, let us consider the following analogy: Imagine a large hall with many chairs. Imagine also that most of the chairs are occupied—just very few of them are free. Now if one of the persons gets up from his chair at position A, and sits on a free chair, at position B—it looks as if the free chair has moved from B to A. The free chairs correspond to the holes and the persons correspond to the electrons.

The highest energy level of the valence band is given by the edge of the valence band. It is designated by the symbol E_v.

(d) There exists a "***conduction band***". Here, the electrons are free to move around because the band is not fully occupied. The lowest energy level of the conduction band is given by the edge of the conduction band. It is designated by the symbol E_c

(e) Between valence band and conduction band, there is a "forbidden zone", also called "gap" or, to be more precise "***bandgap*** $\boldsymbol{E_g}$".

3. **Elementary particles in quantum physics**

(a) ***Electron*** → elementary particle of electric current;
   i. Charge of the electron: $\approx 1.6 \times 10^{-19}$ C
   ii. Mass of the electron: $\approx 9 \times 10^{-31}$ kg
   iii. Energy of the electron: depends on the state of the electron.

(b) ***Photon*** → elementary particle of light
   i. Charge of the photon: 0
   ii. Mass of the photon: 0
   iii. Energy of the photon: depends on the wavelength λ of the light according to the expression $E_{photon} = (h \times c)/\lambda$, with h Planck's constant ($\approx 6.6 \times 10^{-34}$ kg m^2 s^{-1}) and c the speed of light ($\approx 3 \times 10^8$ m s^{-1}).

(c) ***Phonon*** → it is not really an elementary particle, but an "excitation quantum" given by the mechanical vibrations within a crystal. However, for the purposes of this book, we can treat it as if it were an elementary particle
   i. Charge of the phonon: 0
   ii. Energy of the phonon: depends on the state of vibrations within the crystal.

4. **Momentum**

This is a term, which is often difficult to grasp by the beginner. A good way to understand the concept of momentum is to think of a ball, which is being thrown by a child. The ball has a certain mass (corresponding to its weight), say 1 kg. The ball has also a certain velocity, say 2 m per second. Velocity is a quantity with a direction—in the case of our ball, the direction may be horizontal, if the child is throwing the ball forward, OR it may be vertical, if the child is throwing the ball upwards. Now the momentum of the ball is simply the product of velocity times mass.

5. **The dual nature of light**

(a) Light is a propagating wave. Its characteristic properties are the wave length and the amplitude of the wave—the latter being related to the intensity of the light.

Interference effects can be explained by the wave properties of light.

(b) A light wave is also composed of particles, called photons. Their energy depends on the wavelength of the light. The number of photons in a light wave corresponds to its amplitude and, thus, to its intensity.

Interaction of light with matter, such as the generation of electron-hole pairs in solar cell, is treated on the based on the representation of light as composed of particles.

6. **Relevant constants**
   Charge of the electron $q \approx 1.602 \times 10^{-19}$ C
   Mass of the electron $m_e \approx 9.105 \times 10^{-31}$ kg
   Constant of Planck $h \approx 6.626 \times 10^{-34}$ kg m^2 s^{-1}
   Speed of light $c \approx 2.998 \times 10^8$ m s^{-1}
   Boltzmann Constant $k \approx 1.3807 \times 10^{-23}$ m^2 kg/(s^2 K).
7. **Bandgaps and equivalent wavelengths at 25 °C**

Material	Bandgap E_g [eV]	Equivalent wavelength λ [nm]	Comments (nature of gap; see Sect. 3.2.2)	Use
Ge	0.67	1850	Indirect	Substrates for space solar cells
Si	1.14	1088	Indirect	Used today for >90% of solar cells
CdTe	1.50	827	Direct	Used today for thin-film solar cells
GaAs	1.43	867	Direct	Basis of space solar cells
Amorphous silicon	1.75	710	$E_g \approx$ 1.7–1.85 eV Non-direct	Used today for solar cells within pocket calculators

8. **Range of visible light**
   The range of visible light extends from 380 to 760 nm. Expressed in photon energies, this range extends from 1.63 to 3.27 eV.
9. **Quick aid to convert wavelengths into Photon energies and back**
   The easiest way is to use the following expressions:

$$E[\text{eV}] \approx \frac{1.24}{\lambda[\mu\text{m}]}$$

$$\lambda[\mu\text{m}] \approx \frac{1.24}{E[\text{eV}]}$$

3.1 The Photovoltaic Effect: Interaction of Light and Matter

The photovoltaic effect is the operating principle of the solar cell: it is the creation of voltage or electric current in a material upon exposure to light. A young French physicist, Alexandre-Edmond Becquerel (24 March 1820–11 May 1891) was the first to experimentally demonstrate the photovoltaic effect: In 1839, at the age of 19, whilst working in his father's laboratory, he was able to demonstrate, for the first time, the photovoltaic effect, i.e. the generation of an electric current, when light is absorbed in a material [1, 2].

The theoretical basis to understand the photovoltaic effect was given by Albert Einstein in 1905: to obtain an idea of Einstein's arguments, let us now look, at the nature of light:

Light has, in fact, a dual nature; on one hand, it can be described as consisting of electromagnetic waves, with a given wavelength λ, and a given frequency $\nu = c/\lambda$, where c is the speed of light. On the other hand, it can also be viewed as consisting of "particles of light", or "quanta of light", or (as we call them nowadays), of "photons", with a given energy E_{ph}. Einstein was the first to use this second nature of light—i.e. to view light as composed of particles, or "photons". Based upon Max Planck's theory of black-body radiation, Einstein postulated that the energy E_{ph} in each quantum of light was equal to its frequency ν multiplied by a constant: $E_{\mathrm{ph}} = h\nu$ (where h was later called "Planck's constant"). Einstein was trying to explain experiments carried out by Hertz in 1887: Hertz had accidentally stumbled upon the "photoelectric" effect,[1] while doing experiments with radio waves. His observations on the interaction of light and matter could not be explained until Einstein published his paper [3], thereby postulating that only photons with energy E_{ph} above a certain threshold (given by the material used) possess the required energy to eject a single electron, thus creating the effect observed by Hertz. This discovery led to the quantum revolution in physics and earned Einstein the Nobel Prize in Physics in 1921.

[1] The "photoelectric" effect is similar to the "photovoltaic" effect: In both cases light is absorbed by a material, leading to the excitation of electrons within the material. In the "photovoltaic" effect, the electrons are collected via electrodes connected to the material and create an external current: this is the case of solar cells. In the "photoelectric" effect, the material is located in a vacuum chamber; the electron is emitted out of the material and collected on an electrode placed within the vacuum chamber.

3.2 Conversion of Light into Electrical Carriers by a Semi-conductor Diode

3.2.1 Absorption and Energy Conversion of a Photon

When light illuminates a solar cell, the semiconductor material absorbs photons; thereby, pairs of free electrons and holes are created (see Fig. 3.1). However, in order to be absorbed, the photon must have an energy $E_{ph} = h\nu$ (where h is Planck's constant and ν the frequency of light) higher or at least equal to the bandgap energy E_g of the semiconductor. The bandgap energy is the difference in energy levels between the lowest energy level of the conduction band (E_C) and the highest energy level of the valence band (E_V) (Fig. 3.2). For a given semiconductor E_g is a constant, which only slightly depends on temperature. Table 3.1 gives the values of E_g for various amorphous and crystalline semiconductors at $T = 25$ °C.

Depending on the energy of the photon and on the bandgap energy of the solar cell material, three cases can occur:

1. $E_{ph} = E_g$: in this case, the photon can be absorbed and will then generate a single electron-hole pair (Fig. 3.3), without loss of energy.
2. $E_{ph} > E_g$: in this case, the photon can be easily absorbed and will then create a single electron-hole pair. The exceeding energy $E_{ph} - E_g$ is rapidly transformed into heat (thermalisation; Fig. 3.4).
3. $E_{ph} < E_g$: the energy of the photon is not high enough to be absorbed. The photon will be either reflected or absorbed elsewhere, and its energy is lost.

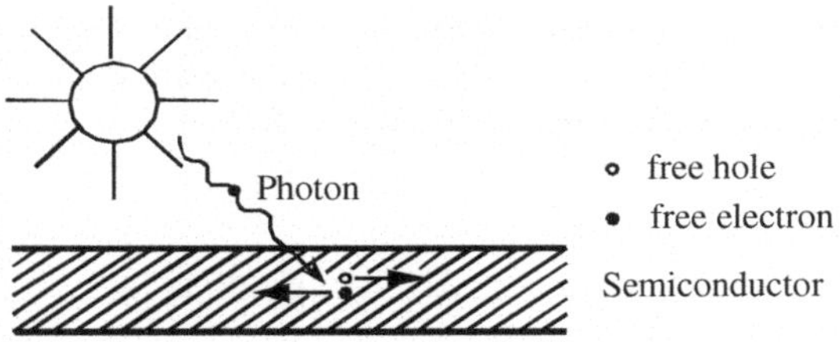

Fig. 3.1 Creation of an electron-hole pair through absorption of a photon of energy $E_{ph} = h\nu$. Reproduced from [4] with the kind permission of the EPFL Press

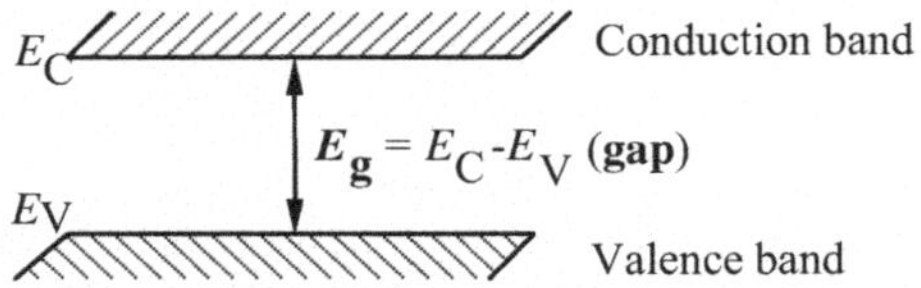

Fig. 3.2 Bandgap energy E_g of a semiconductor. Reproduced from [4] with the kind permission of the EPFL Press

Table 3.1 Bandgap energy E_g of commonly used semiconductors at temperature $T = 25$ °C; see also [5]

Material	Bandgap		References
	Value	Nature[a]	
Crystalline silicon (c-Si)	1.12	Indirect	[6, 7]
Hydrogenated microcrystalline silicon (μc-Si:H)	1.12	Indirect	[4]
Hydrogenated amorphous silicon (a-Si:H)	1.7–1.9	Non-direct	[4]
Crystalline germanium (c-Ge)	0.67	Indirect	[6, 7]
Hydrogenated amorphous germanium (a-Ge:H)	1.1	Non-direct	[4]
Hydrogenated amorphous Si-Ge alloys (a-Si,Ge:H)	1.1–1.8	Non-direct	[4]
Indium arsenide	0.36	Direct	[7]
Cadmium telluride (CdTe)	1.49	Direct	[6]
Copper indium selenide $CuInSe_2$	1	Direct	[8]
Copper-indium-gallium-selenide (CIGS) ($Cu_xIn_yGa_zSe_2$)	1–1.7	Direct	[8]
Kesterite {$Cu_2ZnSn(S,Se)_4$}	1–1.5	Direct	[8]
Cuprous oxide (Cu_2O)	2	Direct	[9]
Zinc telluride (ZnTe)	2.25	Direct	[6]
Gallium arsenide (GaAs)	1.43	Direct	[6]
Pyrite (FeS_2)	0.95	Direct	[10]

[a]See Sect. 3.2.2 for the description of direct, indirect and non-direct bandgaps

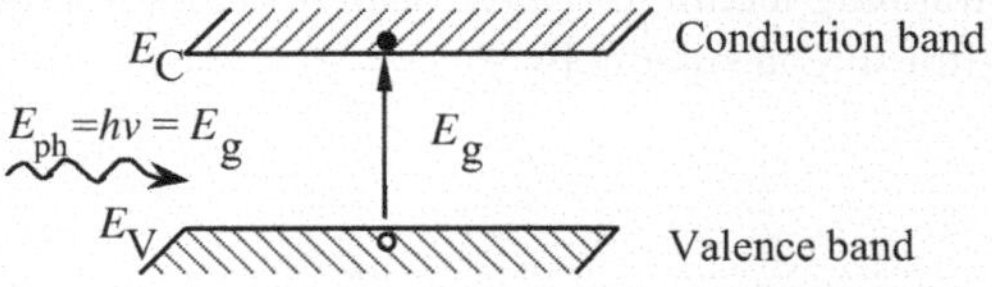

Fig. 3.3 Absorption of a photon when $E_{ph} = E_g$. Reproduced from [4] with the kind permission of the EPFL Press

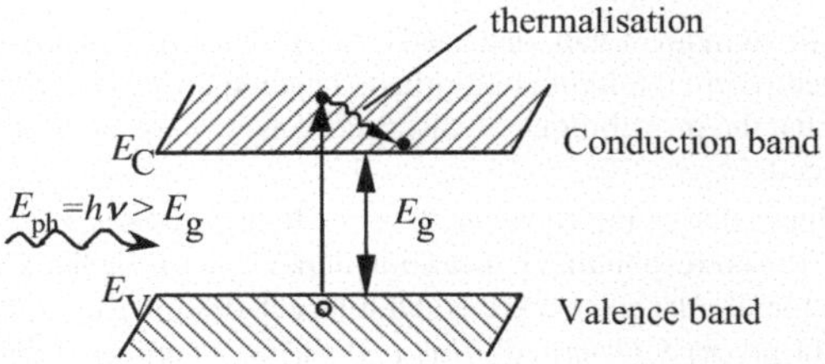

Fig. 3.4 Absorption of a photon when $E_{ph} > E_g$; thermalisation (conversion of excess energy $E_{ph} - E_g$ into heat) takes place very rapidly. Reproduced from [4] with the kind permission of the EPFL Press

3.2.2 Direct, Non-direct and Indirect Band-Gaps

The terms "direct", "non-direct" and "indirect" refer here to the interaction between incoming light and the semiconductor itself:

The semiconductor is considered to have a direct gap, when the absorption of an incoming photon is directly possible, without the interaction of a phonon[2] within the semiconductor. GaAs, CdTe and $CuInSe_2$ are typical examples of semiconductors with a direct gap.

The semiconductor is considered to have an indirect gap, when the absorption of an incoming photon is only possible with the simultaneous interaction of a phonon within the semiconductor. Here, the probability of absorption is significantly reduced, as compared to the case of a semiconductor with a direct bandgap.[3] Si, Ge and C are typical examples of semiconductors with indirect gaps. These materials have absorption coefficients that are relatively low (see Fig. 3.5[4]).

The semiconductor is considered to have a non-direct gap, if it is an amorphous semiconductor, and if it has an indirect bandgap in its crystalline form. Due to its amorphous structure, the interaction with a suitable phonon becomes "easier" (i.e. it has a higher probability), and therefore the absorption coefficient is increased, by about a factor of ten, with respect to the crystalline form. On the other hand, the values of the bandgap are modified (increased) by the amorphous structure.[5]

Therefore, if silicon is used as a solar cell material, we basically have to use relatively thick silicon wafers; this is in contrast to the case of GaAs, CdTe and $CuInSe_2$, where thin photo-absorbing layers are already sufficient. For this reason, sophisticated light-trapping techniques have been developed for silicon solar cells; especially for thin-film silicon solar cells.

---

[2]A phonon is a quantum of mechanical energy, given by the vibration of the atoms within the material.

[3]In semiconductors with an indirect gap, one needs the presence of a specific, suitable phonon, at the time and at the place where the light is absorbed. This requirement reduces the probability of absorbing the light. Thus, the probabilities of absorption, as well as the absorption coefficient are both significantly reduced.

[4]One notes from this figure that as the wavelength of the light increases, the absorption coefficient becomes smaller—and the corresponding penetration depth d_{pen} of the light becomes higher: This means that, in all cases (for all solar cell materials and solar cell designs), light with long wavelengths, i.e. infrared light, is not properly absorbed: This part of the incoming light spectrum penetrates deeply into the solar cell, with only a small part of it being absorbed and used by the solar cell. The majority of such light just passes through the solar cell, without rendering any contribution to the solar cell current. Take, as an example a crystalline silicon wafer with a thickness $d_{wafer} = 100\ \mu m$; Light with a wavelength over about 1000 nm will only be very insufficiently absorbed, as here $d_{pen} > d_{wafer}$.

[5]This effect is up to now not fully understood.

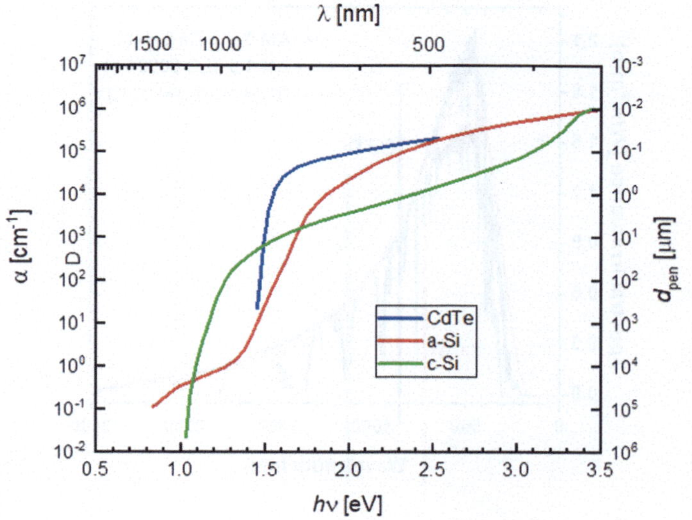

Fig. 3.5 Absorption coefficient α and penetration depth d_{pen} (of monochromatic light), as a function of wavelength λ and of photon energy $h\nu$, for three semiconductor materials commonly used in solar cells. The penetration depth d_{pen} is thereby defined as the depth at which the light has decreased to $1/e$ of its original value; e being Euler's number (The number $e = 2.718\ldots$ is a mathematical constant that is the base of the natural logarithm: the unique number whose natural logarithm is equal to one). Adapted from [11]

3.2.3 *Spectrum of the Incoming Light*

The quantity of light absorbed by a semiconductor depends on the bandgap energy (as previously discussed), but also on the spectrum of the light (i.e. the energy distribution of the incident radiation as a function of wavelength). Figure 3.6 is a reproduction of Fig. 2.1 in Chap. 2: It shows the spectrum of sunlight outside the earth's atmosphere (AM 0) and that of sunlight on the surface of the earth (AM 1.5), under the precise conditions defined in Chap. 2. The AM 1.5 spectrum is considered to be the reference spectrum for all terrestrial solar modules, whereas the AM 0 spectrum is used for space applications of solar cells.

3.2.4 *Relationship Between Light Spectrum and Semiconductor Bandgap*

A semiconductor with a lower value of bandgap energy will be able to absorb a broader range of the solar spectrum (i.e. more low-energy photons) compared to a semiconductor with a higher bandgap. However, in the first case, a substantial part of the incident energy will be lost by thermalisation (i.e. by the energy difference

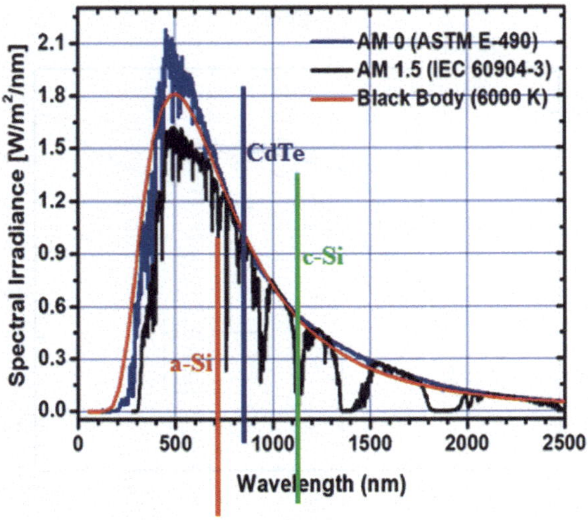

Fig. 3.6 (Same as Fig. 2.1) Spectral distribution of the solar spectrum received on the earth's surface (AM 1.5) and outside the atmosphere (AM 0); given here are the standardized spectra according to standards IEC 60904-3 (for AM 1.5) and ASTM E-490 (for AM 0); the AM 0 spectrum is compared with the radiation of a blackbody at 6000 K. The bandgaps of crystalline silicon (c-Si, green), cadmium telluride (CdTe, blue) and amorphous silicon (a-Si, red) are also indicated in the plot

between the energy of each photon and the bandgap energy). On the other hand, a semiconductor with a higher value of bandgap energy will only be able to absorb a narrower part of the solar spectrum (i.e. a relatively small amount of high-energy photons), but less energy will be lost through thermalisation. One can thus intuitively understand that to maximize the spectral conversion efficiency, one must choose an intermediate value for the bandgap, corresponding approximately to the bandgaps of crystalline silicon (1.12 eV) or of GaAs (1.43 eV).

The actual latent energy that each electron-hole pair (generated by one photon) possesses is the bandgap energy E_g. Assuming that ϕ photons are absorbed per second and unit area, ϕ electrons-hole pairs are generated, having a corresponding total latent energy of $\phi \cdot E_g$ per second and unit area. The *spectral conversion efficiency* η_S can be defined as:

$$\eta_S = \frac{\phi \cdot E_g}{P_{sun}}, \tag{3.1}$$

where P_{sun} is the "power of the sun", i.e. the spectrally resolved solar irradiation per unit of time and area.

Here, the first term (ϕ) in the numerator corresponds to an electrical current ($\phi \cdot q$), whereas the second term (E_g) corresponds to an electrical voltage (E_g/q). Thereby, q is the elementary charge, i.e. the charge of an electron: $q = 1.602 \times 10^{-19}$ [C].

Figure 3.7 shows, for AM 1.5 illumination, the theoretical upper limit obtained for the spectral conversion efficiency η_S—i.e. for the first step of the conversion of light into electrical charge carriers—as a function of the bandgap energy of the semiconductor. Illustrative examples are given in Table 3.2. η_S is almost 50% for bandgaps between 0.8 and 1.5 eV.

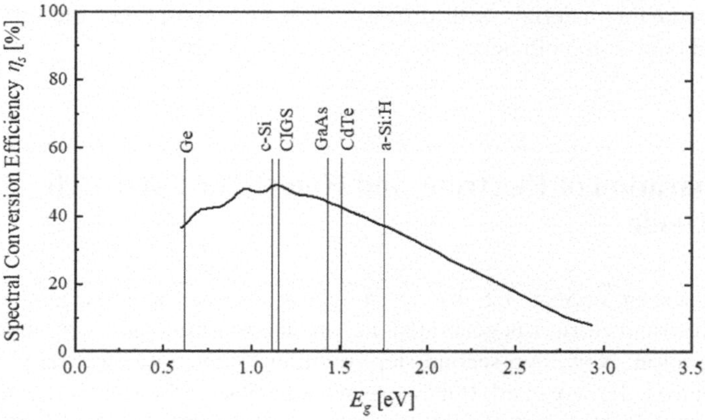

Fig. 3.7 Maximum limit for the spectral conversion efficiency η_S ('first step') as a function of the bandgap energy E_g for AM 1.5 illumination. Values of bandgap energy for different semiconductor materials commonly used in solar cells are also shown

Table 3.2 Comparison of the spectral conversion efficiency η_S of 5 materials: InAs, Ge, Si, GaAs and ZnTe

Material	Bandgap	Absorption	Voltage	Current	Spectral conv. eff.
	E_g [eV]		E_g/q	$\phi \cdot q$	η_S
InAs	0.36 (very small)	All photons of sunlight are absorbed	Very small	Large	Small (≈ 25%)
Ge	0.66 (small)	Almost all photons are absorbed	Small	Large	Average (≈ 40%)
Si	1.12 (average)	Most infrared photons not absorbed	Average (on low side)	Average (on high side)	Large (≈ 48%)
GaAs	1.43 (average)	Infrared photons not absorbed at all	Average (on high side)	Average (on low side)	Large (≈ 47%)
ZnTe	2.25 (large)	Green, red + infrared photons not absorbed	Large	Very small	Small (≈ 25%)

Notes

1. These values apply to the first step only. The second step will result in further losses. The plot for the overall efficiency (first and second step) will be given later, in Sect. 3.5.1 (Fig. 3.19).
2. Furthermore, these values correspond to the case of a solar cell with a single junction, as we have only considered one single bandgap. When two (or more) semiconductor materials with different gaps are superposed, the limit for the spectral conversion efficiency η_S can indeed be increased.

3.3 Separation of Electrons and Holes: The Solar Cell as Diode

We have just seen, in Sect. 3.2, that when light is absorbed within a semiconductor, a pair of electrical carriers is generated, i.e. an electron-hole pair is created for each absorbed photon. Now, in a second step, in order to generate electricity, electrons and holes have to be separated. If this does not take place, electrons and holes simply recombine again—thereby heating up the semiconductor.
Drift transport: (Fig. 3.8) In a solar cell, there is only one way of separating electrons and holes—by the action of an electric field E. Indeed, under the influence of an electric field E, electrons, whose charge is negative, will travel in a direction opposite to that of the electric field, whereas holes, whose charge is positive, will travel in the same direction as the electric field. This phenomenon is called "***drift***" transport and is shown schematically in Fig. 3.8.

We now need a semiconductor structure, which has an internal electric field. The simplest such structure is the *diode*. Therefore (almost) all solar cells are just semiconductor diodes.

In the case of most solar cells (crystalline silicon solar cells, CIGS, CdTe, GaAs and other similar solar cells), we use a *p-n diode*, i.e. a diode with a *p*-doped region immediately adjacent to an *n*-doped region. Here, the internal electric field is at the border of the *p*-doped and *n*-doped regions, in what is called the "depletion region". In such a *p-n* type solar cell, holes and electrons are photo-generated in the bulk of the *p*- and *n*-regions; they then travel by diffusion to the depletion region, where they are separated by the action of the internal electric field. (Note that diffusion is

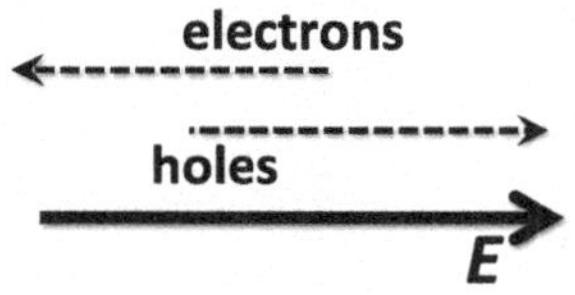

Fig. 3.8 Drift transport of electrons and holes under the influence of an electric field E

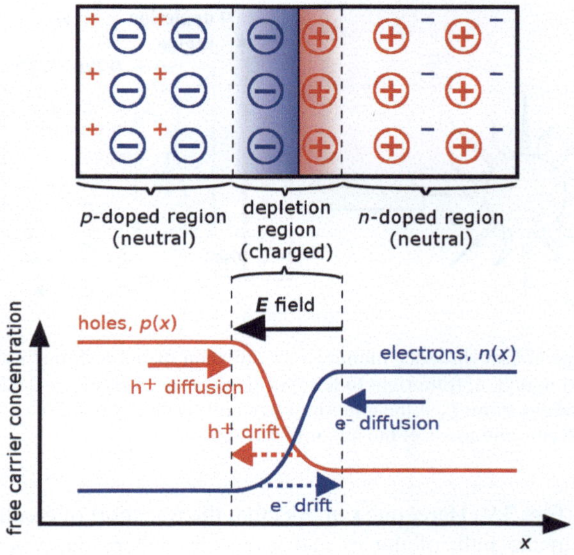

Fig. 3.9 *p-n* Type solar cell in the "dark", i.e. without illumination: transport of electrons (e) and holes (h) by diffusion in the bulk of the *p*- and *n*-regions; and their subsequent separation by drift under the influence of the electric field *E*, in the depletion region

a transport mechanism for charge carriers within a semiconductor, whereby carriers move from a zone, where their density is high, to a zone where their density is lower.)

This situation is shown schematically in Fig. 3.9.

In amorphous and microcrystalline silicon solar cells, we use *p-i-n diodes*. Thereby the letter "*i*" stands for "intrinsic". As this type of solar cells is very rarely used, at the present moment, except for "niche applications" (such as power supplies for watches, calculators and other small-size devices), we shall in this chapter not give any further development for this case. The reader is referred to Chap. 6, Sect. 6.2.1.

The mechanism of (charge) carrier separation and transport is called "carrier collection $\boldsymbol{C}$". During the transport of electrons and holes, a part of these charge carriers, as obtained by photo-generation $\boldsymbol{P}$ are lost, through recombination loss $\boldsymbol{R}$, and the balance is collected. Therefore:

$$\boldsymbol{C} = \boldsymbol{P} - \boldsymbol{R}$$

The ratio between recombination loss $\boldsymbol{R}$ and carrier collection $\boldsymbol{C}$ is, very generally speaking, proportional to the ratio between the "average" or "effective" transport distance d_{cell} for carriers and the average length $l_{\mathrm{transport}}$ of carriers before recombining:

$$(\mathrm{R/C}) \text{ proportional to } (d_{\mathrm{cell}}/\, l_{\mathrm{transport}})$$

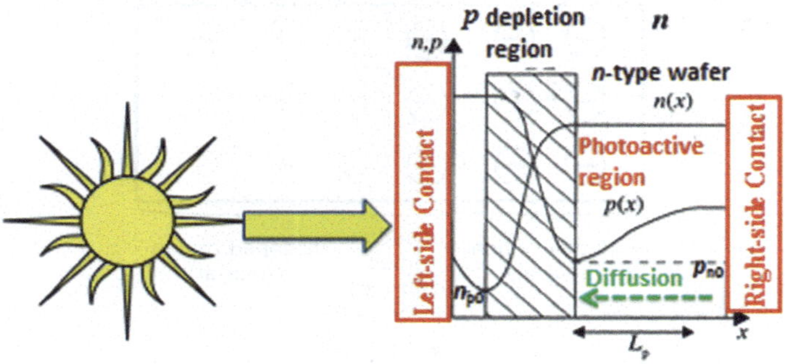

Fig. 3.10 *p-n* Type solar cell under illumination: diffusion of holes through the ***n***-type wafer, towards the depletion region, from right to left; $p(x)$ is the carrier profile of the holes; $n(x)$ is the carrier profile of the electrons; L_p is the minority-carrier diffusion length (of the holes); n_{p0} and p_{n0} are the equilibrium concentrations of the minority carriers

p-n solar cells: (Fig. 3.9) Here, one remarks that the transport of electrons and holes occurs mainly in the bulk of the *p*- and *n*-regions, where there is no significant electric field; this transport is governed by diffusion. Thus, the *p-n* solar cell is called a "diffusion-controlled device". In order to minimize the recombination loss ***R*** and to maximize the collection ***C***, it is necessary, in a *p-n* solar cell, that the minority-carrier diffusion lengths be much larger than the corresponding dimensions of the *p*- and *n*-regions [12].

Diffusion transport: (Fig. 3.10) Diffusion is very generally a mechanism, where particles diffuse from a zone of high concentration to a zone of lower concentration. In semiconductors the particles, which are of interest to us, are the (free) holes and electrons.

Let us look the case of free holes, within a *p-n* type solar cell, as shown schematically in Fig. 3.10: Holes have to diffuse from right to left; they are separated from the electrons by the electric field in the depletion region, and continue their path up to the very left side of the device.

In Fig. 3.10, the sun is shining on the solar cell from the left. Photo-generation of holes and electrons take place throughout the photoactive region. The electrons leave the solar cell on the right side. They have to be transported from their point of generation to the right side of the solar cell. As they are here the majority carriers, their density is very high and their transport is consequently not a problem. The limiting factor are the minority carriers, here the holes. Some of the holes will be generated near the right side contact: The only method, by which they can reach the depletion region, is by diffusion. Indeed, in this example diffusion will take place, because the density of holes just behind the right-side contact is relatively high due to photo-generation. On the other hand, as one approaches the depletion region, the density will be lower, as most of the holes there have been siphoned off by the electric field. In this zone, at the right of the depletion region, the holes are the minority carriers, because this is an *n*-type wafer, where the density of holes is very low. The electrical current carried by these holes is a diffusion current; its current density $J_{\text{p.diff}}$ is given by the equation:

$$J_{\text{p.diff}} = -D_p \frac{\partial p(x)}{\partial x}, \quad (3.2)$$

where D_p is the diffusion constant of the holes and $p(x)$ the carrier profile of the holes.

On the other hand, L_{p} is the minority-carrier diffusion length (of the holes). L_{p} is, in its turn, given by the expression

$$L_{\text{p}} = \sqrt{D_p \times \tau_p}, \quad (3.3)$$

where τ_p is the lifetime of the holes.

The minority carrier diffusion length L_{p} governs the carrier transport on the right side of the device.

p-i-n solar cells: In *p-i-n* solar cells the transport of the photo-generated carriers is mainly by drift (and not by diffusion). Thus, *p-i-n* solar cells are drift-controlled devices. The transport length that now intervenes is the drift length (for electrons and holes)

$$L_{\text{drift}} = \mu \times \tau \times E, \quad (3.4)$$

where μ is the carrier mobility, τ the carrier lifetime and E the electric field.

The link between diffusion and drift is given by the so-called «Einstein relation»:

$$D = (\mu k T)/q, \quad (3.5)$$

where k is the Boltzmann constant, T the absolute temperature and q the charge of an electron.

For further details, the reader is referred to Chap. 6, Sect. 6.2.1.

We will now:

Discuss in more detail, whilst using the corresponding equations, the functioning of a solar cell; the goal here is to look at the main parameters of the solar cell: short-circuit current density J_{sc}, open-circuit voltage V_{oc}, Fill Factor *FF* and efficiency η. We will also give approximate relationships for the dependence of η on temperature and intensity of the incoming light.

3.4 Solar Cell Characteristics, Equivalent Circuits and Key Parameters

We have just stated, in Sect. 3.3, that, in general, solar cells are simply semiconductor diodes, which are exposed to light. We will now look at the behaviour of a diode in the dark (without illumination), and thereafter see how this behaviour is modified, when the diode is illuminated.

3.4.1 Dark Characteristics

The current-voltage characteristic of a semiconductor diode in the dark (without illumination) is given by an exponential function; in (3.6) hereunder, we give the current density[6] J, in function of the applied voltage V:

$$J = J_{\text{dark}} = J_0 \left[\exp\left(\frac{qV}{nkT} \right) - 1 \right], \tag{3.6}$$

where

$q = 1.6022 \times 10^{-19}$ [C] is the elementary charge,
$k = 1.3807 \times 10^{-23}$ [m^2 kg/(s^2 K)] is the Boltzmann constant,
n is the so-called "diode ideality factor",
T [K][7] is the absolute temperature,
J_0 is the reverse saturation current density.

Figure 3.11 shows schematically the current-voltage characteristic of a diode without illumination.

The reverse saturation current density J_0 depends on recombination within the diode; it also depends on the bandgap energy E_{gap} of the semiconductor material used. The higher the bandgap is, the lower J_0 will be. A large value of J_0 is "bad" for the diode, because it means that the diode cannot block negative currents effectively enough, and consequently it is also "bad" for the solar cell: It leads e.g. to a low value of V_{oc}. According to a large amount of experimental data analysed by Martin Green, the minimum value $J_{0\text{min}}^{\text{Green}}$ of the reverse saturation current density J_0 is given by the following semi-empirical limit [12]:

$$J_{0\text{min}}^{\text{Green}} = 1.5 \times 10^8 \times \exp\left(-\frac{E_g}{kT} \right) \text{mA/cm}^2 \tag{3.7}$$

Later in this chapter we will write (3.7) in the form $J_{0\text{min}}^{\text{Green}} = J_{00} \times \exp\left(-\frac{E_g}{kT}\right)$, where J_{00} is a "true constant", equal to 1.5×10^8 mA/cm^2.

J_0 varies over several orders of magnitude. In reality J_0 can be very much larger than $J_{0\text{min}}^{\text{Green}}$, if the diode/solar cell has shunts and other types of defects.

J_0 can be easily determined from the dark J-V characteristics of the diode/solar cell; it turns out to be a very useful diagnostic parameter.

---

[6] To obtain the total current I, one simply has to multiply the current density J, with the cross-sectional area A of the device: $I = A \times J$: it will be often more convenient in this chapter to talk of current **densities**, rather than of currents.

[7] "K" stands for "degrees Kelvin".

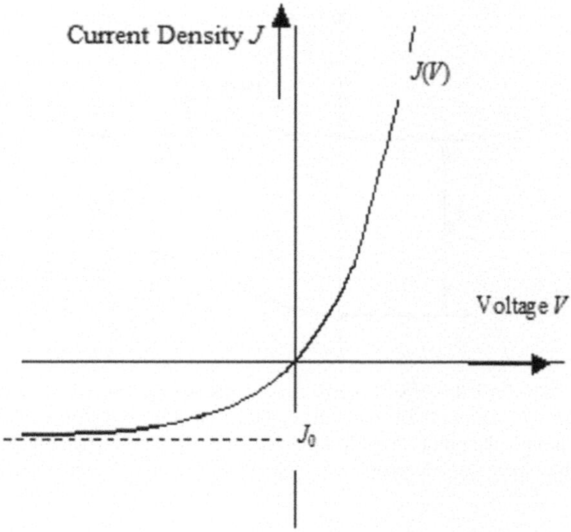

Fig. 3.11 Typical $J(V)$ characteristic for a diode. J_0 is the reverse saturation current density. Reproduced from [4] with the kind permission of the EPFL Press

The diode ideality factor n varies only between 1 and 2. The lower the value of n is, the higher are the solar cell parameters V_{oc} and FF (as defined in Sect. 3.4.2), for a given bandgap value E_g of the semiconductor material used. For the "ideal" *p-n* diode, $n = 1$; for the "ideal" *p-i-n* diode, $n = 2$. In practice, *p-n* diodes have values of n between 1.2 and 1.4 and *p-i-n* diodes values of n around 1.8. The ideality factor n is not a very useful diagnostic parameter.

3.4.2 Characteristics Under the Influence of Light

As soon as light hits the solar cell, a photo-generated current density J_{ph} will be superimposed on the dark current density J_{dark}. Equation (3.7) then becomes:

$$J_{illum} = J_{ph} - J_{dark} = J_{ph} - J_0\left[\exp\left(\frac{qV}{nkT}\right) - 1\right], \tag{3.8}$$

where J_{illum} is the current density of the illuminated solar cell.

There is now a minus sign in front of J_{dark}, because the photo-generated current and the diode forward current flow in opposite directions. The photo-generated current is composed of holes flowing to the *p*-side and electrons flowing to the *n*-side, whereas in the forward operation of a (dark) diode, the flow of carriers is just the opposite.

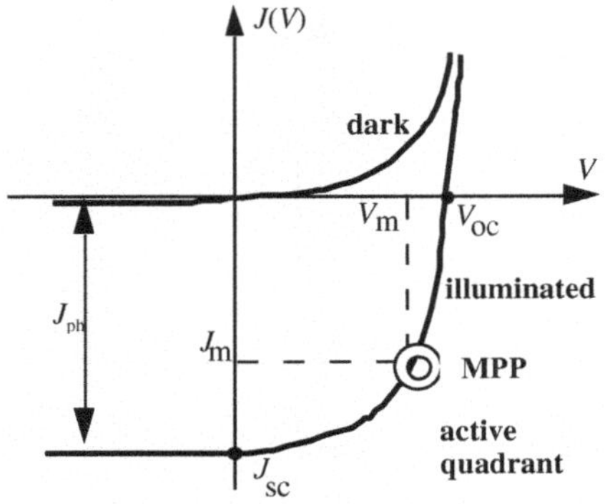

Fig. 3.12 Typical characteristics of solar cells: dark characteristics and illuminated characteristics. The "active quadrant" is the quadrant, where the solar cell can furnish power to a load; MPP is the "maximum power point", the point on the illuminated characteristics, where the power furnished to the load is a maximum (see text). Reproduced from [4] with the kind permission of the EPFL Press

We can now proceed to draw the *J-V* curve defined by (3.8). This is done schematically in Fig. 3.12.

An interesting point on the illuminated *J-V* curve is the Maximum Power Point (MPP), where the power furnished by the solar cell to an external load reaches a maximum. To obtain this point by calculations, we must maximize the product of current density times voltage, i.e. the product

$$J \times V$$

One strives, in all practical situations to keep the solar cells/modules operating at this point (Fig. 3.13). This is obtained by the use of an electronic device called an "MPP-Tracker". The Maximum Power Point (MPP) defines an important key parameter of the solar cell/module, namely the Fill Factor (*FF*). The Fill Factor is given by the following equation:

$$FF = (J_{\mathrm{m}} \times V_{\mathrm{m}})/(J_{\mathrm{sc}} \times V_{\mathrm{oc}}), \tag{3.9}$$

where J_{m} and V_{m} are the current density and the voltage at the MPP, respectively; J_{sc} is the short-circuit current density and V_{oc} the open-circuit voltage.

3.4.3 A Remark About the Theoretical Fundaments of the Basic Solar Cell Equations

Equations (3.6) and (3.8) are the very basis of solar cell theory. Therefore, one may ask oneself: On which theoretical foundations are they based?

Fig. 3.13 One strives, in all practical situations, to harvest a maximum amount of sunrays. Courtesy Dji-illustrations, Neuchâtel

Equation (3.6) is the basic diode equation; it can, for the case $n = 1$, be more or less rigorously derived from basic semiconductor device theory, i.e. from the drift-diffusion equation and from Poisson's equation, albeit in a somewhat "round-about" way, with many assumptions and approximations (see [4], or any textbook on semiconductor device physics) (Fig. 3.14).

The case $n \neq 1$ is based on a very crass approximation, whereby two exponential functions, each with a different argument are combined into a single exponential function with its argument lying somewhere between the first two ones.

Equation (3.8) follows from (3.6) by using the so-called "*superposition principle*", i.e. by postulating that under illumination, the total solar cell current J_{illim} is simply given as superposition (addition) of the diode dark current J_{dark} and the photo-generated current J_{ph}. This is not at all obvious, if one looks in detail at the applicable semiconductor device equations within the solar cell. In fact, the addition of light, i.e. of photo-generated electron-hole pairs, completely changes the entire profiles $n(x)$ and $p(x)$ of electrons and holes within the photo-absorbing layers of the solar cell. However, as the equations used to derive the diode characteristics are all linear, the superposition principle can be intuitively justified, but only for *pn*-type solar cells and under restrictive assumptions (see [12, 13]).

Fig. 3.14 To calculate the basic equations governing the operation of a solar cell is, indeed, no simple task. Courtesy Dji-illustrations, Neuchâtel

3.4.4 Equivalent Circuits for the Solar Cell

1. ***Basic equivalent circuit***

The basic equivalent circuit for all solar cells is given in Fig. 3.15a. It corresponds to (3.8) and consists merely of a diode **D** and a photo-current (density) source J_{ph}.

This equivalent circuit is not of any practical use.

2. ***Standard equivalent circuit***

To arrive at the standard solar cell equivalent circuit, which is used universally for (almost) all solar cell work, one has to add two elements to the basic equivalent circuit of Fig. 3.15a:

(a) A series resistance R_{series}, which stands mainly for the Ohmic losses in the contacts and wiring;
(b) A parallel resistance R_p, which represents two very different effects:

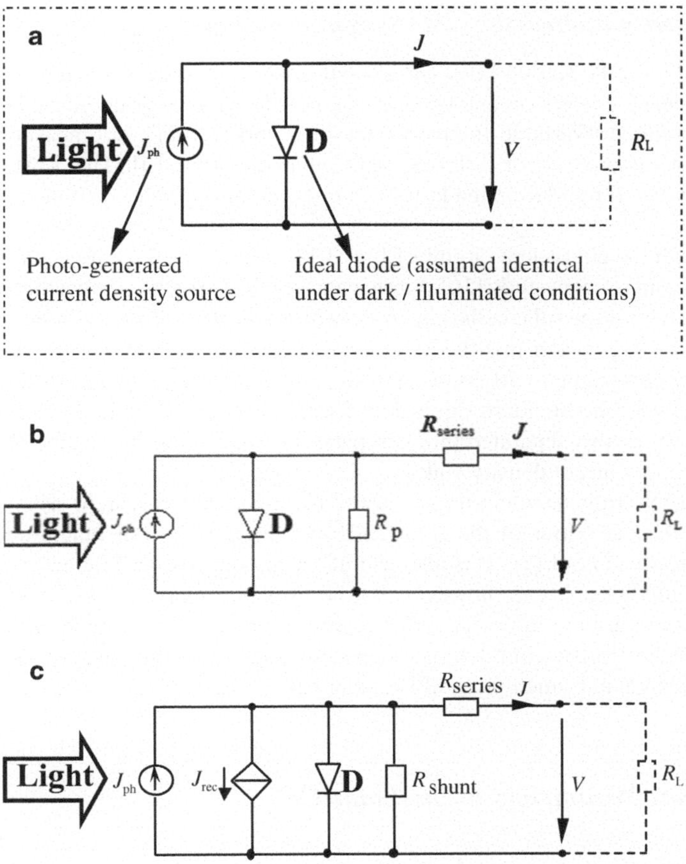

Fig. 3.15 **a** Basic equivalent circuit, for an "ideal" solar cell; an external load resistance R_L has also been drawn. **b** Standard equivalent circuit, for the "real" solar cell; parallel resistance R_p and series resistance R_{series} have been added. **c** Merten-Andreu-Shah (MAS) equivalent circuit, for use in solar cell analysis and diagnosis; the parallel resistance R_p of the standard equivalent circuit has now been replaced by a "true" shunt resistance R_{shunt} and a recombination current density sink J_{rec}. Reproduced from [4] with the kind permission of the EPFL Press

i. Actual, physical shunts, which can be either Ohmic or have the character of diodes.
ii. Recombination losses.

One obtains thereby the equivalent circuit shown in Fig. 3.15b. The reader should take good note of **this** equivalent circuit as she/he will find it generally useful for dealing with almost all practical problems.

This equivalent circuit is indeed recommended for all our readers, who are concerned with the planning, installation and operation of solar cells and modules. It can directly be used for the considerations in Chap. 12.

3. ***The Merten-Andreu-Shah (MAS) equivalent circuit***

In his Ph.D. thesis Merten [14], carried out, under the direction of Professor Jordi Andreu, measurements on amorphous silicon solar cells and modules. These were mainly so-called "Variable Intensity Measurements (VIM)", i.e. measurements of the *J*-*V* characteristics for different light intensities [14, 15]. The VIM measurements conducted by Merten led him to a third type of equivalent circuit as shown in Fig. 3.15c.

Although the equivalent circuit of Fig. 3.15c was derived for amorphous silicon solar cells—in a recent study [16], Merten and co-workers have shown that it can also be applied to *pn*-type solar cells, namely to wafer-based crystalline silicon solar cells. One may, thus, consider that the MAS Equivalent Circuit as shown in Fig. 3.15c is universally applicable to different types of solar cells. The MAS Equivalent Circuit is identical with the Standard Equivalent Circuit, except for the fact that the parallel resistance R_p is now separated into two parts: A "true" shunt resistance R_{shunt} and a recombination current density sink J_{rec}.

The MAS equivalent circuit is useful for solar cell research, as it correctly describes the behaviour of the solar cell over many orders of magnitude of illumination levels, i.e. of J_{ph}. It is also useful for the diagnosis of faults in solar cells and solar modules. It can, however, not be used for analysing and designing the electric circuit around the solar cell, because J_{rec} is not an element, which can be used within an electric circuit diagram, as it depends, according to [15], (3.3), on the ***internal*** design and functioning of the solar cell.

3.4.5 Key Parameters of the Solar Cell

1. ***Short-circuit current density J_{sc}***

In short-circuit conditions, $R_L = 0$. The short-circuit current density J_{sc} is, therefore, for the "ideal" case equivalent to the photo-generated current density source J_{ph} as shown in Fig. 3.15a–c. In Fig. 3.16 one sees now the maximum value of J_{sc}, which can be attained for input illumination corresponding to AM 1.5. This maximum value is obtained when all photons with $E_{ph} = h\nu > E_g$ are usefully absorbed by the solar cell (where h is Planck's constant and ν the frequency of light).

2. ***Open-circuit voltage V_{oc}***

Under open-circuit conditions, $R_L = \infty$ and $J = 0$ (Fig. 3.15a–c). The open-circuit voltage V_{oc} can, therefore, be derived by setting $J_{illum} = 0$ in (3.8) (meaning that no current flows out of the solar cell), and solving the resulting relationship for *V*, rendering:

$$V_{oc} = \frac{nkT}{q} \ln\left(\frac{J_{ph}}{J_0} + 1\right) \approx \frac{nkT}{q} \ln\left(\frac{J_{ph}}{J_0}\right) \quad (3.10)$$

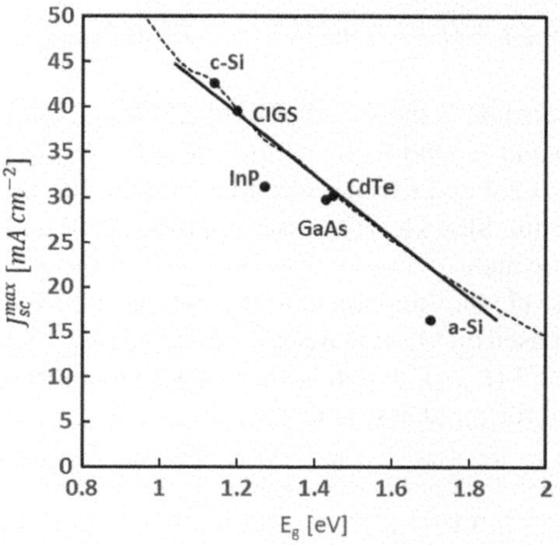

Fig. 3.16 Maximum value of the short-circuit current density J_{sc} as a function of the bandgap E_g. Dots show the maximum values obtained for different solar cell materials in Research Laboratories [17]. The straight line given is merely a very coarse approximation, used here to guide the eye; it corresponds to the equation $J_{sc} \approx 80\,\text{mA/cm}^2 - 34\,\text{mA/}(\text{cm}^2\,\text{eV}) \times E_g;\ E_g$ in eV

One sees here that V_{oc} depends on J_{ph}, i.e. on the illumination level. By decreasing the incoming light by a factor of 10, i.e. by a factor of approximately $e^{2.3}$, V_{oc} will decrease by roughly n times (26 mV × 2.3 ≈ 60 mV), because kT/q is approximately 26 mV at room temperature.

V_{oc} also depends on temperature; this is, one hand, given by the pre-factor (nkT/q) in (3.11), but it is also given by the quantity J_0, which has a pronounced temperature dependence. As a net result, V_{oc} always decreases with increasing temperature T. In a similar manner, V_{oc} always decreases, if the diode quality factor n is increased (indicating a diode with a *lower* quality, i.e. a diode with increased recombination!).

In practice, one of the key quantities determining V_{oc} is the diode reverse saturation current J_0.

By taking Martin Green's semi-empirical limit [12] for J_0 (3.7, above), and by setting at the same time $n = 1$, in the right side of (3.10), one obtains for V_{oc} the *limit*:

$$V_{oc} = \frac{kT}{q} \ln\left(\frac{J_{ph}}{J_0}\right) \approx \frac{kT}{q} \ln\left(\frac{J_{ph}}{J_{00}}\right) + \frac{E_g}{q}. \quad (3.11)$$

where $J_{00} = J_{00}^{\text{Green}} = 1.5 \times 10^8\ \text{mA/cm}^2$.

Note that $\ln\left(\frac{J_{ph}}{J_{00}}\right)$ is always negative, so that $V_{oc} < E_g/q$.

On the other hand, there exists a fundamental thermodynamic limit for V_{oc}, as given e.g. by Shockley and Queisser [18] and by Kiess and Rehwald [19]. The latter derive the following expression for V_{oc}:

$$V_{oc} = E_g/q + (kT/q)\ln\Big\{\big(h^3c^2/2\pi kT\big)\Big(N_{incident}/E_g^2\Big)\Big\}, \tag{3.12}$$

where h is Planck's constant, c the speed of light and $N_{incident}$ the number of photons incident per unit area and second with energies $h\nu > E_g$. $N_{incident}$ corresponds to our quantity ϕ (see Sect. 3.2.4 and (3.1)). Note that Meillaud-Sculati has shown in her Ph.D. thesis [20] that the Shockley-Queisser approach [18] and the Kiess-Rehwald approach [19] are equivalent.

Figure 3.17 shows: (1) the fundamental, theoretical limit for V_{oc} (3.12),[8] (2) the semi-empirical limit based on Martin Green's work ([12] and (3.11)), and (3) a simple rule of thumb $V_{oc} \approx 2/3$ (E_g/q), which is very easy to memorize and is therefore an useful approximation, for rapid assessments.

3. ***Fill Factor FF***

The limit values for the Fill Factor FF are found by taking (3.8), and by computing the product of voltage × current density, i.e. by evaluating

$$P_{ideal} = (J_{illum} \times V),$$

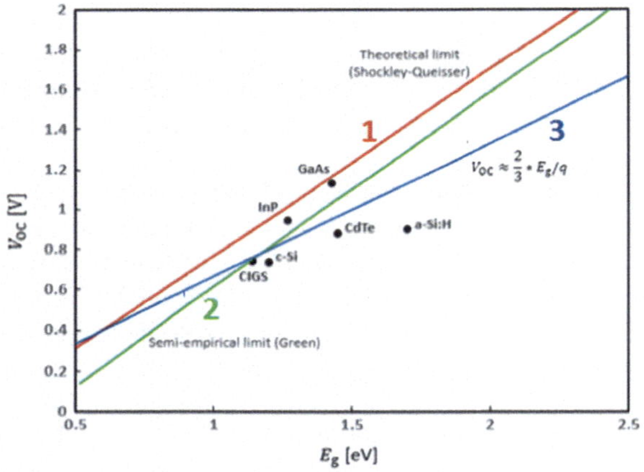

Fig. 3.17 Maximum value of the open-circuit voltage V_{oc} as a function of the bandgap E_g (see text). Dots show the maximum values obtained for different solar cell materials in various Research Laboratories [17]

[8]Note that in Fig. 3.17 the fundamental, theoretical limit is referred to as the "Shockley-Queisser limit", although the line depicted in the figure is based on (3.12), which is taken from the work of Kiess and Rehwald [19]. Shockley and Queisser [18] do not give any expressions for V_{oc}, which would be easily usable here, but as developed in [20], the two approaches are equivalent.

where P_{ideal} denotes the limit value for the output power density[9] that the solar cell can deliver in the "ideal" case, corresponding to Fig. 3.12. We then mathematically maximize P_{ideal} as a function of V, to find the maximum power point MPP (in this ideal case) and compute the quantity

$$FF = P_{max}/(V_{oc} \times J_{sc}) = (V_m \times J_m)/(V_{oc} \times J_{sc}).$$

This gives us the expression [12]:

$$FF = \frac{\tilde{v}_{oc} - \ln(\tilde{v}_{oc} + 0.72\text{V})}{\tilde{v}_{oc} + 1} \tag{3.13}$$

with:

$$\tilde{v}_{oc} = \frac{V_{oc}}{nkT/q}$$

Inserting the semi-empirical limit value for V_{oc}, from the preceding paragraph (3.11), we obtain the two lower curves ($n = 1$ and $n = 2$) in Fig. 3.18 representing the semi-empirical limit for the Fill Factor *FF*.[10] In the same figure the fundamental limit for *FF* is also given, based on the theory of Shockley and Queisser.

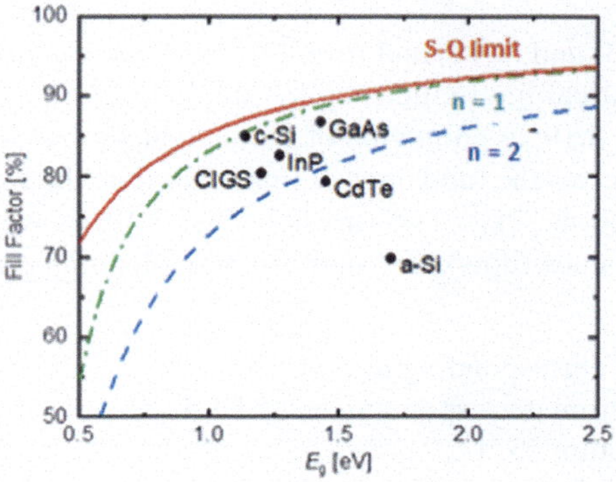

Fig. 3.18 Maximum value of Fill Factor *FF*, as a function of the bandgap E_g (see text). The red curve is the Shockley-Queisser (S-Q) limit. To compute this curve, data from [21] has been used. The green and the blue curves are semi-empirical limits according to the work of Green [12]. Dots show the maximum values obtained for different solar cell materials in Research Laboratories [17]

[9]Note that in this section we use the symbol P to denote power **density**, whereas in Sect. 3.5 we will be using the same symbol P to denote power.

[10]The curves for V_{oc} in Fig. 3.17 have been computed for the case $n = 1$; in fact, curve (2) in Fig. 3.17 was obtained from (3.11), where one has set $n = 1$; in order to obtain the corresponding curves for V_{oc} for the case $n = 2$, we would need to know more about the recombination within the depletion region of the solar cell. As such information is not available, we have here simply combined (3.11) for V_{oc} and $n = 1$ with (3.13) for $n = 2$. The result, as presented in Fig. 3.18 under the heading $n = 2$, is a very coarse approximation.

3.5 Solar Cell Efficiency Limits

3.5.1 Limits at STC (Standard Test Conditions)

To calculate the *overall energy conversion efficiency* η of the solar cell, we must now divide the electrical output power P_{out} (at MPP) by the input (solar) power $P_{in} = P_{sun}$. One has:

$$P_{out} = V_m \times J_m \times A = J_{sc} \times V_{oc} \times FF \times A \tag{3.14}$$

Here J_{sc}, V_{oc} and *FF* have been determined above, and A is the surface area of the solar cell. On the other hand, $P_{in} = P_{sun}$ is the "power of the sun", i.e. the total energy within the solar spectrum per unit of time, for a unit surface. P_{sun} was already used to calculate the *spectral conversion efficiency* η_S (3.1).

Using this formula for η_S we can now write:

$$\eta = \eta_S \times (J_{sc} \times V_{oc} \times FF)/(\phi E_g) \tag{3.15}$$

The relationship between J_{sc} and E_g is given in Fig. 3.16 for AM 1.5 illumination. Thereby one assumes that the solar cell absorbs <u>all</u> incoming light with wavelengths shorter than the absorption edge of the semiconductor, and that collection within the solar cell is ideal.

From (3.11), (3.13) and (3.14) and from Fig. 3.16 one can now compute, numerically, the semi-empirical limit efficiency for single-junction solar cells. The result is represented in Fig. 3.19, as a function of semiconductor bandgap energy. Furthermore, Fig. 3.19 also gives the fundamental limit, based on the work of Shockley and Queisser (as tabulated in [21]). The curves in Fig. 3.19, as well as the practically obtained solar cell values (dots), are valid for so-called <u>*Standard Test Conditions*</u>, i.e. for:

- Illumination level corresponding to AM 1.5
- Illumination spectrum corresponding to AM 1.5
- Solar cell temperature 25 °C.

3.5.2 Variation of Efficiency η in Function of Temperature

In practice, solar cells are rarely operated at a temperature $T = 25$ °C (STC)—they are in most cases operated at higher temperatures $T > 25$ °C. This leads to a drop in their efficiency η. Therefore, we are going to derive an approximate relationship between solar cell efficiency η and solar cell operating temperature *T*.

Now: η is proportional to the product ($J_{sc} \times V_{oc} \times FF$), as given in (3.15). However, one may state:

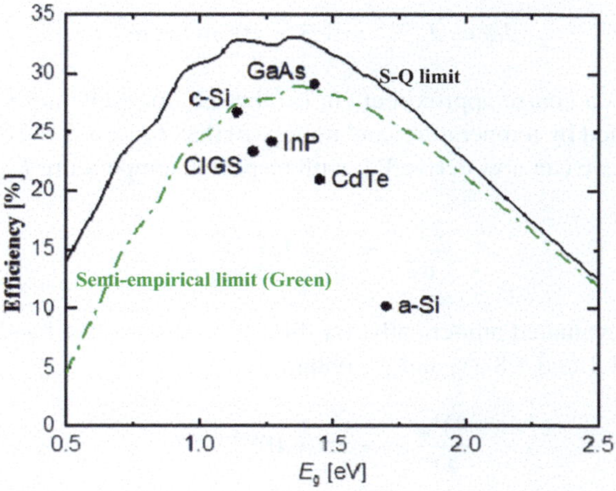

Fig. 3.19 Maximum value of the overall (The conversion efficiency for the *first* step of energy conversion was given in Sect. 3.2.4, Fig. 3.7.) energy conversion efficiency η of the solar cell as a function of the band gap E_g. Dots show the maximum values obtained for different solar cell materials in practice [17]. (At Standard Test Conditions, STC, see text)

- The maximum (limit) value of J_{sc} hardly varies with temperature
- The maximum (limit) value of FF varies very little with temperature
- The maximum (limit) value of V_{oc} varies strongly with temperature.

Therefore, we will, in this section, only look at the variation of V_{oc} with temperature, and in a rather rough approximation, assume that the variation of η with temperature is similar.

We are here interested in the *relative* temperature coefficient

$$TC = \frac{\partial \eta}{\partial T} / \eta \approx \frac{\partial V_{oc}}{\partial T} / V_{oc}$$

applicable to the efficiency of solar cells and modules.

1. **Fundamental considerations**

Fundamental considerations, should, in principle, be based on Thermodynamics, like the approach of Shockley and Queisser. However, the development of Shockley and Queisser leads for V_{oc} to mathematical expressions which are cumbersome; therefore, we will use here for V_{oc} the semi-empirical limit of Martin Green, i.e. (3.11):

$$V_{oc} \approx \frac{kT}{q} \ln\left(\frac{J_{ph}}{J_{00}}\right) + \frac{E_g}{q} \quad \text{with}$$

$$J_{00} = J_{00}^{\text{Green}} = 1.5 \times 10^8\ \text{mA/cm}^2$$

Assuming, as a coarse approximation: (a) that E_g does not vary with T; (b) that J_{ph} is not modified by temperature; and noting (c) that $J_{00} = 1.5 \times 10^8$ mA/cm^2 is a "true constant", we can now derive V_{oc} with respect to temperature T, and we obtain:

$$\frac{\partial V_{oc}}{\partial T} = \frac{k}{q} \times \ln\left(\frac{J_{ph}}{J_{00}}\right)$$

This can be evaluated numerically, for different values of the bandgap E_g, in the range between 1.1 and 1.8 eV, and we obtain:

$$\frac{\partial V_{oc}}{\partial T} \approx -15 \times 10^{-4}\ \text{V/°C} \tag{3.16a}$$

Thereafter, we can divide by the corresponding value of V_{oc}, rendering:

$$TC = \frac{\partial \eta}{\partial T}/\eta \approx \frac{\partial V_{oc}}{\partial T}/V_{oc} \approx -0.2\%/\text{°C} \times (1\ \text{eV}/E_g);\ \ E_g \text{ in eV}, \tag{3.16b}$$

Note that this relationship is only valid for temperatures around 25 °C.

The value of TC in (3.16b) is negative—this means that the solar cell efficiency η will drop as we increase the operating temperature above 25 °C (STC).

Equation (3.16b) is just a very simple "rule of thumb", which enables us to roughly assess the order of magnitude for the temperature coefficient of the efficiency of solar cells and modules.

Equation (3.16b) means also, that the higher the bandgap of the semiconductor material is, the lower the magnitude of the relative temperature coefficient $TC = (\partial\eta/\partial T)/\eta$ will be. From a practical point of view, it is therefore advantageous to use solar cell absorber materials with high bandgaps, if the actual operating temperature of the solar cell is going to be high.

Table 3.3 gives the values of TC, based on (3.16b), for the most common solar cells, listed according to the absorber material they employ.

2. **Practical, experimental results**

In practice, the values of TC are substantially higher (in magnitude) than those suggested by (3.16b). This will be developed now: Table 3.4 gives values found for actual, commercial solar cells and modules: In this table, the lowest temperature dependence is found for amorphous silicon modules ($TC \approx -0.2\%$/°C), and for some CdTe modules ($TC \approx -0.25\%$/°C); and the strongest temperature dependence for some CIGS[11] modules ($TC \approx -0.39\%$/°C) and for certain wafer-based crystalline

[11] CIGS cells and modules have a large variety in the chemical composition of their absorption layer.

Table 3.3 Comparison of the "theoretical" limit values of the relative temperature coefficient $TC = (\partial\eta/\partial T)/\eta$ for different solar cell types: CIGS, a-Si, CdTe, and c-Si

	Bandgap E_g [eV]	"Theoretical" value of *TC* (3.16b) [%/°C]
CIGS	1.2	≈ −0.17
Wafer-based crystalline silicon (c-Si)	1.12	≈ −0.18
Amorphous silicon (a-Si)	1.75	≈ −0.11
CdTe	1.49	≈ −0.13

Table 3.4 Comparison of the relative temperature coefficient $TC = (\partial\eta/\partial T)/\eta$ for different solar cell/module types: homojunction c-Si, heterojunction c-Si, a-Si, micromorph, CdTe, and CIGS (based on data and references in [4] for a-Si and micromorph; and on datasheets for all other values)

Type of solar cell/module	Range of temperature coefficient $TC = (\partial\eta/\partial T)/\eta$
Crystalline silicon homojunctions (c-Si)	≈ −0.29%/°C (TOPCON, PERC) to −0.38%/°C (Al-BSF) (see Chap. 5)
Crystalline silicon heterojunctions (c-Si/HJT)	≈ −0.3%/°C (see Chap. 7)
Amorphous silicon (a-Si)	≈ −0.2%/°C (see Chap. 6)
Micromorph tandems (μc-Si/a-Si)	≈ −0.3%/°C to −0.35%/°C (see Chap. 6)
CdTe	≈ −0.25%/°C to −0.32%/°C (see Chap. 8)
CIGS	≈ −0.32%/°C to −0.39%/°C (see Chap. 8)

silicon modules (*TC* ≈ −0.38%/°C); c-Si heterojunction modules [22], as well as some recently developed homojunction c-Si modules are interesting module implementations: they have high efficiencies and relatively low temperature dependence (*TC* ≈ −0.3%/°C).

This situation is illustrated in Fig. 3.20.

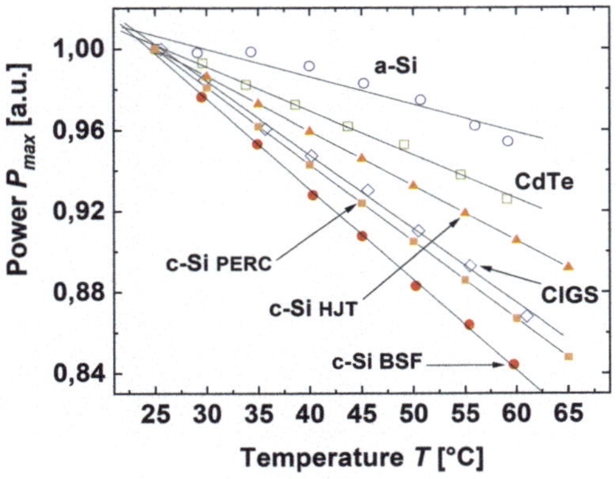

Fig. 3.20 Effect of operating temperature T on the normalized value of the output power P_{max}, at MPP, of typical solar modules, for various cell technologies, based on [23] for CIGS, CdTe and a-Si; and on datasheets for c-Si modules

The **case of c-Si (wafer-based crystalline silicon)** merits, for two reasons, particular attention: (1) these cells account for more than 90% of all cells produced worldwide; (2) they have a large span of practical values of TC; in datasheets of c-Si cells, we find values running from −0.29 up to −0.38%/°C.

When one looks at the whole picture, which results from studying datasheets and published experimental data, one finds that there is a remarkable relationship between the value of V_{oc} and the magnitude of TC.

As V_{oc} approaches its "theoretical" limit value of approximately 800 mV (see Fig. 3.17, line 2, green), the magnitude of TC becomes smaller; thereby, TC tends to approach its "theoretical" limit value of ≈ −0.18%/°C.[12]

Conversely, cells with low values of V_{oc}, have high magnitudes of TC, i.e. a strong temperature dependence.

This relationship would fully explain why crystalline silicon heterojunction modules have a remarkably low magnitude of TC.

Tijmen Slikker of Eternalsun Spire has recently done a series of measurements on different c-Si modules and their temperature behaviour—his results are presented here as Fig. 3.21a, b.

3. **Concluding remarks**

- The relative temperature coefficient TC is a parameter, which is given in the Datasheet of the Module. It varies strongly:
  - Between modules using different absorber materials, i.e. the magnitude of TC is much larger for c-Si modules than for CdTe modules.

[12]We are referring here not to the thermodynamical limit values according to Shockley and Queisser [18], but to the semi-empirical limit values according to Green [12].

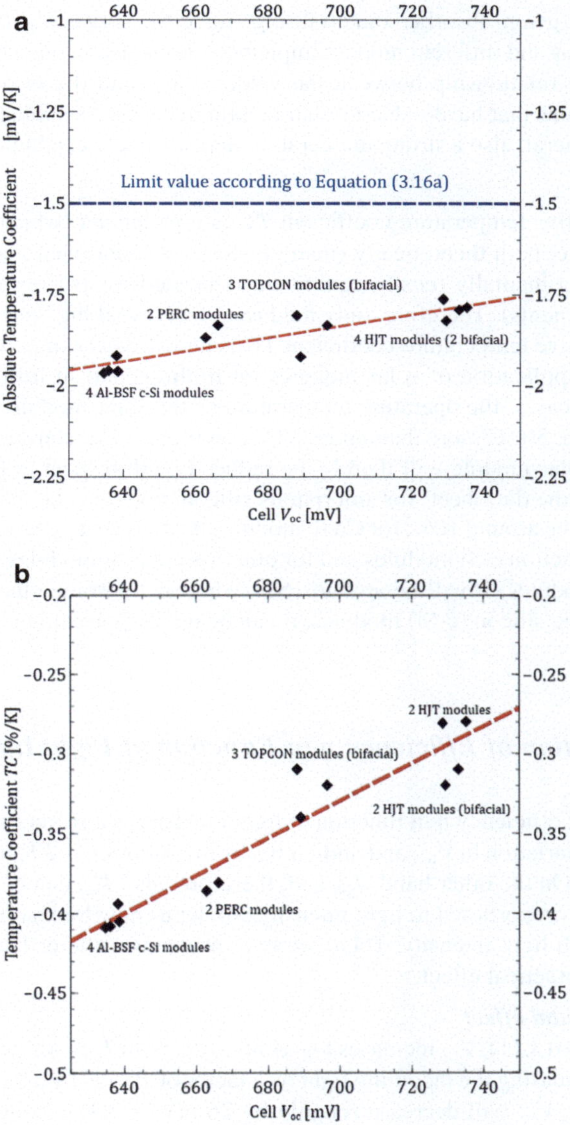

Fig. 3.21 **a** Representation of the relationship between the value of V_{oc} and the "**absolute**" temperature coefficient $\frac{\partial V_{oc}}{\partial T}$ (Courtesy of Eternalsun Spire). **b** Representation of the relationship between the value of V_{oc} and the **relative** temperature coefficient TC (Courtesy of Eternalsun Spire)

  - For a given absorber material, e.g. for c-Si, between different module designs and different module implementations. Here, one notices an interesting relationship between the value of V_{oc} and the magnitude of TC: Modules that have—due to high recombination—low values of V_{oc} have, in general, also a strong temperature dependence, i.e. a high magnitude of TC.[13]

- The relative temperature coefficient TC is a parameter, which is difficult to determine, both theoretically (many factors not mentioned here play a role) and experimentally (costly equipment is needed for performing laboratory measurements). Therefore, only field tests will reveal the "full truth".
- The relative temperature coefficient TC is an extremely important parameter for the application of solar modules on rooftops and in tropical countries. In these cases, the operating temperature of the solar module easily reaches 75 °C, i.e. 50 °C more than under STC conditions. The conversion efficiency of the solar module will thereby be reduced, with respect to the STC value given in the datasheet: for amorphous silicon modules, the efficiency reduction will be around 10%, for CdTe modules it will be between 12 and 15%, for heterojunction c-Si modules and for other recent c-Si modules (such as TOPCON modules) it will be around 15%, whereas for some other wafer-based crystalline silicon (c-Si) modules, it can be up to approximately 20%.

3.5.3 Variation of Efficiency η in Function of Light Intensity

The variation of efficiency η in function of light intensity is governed directly by the corresponding variation in V_{oc}, and, indirectly, by the variation in FF, mainly through V_{oc} (see 3.13). On the other hand, J_{ph} and, therefore, also J_{sc}, can be considered to remain directly proportional to light intensity. So, let us now look at the variation of V_{oc} (and η) with light intensity. There are two mechanisms here: 1. a fundamental effect and 2. a practical effect:

1. ***A fundamental effect***
   According to (3.11), V_{oc} increases logarithmically with J_{ph}—i.e. with light intensity. By decreasing the incoming light by a factor of 10, i.e. by a factor of approximately $e^{2.3}$, V_{oc} will decrease roughly by 26 mV $\times$ 2.3 $\approx$ 60 mV. This is so,

[13]There are two reasons for this: (1) If we have a higher value of V_{oc} (and of η), we will be obtaining a higher magnitude of TC, as TC is the relative temperature coefficient, and we have to divide by V_{oc} (or by η), in order to obtain it; (2) the difference in TC between "theoretical" and experimental values (i.e. between Tables 3.3 and 3.4) is so large that it cannot be explained only by (1). We therefore postulate that Surface Recombination, which is mainly responsible for the difference in V_{oc} obtained for c-Si modules, is strongly temperature dependent. This postulate remains to be theoretically justified. Note that the slope of the red line in Fig. 3.21a is due to (2).

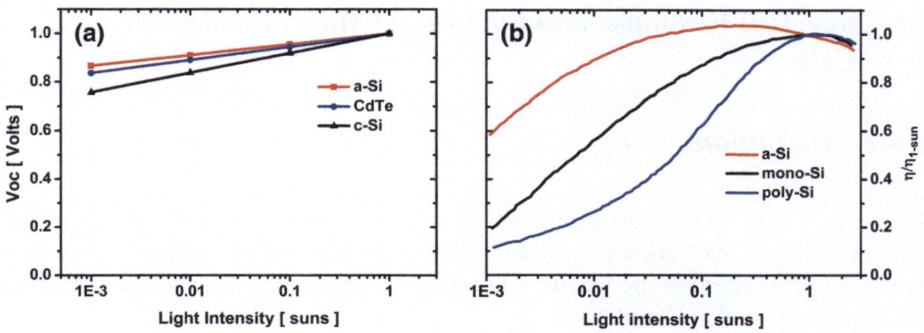

Fig. 3.22 **a** V_{oc} as a function of light intensity: theoretical curves for three materials (amorphous silicon, CdTe, and crystalline silicon); **b** normalized cell efficiencies $\eta/\eta_{1-\text{sun}}$: experimental curves for three types of silicon solar cells (amorphous silicon, polycrystalline silicon and monocrystalline silicon)—measurements done at IMT, PV-Lab, Neuchâtel

because kT/q is approximately 26 mV at room temperature.[14] The overall result is shown in Fig. 3.22a, for amorphous silicon (a-Si), crystalline silicon (c-Si) and cadmium telluride solar cells.

2. ***A practical effect, depending on the imperfections of the solar cell***
According to Fig. 3.15b, c, FF (and therefore also η) will decrease with increasing light intensity, under the influence of the series resistance R_{series}, which is responsible for a voltage drop under all operation conditions, except for open-circuit conditions.
On the other hand, V_{oc} and FF will decrease, with decreasing light intensity, under the influence of the shunt resistance R_{shunt}, which siphons off, at lower light intensities, a correspondingly larger fraction of the photo-generated current J_{ph} (see Fig. 3.15b, c) leaving less current to build-up V_{oc}.
The phenomena described here (under point 2), depend very much on solar cell type, solar cell design, and solar cell fabrication process. A practical **example** is given in Fig. 3.22b for three cells, which were measured at IMT PV-Lab Neuchâtel. By comparing Fig. 3.22a, b, one can deduce that at low light intensities, the imperfections of the solar cell (in this case, the shunts within the solar cell) are the predominant factor responsible for reducing the efficiency.

3. ***Concluding remarks***
The exact behaviour of solar cell efficiency η in function of light intensity cannot be predicted in a general manner, but depends (as stated above) on solar cell type, solar cell design, and solar cell fabrication process. Amorphous silicon solar cells have, in most cases, a better efficiency at very low light intensities than wafer-based crystalline silicon solar cells: for this reason, they are often used for indoor applications, where the light intensity can be very low.

[14]This is only valid for relatively high light intensities, say for light intensities in the range between 0.1 and 1 sun. Remember: 1 sun is equivalent to an irradiation of 1000 W/m^2; STC conditions are based on 1 sun and on the AM 1.5 spectrum (see Chap. 2).

3.6 Spectral Response and Quantum Efficiency in Solar Cells

3.6.1 *Definitions*

We need "tools" to evaluate how a given solar cell or a given solar module will respond to light that reaches it—light which will have different spectral components—depending upon the location, the time of the day, the weather, etc. All this has been explained in Chap. 2.

These "tools" bear the names "Spectral Response" and "Quantum Efficiency".

What exactly is "***Spectral Response***[15]"? This is illustrated schematically in Fig. 3.23.

Spectral Response[15] is defined as the ratio of the electric current coming out of the cell or module[16] over the power of the incoming light. This is plotted as a function of the wavelength of the incoming light.

What exactly is "***Quantum Efficiency***"? This is illustrated schematically in Fig. 3.24.

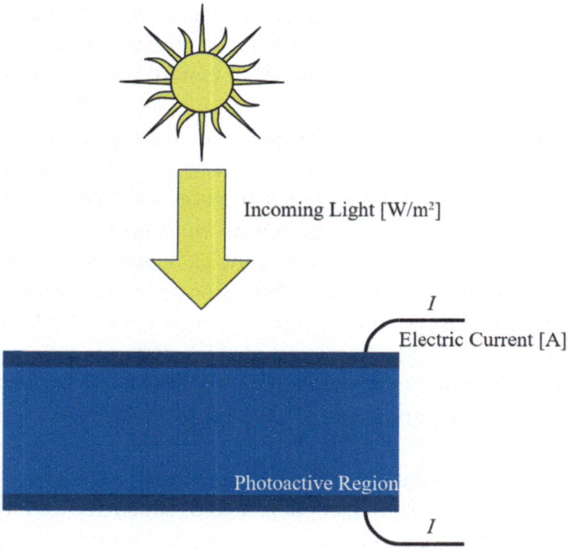

Fig. 3.23 Schematic representation illustrating how spectral response is defined. Courtesy of Peter Fiala, PV-Lab, IMT Neuchâtel, EPFL

[15]Officially called (according to the latest standards) "Spectral Responsivity".

[16]At short circuit conditions (and at STC, of course).

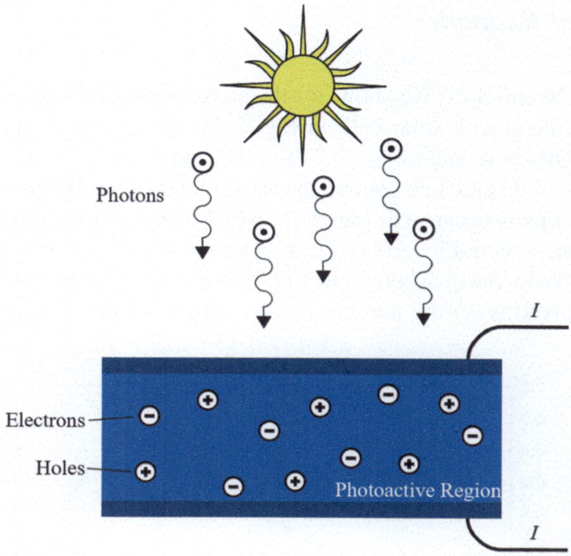

Fig. 3.24 Schematic representation illustrating how quantum efficiency is defined. Courtesy of Peter Fiala, PV-Lab, IMT Neuchâtel, EPFL

Quantum Efficiency is defined as the ratio of the number of electron-hole pairs generated within the photoactive region of the cell over the number of photons reaching the solar cell. This is again plotted as a function of the wavelength of the incoming light. Or, alternatively, it can be plotted as a function of the energy of the photons.

Note that Quantum Efficiencies are always lower than 1, because photons can only be "lost", on their way from entering into the solar cell, until the point, where they generate electron-hole pairs.

Furthermore, one distinguishes between External Quantum Efficiency (EQE) and Internal Quantum Efficiency (IQE). To evaluate EQE, one counts all the photons, which reach the cell; whereas, in IQE one only counts the photons, which reach the photoactive region.

Note that there is very simple relationship between EQE and Spectral Response[15] (SR):

$$\mathrm{SR}(\lambda) = (q\lambda/h \times c)\mathrm{EQE}(\lambda) \tag{3.17}$$

where SR is the spectral response, EQE the External Quantum Efficiency, q the charge of the electron, λ the wavelength of the light, h the Constant of Planck and c the speed of light.

3.6.2 Typical Examples

In Figs. 3.25, 3.26 and 3.27, we show, for the same individual solar cell—a heterojunction crystalline silicon solar cell: in Fig. 3.25, the spectral response curve; in Fig. 3.26, the EQE curve and in Fig. 3.27 the IQE curve.

Actually, Figs. 3.25 and 3.26 contain the same information, but have a completely different form—this is because in Fig. 3.25, which is a spectral response curve, we are looking at the spectrally resolved power of the incoming sunlight, which has the dimension [W/m^2/nm], whereas in Fig. 3.26, we are "just" counting (also in a spectrally resolved way) the number of incoming photons. Now, photons in the

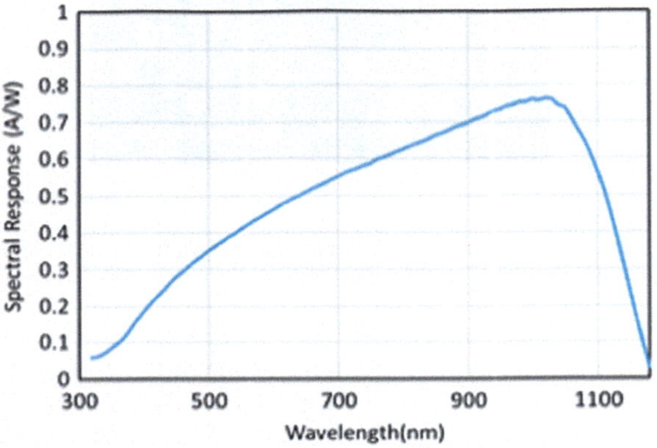

Fig. 3.25 Spectral response curve of a heterojunction crystalline silicon solar cell. Courtesy of Mathieu Boccard, PV-Lab, IMT Neuchâtel, EPFL

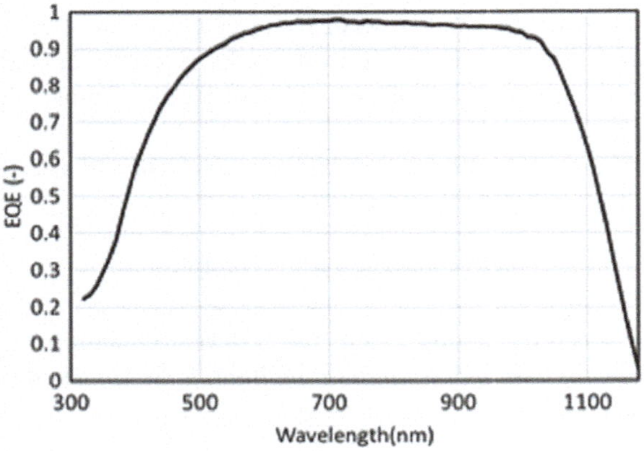

Fig. 3.26 EQE curve of a heterojunction crystalline silicon solar cell. Courtesy of Mathieu Boccard, PV-Lab, IMT Neuchâtel, EPFL

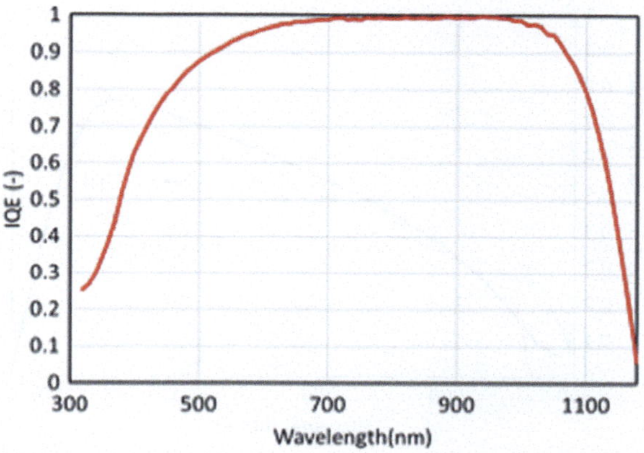

Fig. 3.27 IQE curve of a heterojunction crystalline silicon solar cell. Courtesy of Mathieu Boccard, PV-Lab, IMT Neuchâtel, EPFL

range of long wavelengths have a much lower energy per photon, than the photons in the range of short wavelengths. In fact, according to the Preamble to this chapter, the energy of a photon depends on the wavelength λ of the light according to the expression $E_{photon} = (h \times c)/\lambda$, with h being the constant of Planck ($\approx 6.6 \times 10^{-34}$ kg m^2 s^{-1}) and c the speed of light ($\approx 3 \times 10^8$ m s^{-1}). Thus, for high values of λ (long wavelength region) E_{photon} has a small value, whereas for low values of λ (short wavelength region) E_{photon} has a high value.

If we now compare Figs. 3.26 and 3.27, these are both Quantum Efficiency Curves, so we are, for both figures, counting the incoming photons—however to plot Fig. 3.26, which is the EQE curve, we are counting **all** the photons, which reach the solar cell, whereas in Fig. 3.27, which is the IQE curve, we count only those photons, which reach the photoactive region of the cell.

3.6.3 *Practical Consequences*

In Chap. 2 of this book, various curves of Spectral Irradiance were presented (with the units [W/m^2/nm]) and for different conditions: (a) different times of the day (one noted there, that in the mornings and in the evenings, there was comparatively more red light coming in than at noon); (b) for different weather conditions (one noted there that if there was snow and sun, there was comparatively more red light coming in than if one had a cloudy day); (c) for indoor lighting with various lamps (one noted there that for most LED lamps, there was comparatively more blue light coming in than if one had direct sunlight).

Thus, if the ambient conditions to which the solar cell or module is exposed changes, not only the intensity of the light is modified—even drastically reduced in

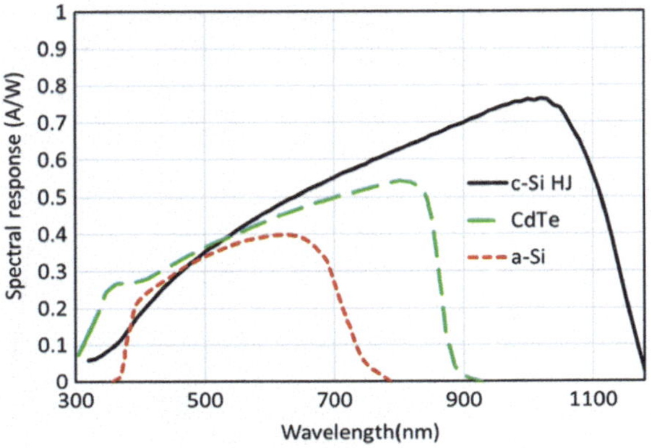

Fig. 3.28 Typical spectral response curves for solar cells: red: amorphous silicon; green: CdTe; blue: crystalline silicon. Courtesy of Mathieu Boccard, PV-Lab, IMT Neuchâtel, EPFL

the case of indoor lighting—this was treated in Sect. 3.5.3 above—but the **spectrum** of the incoming light is also modified: This effect will be treated here.

The correct tool to be used is the spectral response curve of the cell or module, but the EQE curve can also be used, because it contains the same information.

Figure 3.28 shows Spectral Response curves for three different types of solar cells: red: a typical amorphous silicon solar cell; green: a typical CdTe solar cell; blue a typical crystalline silicon solar cell.

From Fig. 3.28 one notes that amorphous silicon solar cells are spectrally well adapted for use in Indoor Light Conditions, where one finds, especially for certain types of LEDs (see Fig. 2.8 in Chap. 2 and look at the curve for "cool white LED") a preponderance of light in the range of wavelengths, between 400 and 600 nm. On the other hand, CdTe and c-Si solar cells and modules are spectrally well adapted for use in Outdoor Light Conditions.

To conclude this section, we can state that Spectral Response curves are generally used when evaluating the performance of modules in practical situations, in the "field", whereas EQE and IQE curves are the proper tools for analysing what is happening inside a given solar cell—they are mainly used for Research on Solar Cells.

To evaluate the efficiency of a solar cell/module, we have to consider both the spectral response curve SR (λ) of the solar cell, as well as the spectrum of the incoming light (see Chap. 2).

1. In outdoor applications, we usually have, on a sunny day, an AM 1.5-like solar spectrum at noon, whereas, we have a red shift of the spectrum (towards what is called an AM 3- or AM 4-like spectrum), during morning and evening. Snow and the vicinity of large water surfaces (lakes, large rivers, the sea) shift, on the other hand, the spectrum towards the blue. The latter (blue shift of incoming light) is also true for cloudy weather. This explains why crystalline silicon solar modules

have a relative advantage during mornings and evenings on sunny days, whereas CdTe modules have a relative advantage on cloudy days and in the vicinity of snow and large water surfaces.

2. In indoor applications, we often have, especially if fluorescent bulbs or light emitting diodes (LEDs) are used to illuminate the room, a spectrum with a larger blue content than the solar spectrum. This effect, together with the variation of efficiency with illumination level, explained in Sect. 3.5.3, underlines the decisive advantage of amorphous silicon solar cells for indoor applications.

References

1. R. Williams, Becquerel photovoltaic effect in binary compounds. J. Chem. Phys. **32**(5), 1505–1514 (1960)
2. E. Becquerel, Mémoire sur les effets électriques produits sous l'influence des rayons solaires. Comptes Rendus **9**, 561–567 (1839)
3. A. Einstein, Über einen die Erzeugung und Verwandlung des Lichtes betreffenden heuristischen Gesichtspunkt. Ann. Phys. **17**(6), 132–148 (1905)
4. A. Shah (ed.), *Thin-Film Silicon Solar Cells* (EPFL Press, Lausanne, 2010)
5. B.G. Streetman, S. Banerjee, *Solid State Electronic Devices*, 5th edn. (Prentice Hall, New Jersey, 1982)
6. S. Kasap, P. Capper (eds.), *Springer Handbook of Electronic and Photonic Materials* (Springer, New York, 2006), pp. 54, 327
7. Ioffe database, http://www.ioffe.ru/SVA/NSM/Semicond/
8. A.V. Shah, A.N. Tiwari (eds.), Thin-film photovoltaic solar cells. Sol. Energy Mater. Sol. Cells **119** (2013)
9. P.W. Baumeister, Optical absorption of cuprous oxide. Phys. Rev. **121**(2), 359 (1961)
10. K. Ellmer, H. Tributsch, Iron disulfide (pyrite) as photovoltaic material: problems and opportunities, in *Proceedings of the 12th Workshop on Quantum Solar Energy Conversion (QUANTSOL 2000)*
11. A. Shah, J. Meier, A. Buechel, U. Kroll, J. Steinhauser, F. Meillaud, H. Schade, D. Dominé, Towards very low-cost mass production of thin-film silicon photovoltaic (PV) solar modules on glass. Thin Solid Films **502**, 292–299 (2006)
12. M.A. Green, *Solar Cells, Operating Principles, Technology and System Applications* (Prentice Hall, New Jersey, 1982)
13. H.J. Möller, *Semiconductors for Solar Cells* (Artech House, Norwood, Massachusetts, 1993)
14. J. Merten, Photovoltaics with amorphous silicon: aspects of technology, physics and application, Ph.D. thesis, University of Barcelona (1996)
15. J. Merten, J.M. Asensi, C. Voz, A. Shah, R. Platz, J. Andreu, Improved equivalent circuit and analytical model for amorphous silicon solar cells and modules. IEEE Trans. Electron Devices **45**, 423–429 (1998)
16. J. Merten, J. Coignus, G. Razongles, D. Muñoz, Novel equivalent circuit for heterojunction cells and diagnostic method based on variable illumination measurements (VIM), in *Proceedings of the 27th European Photovoltaic Conference* (2012), pp. 1268–1271
17. M.A. Green, E.D. Dunlop, D.H. Levi, J. Hohl-Ebinger, M. Yoshita, A.W.Y. Ho-Baillie, Solar cell efficiency tables (Version 54). Prog. Photovolt. Res. Appl. **27**, 565–575 (2019)
18. W. Shockley, H.J. Queisser, Detailed balance limit of efficiency of p-n junction solar cells. J. Appl. Phys. **32**, 510–519 (1961)
19. H. Kiess, W. Rehwald, On the ultimate efficiency of solar cells. Sol. Energy Mater. Sol. Cells **38**, 45–55 (1995)

20. F. Meillaud-Sculati, Microcrystalline silicon solar cells: theory, diagnosis and stability, Chap. 3, Ph.D. thesis, "Faculté des Sciences" of the University of Neuchâtel (2006)
21. S. Rühle, Tabulated values of the Shockley-Queisser limit for single-junction solar cells. Sol. Energy **130**, 139–147 (2016)
22. E. Maruyama, A. Terakawa, M. Taguchi, Y. Yoshimine, D. Ide, T. Baba, M. Shima, H. Sakata, M. Tanaka, Sanyo's challenges to the development of high-efficiency HIT solar cells and the expansion of HIT business, in *Conference Record of the 2006 IEEE 4th World Conference on Photovoltaic Energy Conversion* (IEEE), pp. 1455–1460
23. A. Virtuani, D. Pavanello, G. Friesen, Overview of temperature coefficients of different thin film photovoltaic technologies, in *Proceedings of the 22nd EU PVSEC Conference* (Milano, 2007)

Arvind Shah is the Founder of the Photovoltaics Research Laboratory (PV-Lab), at the Institute of Microtechnology (IMT), in Neuchâtel, Switzerland.

PV Lab Neuchâtel has done pioneering work in the establishment of low-cost production methods for solar cells based on silicon: It introduced a novel plasma-assisted deposition method called "VHF deposition" permitting a significant increase in the deposition rate for thin-film silicon layers. It also introduced microcrystalline silicon, deposited by VHF plasma, and with very low oxygen content, as novel absorber layer, within thin-film solar cells.

From 1979 to 2005, Arvind was Professor at the University of Neuchâtel. From 1987 to 2005, he was additionally part-time professor at the EPFL Lausanne.

In 1975 he founded and co-directed the Centre for Electronics Design and Technology (CEDT) at the Indian Institute of Science in Bangalore. CEDT is now one of India's leading University Centres in the field of Electronics.

Since 2006, Arvind has been active as scientific consultant to the PV Lab and to various Industries, in Europe, India and the USA.

Arvind received the Swiss Solar Prize, together with Johannes Meier in 2005. He received the Becquerel Award in 2007.

Chapter 4
Solar Cells: Optical and Recombination Losses

Sylvère Leu and Detlef Sontag

Abstract In Sect. 4.1 on Optical Losses, we first look at the absorption of light in a semiconductor, distinguishing thereby between semiconductors with direct transition (direct bandgaps), like GaAs, CdTe etc. and semiconductors with indirect transitions (indirect bandgaps) like silicon. Thereafter we introduce various methods for achieving trapping of the light within the semiconductor, such as (a) the introduction of an ARC (antireflection coating layer), (b) texturing of the front and the back surfaces, (c) mirror formation on the back side of the solar cell. In Sect. 4.2 on Recombination, we will introduce the four main types of recombination processes (a) radiative recombination, (b) Shockley-Read-Hall (SRH) recombination, (c) Auger recombination and (d) surface recombination. In Chaps. 5 and 7 we will apply this basic information to design highly efficient solar cells.

4.1 Optical Losses

The losses of a solar cell can be divided into three categories:

1. Optical losses
2. Losses due to recombination
3. Ohmic losses.

In this chapter, we cover the basics of optical losses and recombination losses. Ohmic losses occur mainly when individual solar cells are assembled into entire modules; they will find application in Chaps. 9 and 10.

S. Leu · D. Sontag (✉)
Meyer Burger Technology A.G., Gwatt, Switzerland
e-mail: dsontag@web.de

S. Leu
e-mail: sylvere.leu@ciptec.ch

A. Shah (ed.), *Solar Cells and Modules*, Springer Series in Materials Science 301,
https://doi.org/10.1007/978-3-030-46487-5_4

4.1.1 Preliminary Remarks: Reflection, Refraction, Absorption and Transmission

In Chap. 3, we learned that electricity is generated when a photon is absorbed by a solar cell. The basic materials for solar cells are semiconductors. Semiconductors are characterized by having an energetic bandgap (forbidden band or forbidden zone) between the valence band and the conduction band. By absorbing photons, electrons can be lifted from the valence band into the conduction band. The condition for this is that the photon energy is equal to or greater than the bandgap energy of the semiconductor.[1] If this condition is fulfilled, positively and negatively charged electric charge carriers can be generated by the incoming sunlight. The conductivity of the semiconductor plays an important role in the separation of the charge carriers and letting them flow out of the solar cell. Doping can increase the conductivity. In this Sect. 4.1 we want to learn how we can absorb as many photons as possible.

Light that strikes the surface of a material will partly be reflected; the rest of the light will penetrate into the material. The light, which penetrates into the material, is refracted. By "refraction" one designates the bending of the light rays when they enter into a material. Light rays are refracted at each interface as they transit from one material, to another material that has different optical properties. In photovoltaics, one has very many interfaces, between regions, which have different optical properties—either because they are doped in a different way or because their chemical composition is different.

These are the 4 basic phenomena in Optics: **reflection, refraction, absorption and transmission**. Newton already thought about the phenomenon of partial reflection. Indeed, at the boundary (interface) between two different materials, it is not clear how a photon can "decide" whether to penetrate into the second material or to be reflected back into the first material. Quantum Physics give us the answer: it tells us that we can only state the probability with which photons are reflected. We do not know which photon will be reflected and which one will be transmitted. But we know that always the same proportion of photons is reflected; we also have precise data on the optical properties of different materials.

Another interesting phenomenon of optics, which is important to us, was found by Pierre de Fermat: When light is refracted at a boundary layer or "interface", it will always take the fastest path, not the shortest path. This property also gives rise to puzzles: How can the light at the boundary layer "know" which path is the fastest one? And how can the photons "know" which one of them should be reflected and which one may penetrate into the next material, so that always the same proportion of photons are reflected? Quantum Physics provides the following explanation: the light simply "tries out" all possible ways at the same time. All alternative paths (i.e. all paths except the one that is selected) are cancelled out by destructive overlapping.

[1]The bandgap of a silicon crystal is 1.12 eV.

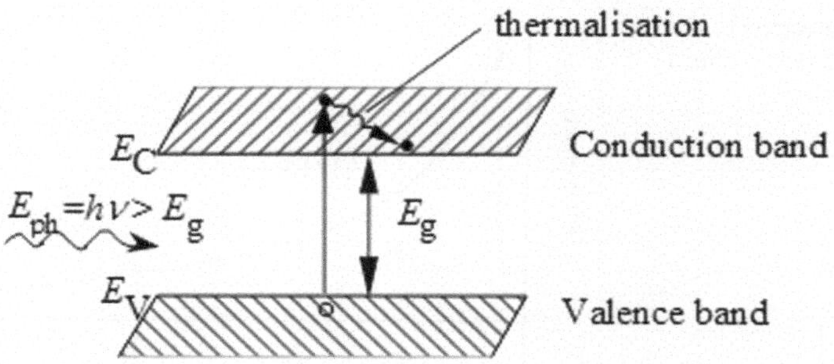

Fig. 4.1 Illustration of the photo-generation of an electron-hole pair according to Sect. 3.2.1

The above questions led to the refraction laws, which were published mainly by Willebrord van Roijen Snell, in 1621. Augustin Jean Fresnel (1788–1827) contributed essential insights into the wave character of light. Both of these distinguished researchers of "past times" will assist us with the question of light trapping within solar cells.

Per m^2 and second, 10^{21} photons hit the Earth from space with an energy between 1 and 5 eV. It is now important to absorb as many of these photons as possible; and to convert a large part of them into electrical energy. This is described in the following sections.

4.1.2 Absorption

Absorption of light in a solar cell means that a photon is absorbed in the semiconductor and gives off its energy to create an electron-hole pair. Thanks to the energy of the photon, a bound electron, which is closely attached to a silicon atom, is released and becomes a "free electron". In semiconductor physics, one says that the electron has been moved from the valence band to the conduction band (see Fig. 4.1). Thereby a "free hole[2]" is left behind in the valence band. Thus, a pair of one "free electron" and one "free hole" is created. This only happens if the energy of the photon is greater than the bandgap energy E_g. If the photon energy is too small, the photons pass unimpeded through the silicon crystal and the energy of the photon is lost for the solar cell. This happens because photons with lower energy cannot produce electron-hole pairs. For such photons, the semiconductor is virtually transparent.

Even with photon energies higher than the bandgap energy E_g, not all photons are immediately absorbed near the surface; in fact, most of them penetrate deeper into the solar cell. The absorption coefficient α determines the penetration of light within the silicon crystal. Here, α is a function of the wavelength λ of the light. As

[2] A "hole" is simply the absence of an electron, where originally there was one. According to semiconductor physics, holes behave just as if they were elementary particles themselves.

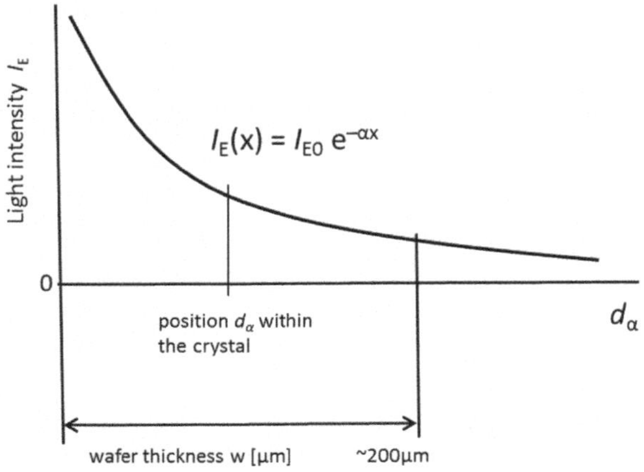

Fig. 4.2 Illustration of the absorption of light within a silicon crystal

illustrated in Fig. 4.2, the light intensity $I_E(x)$ decreases exponentially within the material at the position x. One has the following function for the light intensity I_E:

$$I_E(x) = I_{E0}e^{-\alpha x} \tag{4.1}$$

The absorption depth d_α indicates how deep light of a specific wavelength λ penetrates into the material, before its intensity has fallen to $1/e$, e.g. $\approx$ 36% of its original intensity.[3] In silicon (and in most other semiconductors used for solar cells), d_α increases for increasing wavelengths λ. For light with a wavelength $\lambda = 575$ nm, the absorption depth d_α is 1 µm and for $\lambda = 980$ nm d_α is already 100 µm. At longer wavelengths there is, thus, the danger that some of the photons leak out from the back side of the solar cell.

For semiconductors with direct transitions (like GaAs, CdTe, etc.), an electron, which absorbs the energy of a photon, does not need to change its momentum.[4] The crystal structure is formed in such a way that in the diagram of Energy versus

[3] $e \approx 2.71828\ldots$ is a mathematical constant called "Euler's number"—it is the base of the natural logarithm.

[4] In contrast to energy, the momentum has an amount and a direction. In quantum physics, light, electrical current and mechanical vibration are all represented by "quanta" or "elementary particles".

- for light: photons
- for electrical current: electrons
- for mechanical vibration: phonons.

These particles are not only characterized by their energy E but also by their momentum P.

For a transition, i.e. from valence band to conduction band, it is necessary to consider not only energy but also momentum. In a rough approximation one can say: electrons have both energy and momentum; photons have energy but zero momentum; phonons have very little energy, but considerable momentum.

Since the momentum of a photon is zero it cannot change the momentum of an electron-hole-pair. This requires the additional vibrational energy of the crystal lattice, which is transmitted by a phonon.

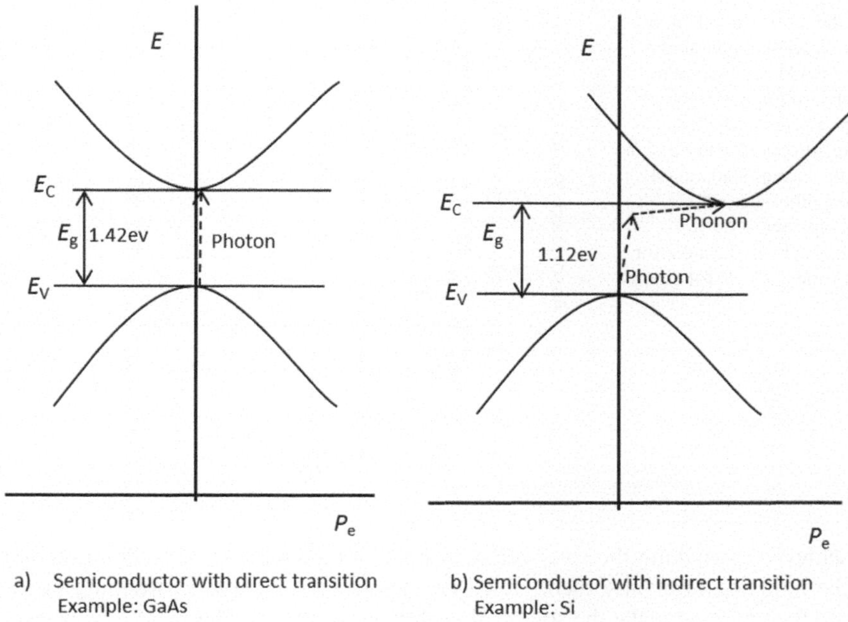

Fig. 4.3 Diagram of Energy versus Momentum for electrons in **a** semiconductors with direct transitions and **b** semiconductors with indirect transitions: the energy level E is plotted on the y-axis and the momentum P_e is shown on the x-axis. E_C stands for the lowest energy level of the conduction band and E_V for the highest energy level of the valence band

Momentum, the minimum of the conduction band lies directly above the maximum of the valence band. This is illustrated in Fig. 4.3a.

On the other hand, in semiconductors with indirect transitions (like silicon), the crystal structure is formed in such a way that in the diagram of Energy versus Momentum, the minimum of the conduction band does not lie above the maximum of the valence band. In a semiconductor with an indirect transition, the electron must change its momentum (Fig. 4.3b). This is only possible with the help of a phonon[5] [1].

Thus, it is also understandable that in a direct semiconductor the absorption coefficient α increases very steeply in function of the wavelength as soon as the band energy E_g is reached. In contrast, in an indirect semiconductor, the absorption coefficient does not increase so steeply. This is because, for the absorption of a photon, a "detour" via third particles, namely phonons, must be carried out. The situation is shown in Fig. 4.4.

In order for a photon to be able to produce an electron-hole pair with high probability, the optical path through the silicon wafer must be long enough. This can

[5]A "phonon" is in quantum physics the elementary particle describing a mechanical vibration. (Here, the mechanical vibration within the semiconductor crystal we are looking at—for example, silicon.) In a similar way, a "photon" is the elementary particle describing light, and an "electron" is the elementary particle describing electric current.

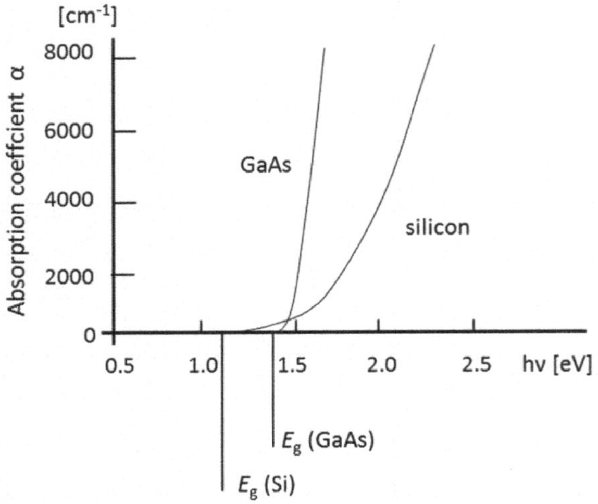

Fig. 4.4 Absorption curve for a direct semiconductor (here: GaAs) compared to the absorption curve for an indirect semiconductor (here: silicon). It can be seen that the curves start exactly at the point where the bandgap energy E_g is reached. This is 1.44 eV for GaAs and 1.12 eV for silicon

be achieved by making the solar cell very thick. At a thickness of 10 mm (almost) all the light would be absorbed. However, material costs are a constraining factor and so the aim is to make the solar cell as thin as possible. Today's solar cells have thicknesses of 160–180 μm with a wafer size between M2 (156 × 156 mm^2) and M6 (166 × 166 mm^2); the thinner the solar cell becomes, the more important it is to increase the absorption of sunlight in ways other than thickness. There are basically three possibilities:

1. Front side: reducing reflection with optimized surface coating
2. Texturing of the front surface
3. Passivation of the back surface; mirror formation at the back.

4.1.3 Front Side: Avoiding Reflection with Optimized Surface Coating

Upon perpendicular incidence, an untreated silicon surface reflects back about 35% of the incident sunlight. This can be easily deduced from the Fresnel equations.

$$R = ((n_1 - n_2)/(n_1 + n_2))^2 \tag{4.2}$$

n_1 refractive index of air ≈ 1
n_2 refractive index of silicon ≈ 3.9

The refractive index n is a dimensionless number. It is a measure of how much the speed of light is reduced in the material we are considering. In vacuum the speed

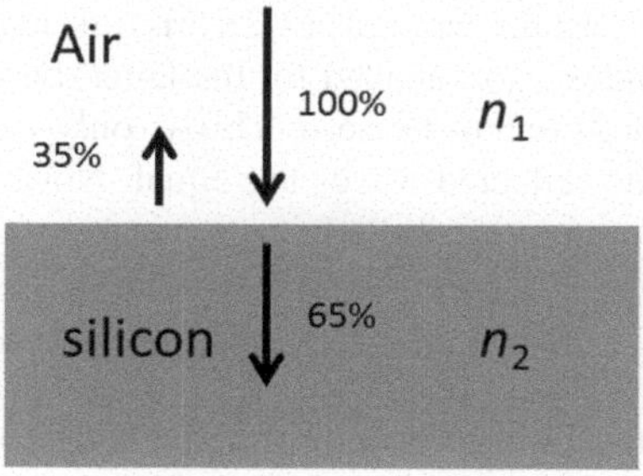

Fig. 4.5 Air-silicon interface: 35% of the incoming light is reflected

of light is $c = 299{,}792\ \mathrm{km\ s^{-1}}$. The refractive index n for vacuum is exactly 1. The refractive index n_{material} for a given material is consequently:

$$n_{\mathrm{material}} = c/c_{\mathrm{material}} \tag{4.3}$$

c	speed of light in vacuum
c_{material}	speed of light in the material considered

In a material that is optically denser than vacuum, the speed of light decreases. Just as in vacuum the refractive index n is exactly 1, the refractive index n_{air} for air is ≈ 1, in a very good approximation. For silicon it is $n_{\mathrm{silicon}} \approx 3.9$. Substituted into (4.2), the reflectance R at the silicon-air interface is given by:

$$R = \left(\frac{(1-3.9)}{(1+3.9)}\right)^2 = 0.35 = 35\% \tag{4.4}$$

This means that silicon reflects 35% of sunlight. This applies to light falling vertically onto the solar cell, as illustrated in Fig. 4.5. If the light falls diagonally onto the solar cell, the reflection is even higher and even more sunlight is returned unused. In order to reduce the reflectance, we have to process the solar cell surface. In optics, this step is also called "applying an anti-reflective coating" (ARC). Ideally, we put a thin layer on top of the solar cell, so that the incident and reflected light waves cancel out. We do this by:

1. Adjusting the thickness of the additional thin layer, by adapting it to the wavelength of the incident sunlight[6] so that the desired phase difference of the incident and reflected light results in a destructive interference, and
2. Choosing the refractive index of the additional layer to lie between the refractive indices of the two adjacent materials—in our case air and silicon.

---

[6]The incident sunlight does not have only a single wavelength. Rather, it has a whole spectrum of wavelengths (see Chap. 2). This question will be addressed later, in Sect. 4.1.4: We will look for the time being only at the "dominant" wavelength—the wavelength where the solar spectrum has a maximum—this is the case at 575 nm (if one takes the absorption of the silicon solar cell also into account).

Again, we apply (4.2); this time to the ARC layer in order to calculate the refractive index n_{ARC} needed for the latter. For complete absorption, the refractive index R in (4.2) has to be zero. This is only possible if the phase shift and the amplitude of the reflected waves are equal. Now, this will happen (see [2]) when the following formula is fulfilled:

$$n_{\text{ARC}} = (n_1 n_2)^{1/2} \tag{4.5}$$

n_{ARC} is the desired refractive index of the ARC layer.

$$n_{\text{ARC}} = (1 \times 3.9)^{1/2} = 1.97 \tag{4.6}$$

Now, it's all about finding a material that has the desired value of refractive index n_{ARC}. Silicon nitride with a refractive index in the range 1.9–2.1 comes very close to having the required value. In addition, depending on the deposition technology chosen, silicon nitride may contain hydrogen. Due to the small size of hydrogen atoms, hydrogen diffuses very rapidly into silicon and can passivate impurities and recombination centres; thereby reducing recombination losses in silicon (see Chaps. 5 and 7). Previously, SiO_x ($n = 1.3$) and TiO_2 ($n = 2.52$) were used, both of which have a refractive index, which is quite far from the required "ideal" value. The coating with silicon nitride fulfils, thus, two functions: it serves as an antireflection coating (ARC) and as a passivation layer.

In the next step we have to determine—in a more precise way—the thickness of the ARC layer. Most of the incoming light on Earth has a wavelength of 550–600 nm, which is in the green part of the solar spectrum (see Chap. 2, Fig. 2.1 and Chap. 3, Fig. 3.6). Silicon, in turn, is only transparent to light with a wavelength larger than 1127 nm, due to its bandgap of $E_g = 1.12$ eV. So, we can optimize the layer thickness for a light wavelength of 575 nm, because it is for a wavelength of 575 nm that we expect the largest quantity of photons, which can be used with a silicon solar cell.

As illustrated in Fig. 4.6, the light waves are reflected at the interface A between air and the antireflective coating—ARC (r_1) and at the interface B between the antireflective coating (ARC) and the substrate (r_2). Thus, the light wave r_2 makes a longer optical path, since it also has to pass twice through the ARC layer. In order to achieve destructive interference and, thus, minimize reflection, the phase difference in the ARC layer must correspond to half a wavelength. Since the ARC layer is passed through twice, its thickness must be—in the simplest form—one quarter of the wavelength.[7] Taking into account the refractive index n_{ARC} of the antireflective coating (ARC) we obtain

$$d = \lambda/(4n_{\text{ARC}}) \tag{4.7}$$

d geometric layer thickness of the ARC

[7] It could also be an odd multiple of one quarter of the wavelength, i.e. $(3\lambda/4)\, n_{\text{ARC}}$ or $(5\lambda/4)\, n_{\text{ARC}}$ etc.

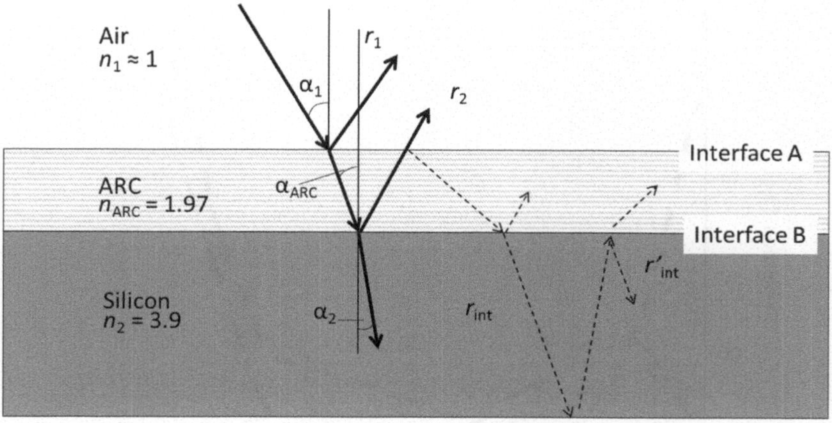

Fig. 4.6 Schematic illustration of the function of an ARC layer. When the refractive index increases from the first material to the second one (e.g. from air to ARC or from ARC to silicon), the light is refracted into the material, e.g. the angles to the vertical become smaller. At the two interfaces A and B, a phase jump of π occurs for the reflected light. (The light rays r'_{int}, which are reflected back into the solar cell at the various layer interfaces are not considered here and are neglected in our calculations.)

λ wavelength
n_{ARC} refractive index of the ARC

In our case, the silicon nitride ARC layer should have a thickness

$$d = 575\,\text{nm}/(4 \times 1.97) = 73\,\text{nm} \tag{4.8}$$

For obliquely incident light (Fig. 4.6), the wavelength leading to destructive interference can be calculated by Snell's law

$$n_1 \sin \alpha_1 = n_{ARC} \sin \alpha_{ARC} \quad \text{and} \quad n_{ARC} \sin \alpha_{ARC} = n_2 \sin \alpha_2 \tag{4.9}$$

Equation (4.9) finally gives the thickness of the ARC layer d as a function of the angle α_1 at which the light hits the solar cell. The derivation can be found in [2]:

$$\lambda = 4d(n_{ARC}{}^2 + n_1 \sin 2\alpha_1)^{1/2} \tag{4.10}$$

where α_1 is the angle between the incident sunlight and the vertical axis. The effect of different ARC layers on the reflection of light is illustrated in Fig. 4.7.

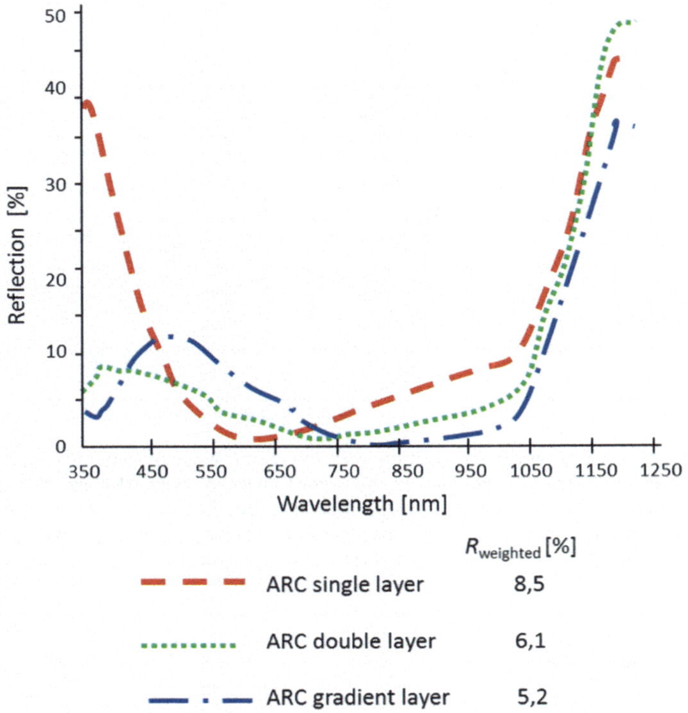

Fig. 4.7 Reflection as a function of wavelength for three different cases of ARC layers: **a** single layer; **b** double layer; **c** gradient layer. (Gradient layers: the deposition process is carried out in such a way that the transition between the two different layers is continuous [3].) Note that today single layers are the ARC layers which are most used. The reflection $R_{weighted}$ is calculated as follows: determine the reflectance R at the considered wavelength (as given by the graph) and multiply it by its share of intensity within the AM 1.5 spectrum. Then integrate over all relevant wavelengths

4.1.4 Additional Considerations: How to Increase Light Trapping Further

1. **Texturization**: A further method to improve light trapping in solar cells is to structure the front and the back surfaces of the solar cell, giving them a suitable texture as will be described in Sect. 5.1.6.
2. **One-axis or two-axis trackers**: The angle of the incidence of sunlight determines the optimal thickness of the ARC layer according to (4.9). If modules are installed on a fixed mounting rack, the angle of incidence of sunlight changes during the day. Since the ARC layer itself and also the whole solar cell is fixed, the light yield is not always optimal. Therefore, single-axis and two-axis trackers are used to optimize the effectiveness of light collection. With single-axis trackers, approx. 10% and with two-axis trackers approx. 20% more energy can be generated over

the year compared to the situation, where the modules are mounted with a fixed angle.

3. **Double layer coating**: If the reflections are to be further reduced over a wide range of wavelengths and angles, it is necessary to stack several layers with different refractive indices and thicknesses, one above the other. Figure 4.7 illustrates the reflection of different ARC stacks. Basically, an attempt is made to keep the reflectance low at the wavelength of light where the greatest amount of light energy hits the earth. This means the reflection should be low between 550 and 900 nm (see Chap. 2, solar spectrum). The double layer for which data is given in Fig. 4.7 uses two different ARC layers SiN_x and SiO_x. Double layers are used in newer cell concepts (see Chap. 5, Sect. 5.4.2).

4.1.5 Wafering and Its Effects on Light Trapping

Texturing is one of the most important ways to improve light trapping. The wafering process has a significant influence on the texture and, thus, on the efficiency of the solar cell. Therefore the different wafering processes will now be described in more detail here. In our discussion here we start out with silicon ingots,[8] which are cut into thin wafers with a thickness of 120–200 μm. Today two processes are used in mass production: the slurry process and the cutting process with diamond wire. Figure 4.8 illustrates the two processes.

1. **Slurry process**: In the slurry process, a brass coated steel wire with a diameter of ~100 μm pushes silicon carbide grains, which are about 20–30 μm in size, through the ingot. The silicon carbide (SiC) grains are harder than silicon. The slurry itself is a liquid containing oil, glycol and SiC grains. This process was mainly used for multicrystalline material until 2017. Process time is about 5 h.[9] The process is illustrated in Fig. 4.8a.
2. **Diamond wire**: In the diamond wire process, a steel wire with a diameter of ~60 μm is covered with small diamonds, which have a size of 5–20 μm, and which are attached to the wire by means of a nickel layer. This wire saws the silicon ingot into wafers. This process is mainly used for monocrystalline material. It is very efficient, process time is 1–2 h, saw damages are less pronounced and kerf loss is smaller than in the slurry process. In the diamond wire process, the wire is run in the so-called Pilgrim Mode.[10] In this way, the service life of the wire can be increased. Wire consumption is about 0.5 m/wafer. The process is

[8]Manufacturing of ingots and silicon crystals will be described in Chap. 5.

[9]In semiconductor industry this type of wafering is still used.

[10]In Pilgrim mode, the diamond wire is fed forward and backward through the silicon, similar to cutting wood with a hand saw. In the forward cut one feeds more wire than in the backward cut. In this way, fresh wire is added. In forward cutting, for example, 600 m of wire is used for cutting and in reverse cutting less wire is used, for example 580 m of wire. So 20 m of fresh wire is fed per cycle into the cutting process.

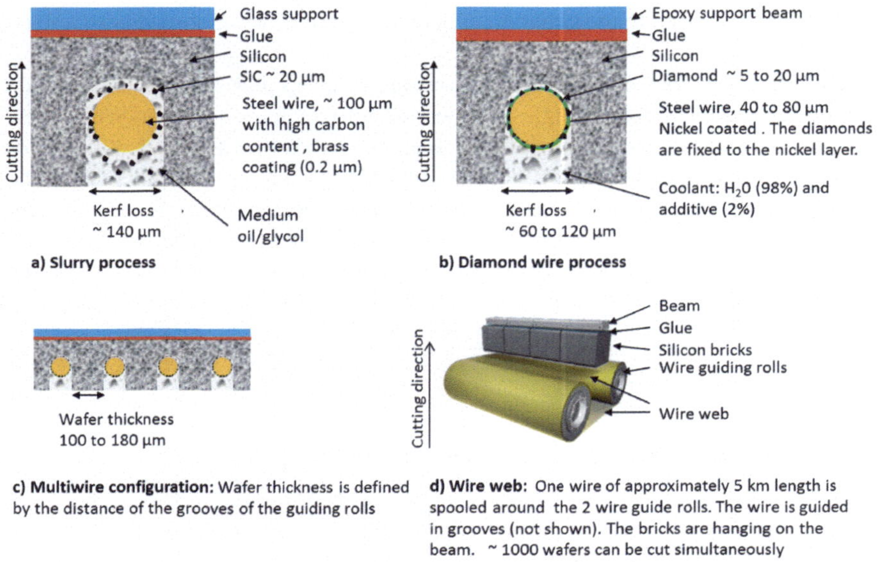

Fig. 4.8 Illustration of **a** the slurry process; **b** the diamond wire process; **c** multi wire configuration; **d** wire web. Courtesy of Meyer Burger Technology AG

illustrated in Fig. 4.8b, c, d. The kerf loss (Fig. 4.8a, b) depends on wire diameter and the size of the SiC grains (Slurry) or the size of diamonds. The glycol/oil in the slurry process and the coolant additives in the diamond cutting process are recycled. The silicon from the kerf can only be recovered in contaminated form and it is therefore not yet economically worthwhile to recycle it. The wire is fed from a first roll (not shown in Fig. 4.8d) onto the so-called wire web consisting of two rolls (Fig. 4.8d) and after the cutting process it is rewound onto a fourth roll (not shown in Fig. 4.8d).

Only since 2015, did one learn how to cut **multicrystalline** material with the diamond wire process. However, unlike the slurry cut wafer surface, the wafer surface cut by diamond wires in this way cannot be textured using traditional wet-chemical methods. New texturing processes such as plasma texturing or metal catalytic texturing must be used.

4.1.6 Texturing of the Front Surface

If the surface of monocrystalline silicon wafers is treated in an alkaline etching bath, one obtains an anisotropic etching effect (see also Chap. 7); this process depends on the crystal orientation.[11] For {100} oriented monocrystalline silicon material,

[11]Details on the crystal orientations can be found in Chap. 7.

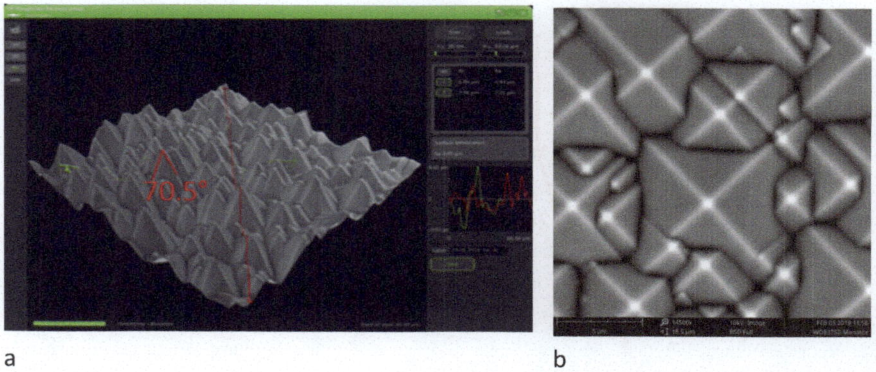

Fig. 4.9 Surface of a monocrystalline wafer after texturing **a** side view; **b** top view. Courtesy Meyer Burger Technology AG

pyramids with a size of 1–7 μm can be etched out at an angle of ~35° to the vertical axis. Figure 4.9 shows the surface of a monocrystalline, textured wafer.

With a refractive index for the ARC layer of 2, we obtain a refraction angle of 24°, if we solve (4.9) for β_2

$$n_1 \times \sin(\beta_1) = n_2 \times \sin(\beta_2) \rightarrow \beta_2 = \arcsin(n_1/n_2 \times \sin(\beta_2)) \\ = \arcsin(0.5 \sin(55^\circ)) = 24^\circ \qquad (4.11)$$

with:

n_1 Refractive index of air ≈ 1
n_2 Refractive index of ARC ≈ 2
β_1 Angles of the pyramids in the textured layers to the vertical:

$$70.5^\circ/2 \approx 35^\circ \quad \beta_1 = 90^\circ - 35^\circ = 55^\circ$$

β_2 Refraction angle of ARC

Vertically incident light strikes the pyramidal texture at an angle of 35° (resp 55°) (see Fig. 4.10). The first part 1 namely about 30%, is reflected, also at 35° (angle of incidence = angle of reflection) and impinges on an adjacent pyramid, whereas the second part 2, namely 70%, is refracted into the antireflexion coating (ARC) at an angle of ~24° and then enters the solar cell at an angle of ~12°. The light that is refracted is, in the ideal case, thrown back into the solar cell from the back side and hits the texture from below. In this way, the optical length of the light is increased and with it, the probability of generating electron-hole pairs.

For multicrystalline wafers, pyramids cannot be etched because the crystal structure is not aligned, as in a monocrystal. Attacking and roughening the surface with

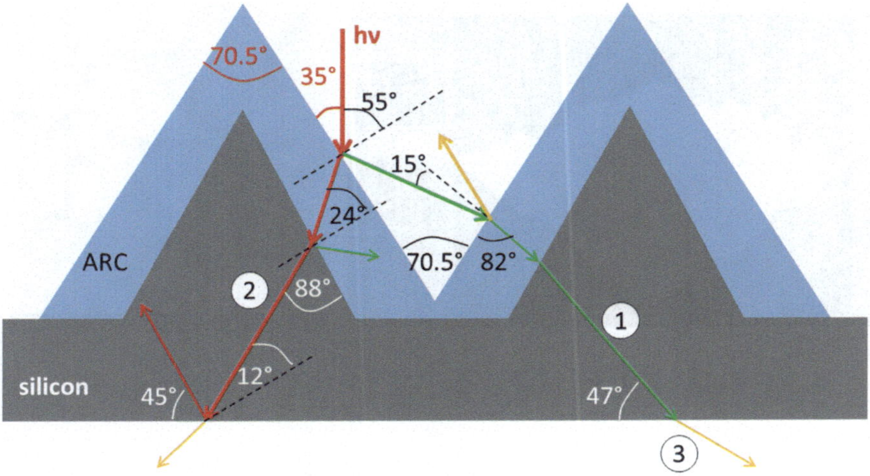

Fig. 4.10 Illustration of light trapping for a textured monocrystalline cell. Glass and encapsulation material are not taken into account

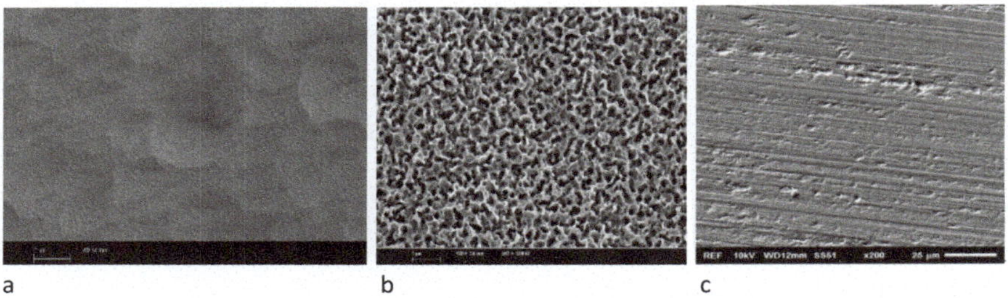

Fig. 4.11 **a** Surface of a multicrystalline cell, cut with **slurry**, after acidic texturing; **b** crystal structure of a multicrystalline wafer cut with **diamond wire** after plasma texturing; **c** multicrystalline wafer cut with diamond wire and wet-chemically textured with traditional texture. It can be seen that the wetting liquid cannot attack the wafer surface and light trapping cannot take place. The wafer surface is rather flat and not rough as shown in figures (**a**) and (**b**) this is the reason for a reflective property, which is too high, and, thus, for a poor cell efficiency. Courtesy Meyer Burger Technology AG

an acidic solution as shown in Fig. 4.11a leads to the roughening of the surface of multicrystalline wafers that are cut with slurry. In this way one obtains a spongy structure with steep flanks. Due to the steep flanks the surface area can be increased, the reflections can be reduced and an increased penetration of light into the solar cell becomes possible.

If the multicrystalline wafer is not cut with slurry as noted above, but with **diamond wire**, the wafer can be cut with less cost. But the wafers cut in this way can no longer be textured by a wet-chemical method, and light trapping cannot be realized in this simple way. Plasma texturing is a possible alternative etching method for such wafers, as well as metal-catalytic processes.[12] Figure 4.11b shows the surface of a multicrystalline cell cut with diamond wire and texturized by plasma. Thanks to these newer methods, diamond-wire-cut, multicrystalline wafers can be processed into solar cells, whose cell efficiency is higher than those of slurry-cut wafers, whilst at the same time the manufacturing costs are reduced.

4.1.7 Passivation of the Back Surface and Mirror Formation at the Back

On the back we can apply two improvements:

- Passivation and light reflection with a metal paste (aluminium paste)
- Texturing the back side.

At the back of the solar cell, a reflector is used. Thus, the light that travels through the cell is reflected there and the optical path is doubled. The light, thus, receives a second chance to be absorbed in the silicon crystal. The back side of a solar cell has to be passivated in all cases, i.e. the recombination centres formed by the surface have to be deactivated; otherwise there is a substantial reduction in current and efficiency (see Chap. 5). The metallization pastes (aluminium pastes) one uses to form electrical contacts enhances the reflection at the back. In this case, a clever arrangement of dielectric layer stacks can achieve practically total reflection and, thus, an "ideal" optical mirror can be implemented (see Chap. 5).

If the back surface itself is textured or roughened, one should ideally obtain a Lambertian reflector. With this type of reflector it is possible to increase the optical path by a factor of

$$4 \times {n_{\text{silicon}}}^2 = 4 \times 3.92^2 \approx 50 \tag{4.12}$$

If the cell thickness is 200 μm, then the Lambertian reflector will result in a 10% increase of the generated electric power. If the cells are very thin, then the increase in electric power is much higher: As an example, if the cell thickness is 10 μm, then the power yield can be almost doubled, thanks to a Lambertian back reflector. Of course, there are no materials, whose surface fully complies with Lambert's Law. An example for partial implementation of the Lambertian effects is given by texturing the back side of solar cells (as e.g. in bifacial HJT cells).

Now, if we have absorbed as much light as possible and created electron-hole pairs, the goal is for the electrons and holes to reach the corresponding metal contacts on the surface of the solar cell, enter from there into the external power circuit and contribute to the generation of electricity for the benefit of an external consumer. For the electrons (and holes) to reach their goal at all, they have to "live" long enough.

---

[12]Silver ions are used here which adhere to the silicon surface, so that the etching liquid has points of attack to structure the surface.

In that way, they reach the metal contacts without recombining. The speed at which the charge carriers (electrons and holes) travel is determined by their mobility μ. The average time until they recombine is given by their lifetime τ. The quantity governing the travel of charge carriers is the diffusion length. The diffusion length L_{diff} is given by

$$L_{\text{diff}} = (D \times \tau)^{1/2} \text{ with } D = \frac{q}{k_B T}\mu \tag{4.13}$$

L_{diff} diffusion length (see also Chap. 3)
μ mobility, D diffusion constant, k_B Boltzmann constant, q charge of the electron, T absolute temperature
τ lifetime

4.2 Recombination Losses

4.2.1 General Concepts

When we illuminate a solar cell, more and more electrons are energetically lifted into the conduction band and become available for the generation of an electric current. They "jump over" the forbidden zone. This process is called "photo-generation". Now, the question arises: how do the electrons come back into the valence band? The semiconductor has a high conductivity under illumination. As soon as we switch off the light, the semiconductor loses its increased conductivity again. The electrons, thus, leave the conduction band after a certain time and recombine with the holes in the valence band. The time between generation and recombination we call the lifetime τ. The holes also have a lifetime.

The lifetime is defined by the equation

$$\tau(\Delta n) = \Delta n / (R(\Delta n)) \tag{4.14}$$

τ lifetime
Δn excess carrier density
R recombination rate

This equation indicates how long the excess carriers Δn exist before they recombine again. The recombination process is governed by the recombination rate R.

For "**direct-bandgap**" semiconductors like GaAs, see Sect. 4.1, this is relatively fast, because after the thermalisation, e.g. after the release of surplus energy to the crystal lattice, the electron jumps directly into the valence band by skipping over the forbidden zone and recombines there with a hole. The momentum of the electron does not have to be changed.

For an "**indirect-bandgap**" semiconductor such as silicon, this is not so easy. In this case, the momentum of the electron must also be changed. Here, a third particle is needed, for example, a phonon from the crystal lattice,[13] which provides the necessary change of momentum. This process has a low probability because now three particles have to interact with each other. Therefore, the lifetime τ of the carriers of "indirect-bandgap" semiconductors is larger than the lifetime for carriers of "direct-bandgap" semiconductors. This is advantageous for silicon solar cells. On the other hand, the mobility of charge carriers is greater in some direct semiconductors like GaAs. Nevertheless, the diffusion length, defined as $\sqrt{D\tau}$ (square root of lifetime times diffusion constant), according to (4.13), is only 30–50 μm for GaAs and approximately 1200 μm for Si.

4.2.2 Recombination

As soon as an electron from the conduction band unites with a hole in the valence band, recombination takes place. We distinguish here between four different types of recombination:

1. Radiative recombination
2. Shockley-Read-Hall recombination
3. Auger recombination
4. Surface recombination

1. ***Radiative Recombination***

In radiative recombination, when an electron recombines with a hole, the released energy is emitted as a photon; this is illustrated in Fig. 4.12.

Therefore, it is the reverse process of absorption. This process is more pronounced, the higher the concentrations of electrons and of holes are:

$$R_{\mathrm{rad}} = B\left(np - {n_{\mathrm{i}}}^2\right) \tag{4.15}$$

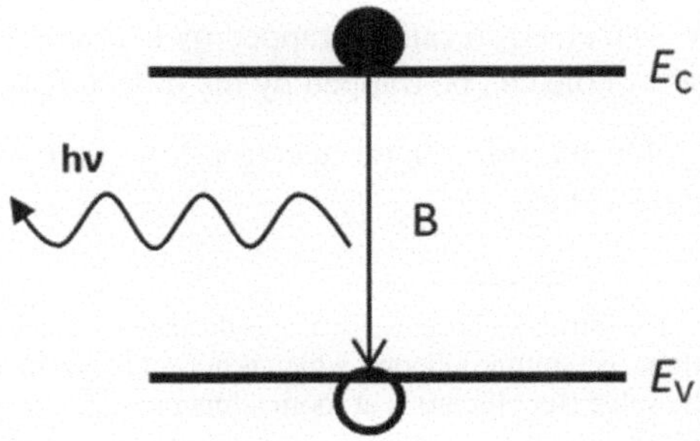

Fig. 4.12 Illustration of the process of radiative recombination. An electron gives its energy during recombination to a photon with the energy $E = \mathrm{h}\nu$ [4]

[13] The crystal lattice is the regular three-dimensional arrangement of the atoms.

n_i denotes the intrinsic density of charge carriers[14] and is in Standard SI units[15]: 10^{10} cm^{-3} for silicon

n, p are the concentrations of electrons and of holes

for electrons: $n = n_0 + \Delta n$
for holes: $p = p_0 + \Delta p$

n_0, p_0 number of electrons, holes in the unexcited (dark) intrinsic state
Δn, Δp excess carriers: electrons, holes
B is the constant of radiative recombination

B is a temperature-dependent and material-dependent constant and characterizes radiative recombination. In silicon it has, at room temperature, the value: $B_{Si} = \sim 10^{-15}$ cm^3 s^{-1}.

In gallium arsenide, we have, at room temperature $B_{GaAs} = \sim 10^{-10}$ cm^3 s^{-1}: this is, thus, a value that is 100,000 times larger than for silicon. This shows that radiative recombination occurs less frequently and is not so important for indirect-bandgap semiconductors such as silicon—when compared to direct-bandgap semiconductors, such as gallium arsenide. Radiative recombination is a material-inherent type of recombination and does not depend on impurities or doping. It is present in all semiconductors.

2. ***Shockley-Read-Hall (SRH) Recombination***

In all semiconductors there are also defects caused by impurities, such as oxygen, carbon or iron. If these impurities (defects) are located on crystal lattice sites of the silicon wafer, they lead to additional energy levels within the bandgap. These energy levels are easier to reach for electrons and holes. They can climb down like on a staircase, which is easier for them, than crossing the whole bandgap in one step. They, thus, act on the charge carriers as recombination centres. Iron introduces an energy state that is located in the middle of the Si bandgap. Iron is, thus, attractive for both electrons and holes and acts as a strong recombination centre. Ni, Cu, Au and oxygen precipitates—also promote recombination and are strong lifetime-killers. SRH recombination is the predominant type of recombination in silicon photovoltaic solar cells. In particular, it will take place in different steps as shown in Fig. 4.13 [5]:

- An electron can be trapped by the defect, Case I
- A hole can be trapped by the defect, Case III.

On the other hand, electron-hole pairs can also be generated starting from the defect states:

[14]The intrinsic carrier density n_i is the density of carriers in an undoped sample of the same material (here: crystalline silicon), which is not subjected to any external field or other activation (e.g. light). Its value (for silicon) is at room temperature approximately 1×10^{10} cm^{-3}.

[15]«SI units» refers to the **International System of Units** (**SI**, abbreviated from the French *Système international* [*d'unités*]); it is the modern form of the metric system and is the most widely used system of measurement.

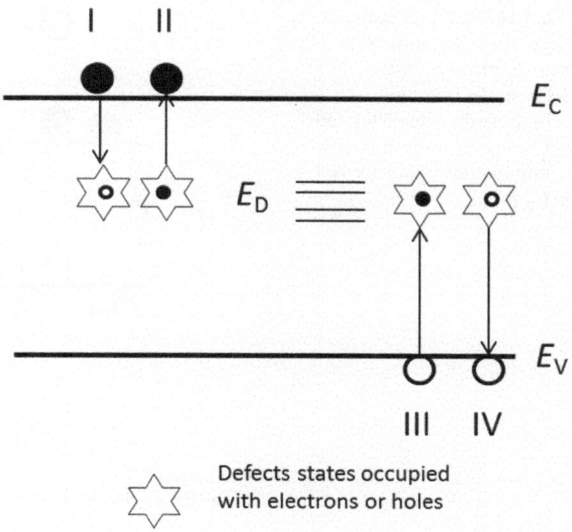

Fig. 4.13 Mechanisms of Shockley-Read-Hall recombination caused by defect states within the bandgap. The electrons or holes stay in the defect state for different durations and finally recombine (Case I for electrons and Case III for holes). Conversely, electrons (Case II) or holes (Case IV) can be emitted from the defect state. E_D denotes the energy levels of defect states (impurities) that represent the recombination centres [6]

- The trapped electron can be lifted back from the defect state into the conduction band, Case II
- Finally, a trapped hole can be returned to the valence band, Case IV.

The SRH recombination rate R_{SRH} is dependent on the number (density) N_D of defect states lying within the bandgap and on their energy levels E_D. It also depends on the capture cross-sections of the defect states for the capture of electrons (σ_n) or for holes (σ_p) and finally it depends on the density n of electrons in the conduction band and the density p of holes in the valence band—and also and on the thermal velocity v_{th} of the charge carriers

$$R_{SRH} = \frac{np - n_i^2}{\tau_n(n + n_{SRH}) + \tau_p(p + p_{SRH})} \tag{4.16}$$

where n_{SRH} and p_{SRH} are auxiliary variables, which in their turn depend on the intrinsic density n_i of free carriers and on the position (energy) E_D of the defect states with respect to the Fermi-level.

n_i is the intrinsic density of charge carriers and is for silicon 10^{10} cm^{-3} (see Footnote 14).

$$n_{SRH} = n_i \exp\left(\frac{E_D - E_F}{k_B T}\right) \tag{4.17}$$

$$p_{SRH} = n_i \exp\left(\frac{E_F - E_D}{k_B T}\right) \tag{4.18}$$

$$\tau_p = (\sigma_p V_{th} N_D)^{-1} \tag{4.19}$$

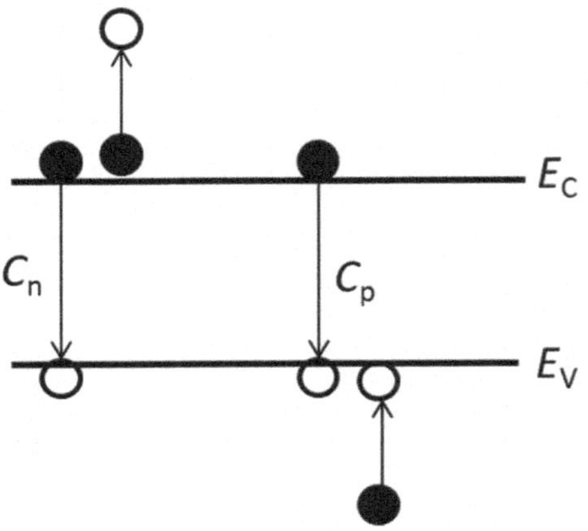

Fig. 4.14 Representation of Auger recombination with the two different cases: electron-electron-hole recombination, characterized by C_n and electron-hole-hole recombination, characterized by C_p [4]

$$\tau_n = (\sigma_n V_{th} N_D)^{-1} \tag{4.20}$$

The lifetime of the electrons in the conduction band is τ_n and the lifetime of the holes in the valence band is τ_p.

SRH recombination is based on impurities and other defects within the semiconductor. In order to reduce SRH recombination, with the goal of increasing solar cell efficiencies, one must minimize the density of defects. This is possible—in the case of silicon—through a skilful design of the crystallization process of silicon, by exploiting the different segregation properties of the impurity atoms. Nevertheless, to take the example of iron impurities, impurity concentrations below 10^{12} cm^{-3} can hardly be achieved. Note that even after crystallization, it is possible to further reduce impurity concentrations by a suitable gettering[16] process (see Chap. 5).

3. ***Auger Recombination***

In Auger recombination, the energy released when one electron jumps from the conduction band into the valence band is transferred to a third particle. The energy can be given over to an electron or to a hole. The first case is called electron-electron-hole recombination and the second case is called electron-hole-hole recombination. The two cases are shown in Fig. 4.14.

In electron-electron-hole recombination, the relationship for the corresponding Auger recombination $R_{aug}(e)$ is given by (4.21)

$$R_{aug}(e) = C_n n^2 \cdot p \tag{4.21}$$

And if an electron and two holes are involved, then we have, by analogy:

$$R_{aug}(p) = C_p p^2 \cdot n \tag{4.22}$$

[16]The gettering process reduces contaminants in a wafer and increases the carrier lifetime.

Since both processes run in parallel, the following relationship holds:

$$R_{\text{aug}} = n \cdot p \cdot (C_{\text{n}} \cdot n + C_{\text{p}} \cdot p) \tag{4.23}$$

For silicon, we have: $C_n \approx C_p \approx 10^{-30}\ \text{cm}^6\ \text{s}^{-1}$.

To fabricate a solar cell, we need doping. This unavoidable doping mainly causes Auger recombination. The higher the doping, the stronger that Auger recombination will be. Doping will lead to many defects.

Today's most common solar cells use *p*-type silicon. *p*-type silicon is, to start with, doped with boron. The *pn*-junction is then realized by overcompensation with phosphorus. Low phosphorus doping can reduce Auger recombination, but in this case, the *n*-region (emitter[17]) has a higher resistivity and it becomes difficult to obtain low contact resistances by metallisation. Emitter sheet resistances above about 100 Ω/□ are difficult to contact (metallisation) without losing cell efficiency.

It is often the case that the resistance and thickness of the emitter are not known. This makes it difficult to determine the properties required for the metallization. With the four point probe[18] one can measure the sheet resistivity. From this the optimal metallisation can then be determined. The sheet resistance depends on both the resistivity and the thickness.

Explanation of Sheet Resistance R_s

The term "sheet resistance R_s" is often used in semiconductor industry when screen-printed pastes or other thin layers are employed. It is a very important parameter for characterizing thin films and plays a decisive role in photovoltaics in the assessment of metallization layers.

When we apply a metal contact to the emitter of a solar cell, we basically should know the resistivity and the thickness of the emitter, in order to optimize the metal contact. It is often the case that the resistance and thickness of the emitter are not known. This makes it difficult to design the metallization. On the other hand, the sheet resistivity of a homogenously doped emitter can be measured very easily.

The sheet resistance resistivity is defined as follows:

$$R_{\text{s}} = \rho / T \tag{4.24}$$

ρ specific resistance (Ω mm);
T (here:) thickness of the layer (mm);
R_s sheet resistance or sheet resistivity (Ω/square) or (Ω/□).

R_s depends both on the specific resistance ρ (Ω mm) or more precisely on ρ (Ω mm²/mm) and on the thickness T (mm). The units used for R_s are (Ω mm²/[mm ×

[17]The term emitter is often used in photovoltaics within the description of a *pn*-junction. In *p*-type silicon the *n*-region is called emitter whereas in *n*-type silicon the *p*-region is called emitter.

[18]Four measuring probes are placed on the cell at a constant distance from each other and on a straight line. A current is driven through the two outer probes and the two inner probes measure the voltage. An alternative is to use the Electrochemical Capacitance-Voltage (ECV) method. This method measures additionally the active carrier concentration profiles in semi-conductor layers.

mm]) or simply Ω. Since the sheet resistance R_s refers to a surface, one writes ohm/□, where □ is dimensionless. R_s gives an indication of how the metallization is to be realized. And as mentioned before it is easy to measure. If, for example, the phosphorous is driven into the p-type silicon for formation of the *pn*-junction[19] a sheet resistivity between 97 and 205 Ω/□ results. 97 Ω/□ can be achieved, for example, with a phosphorus concentration of 8×10^{19}; 205 Ω/□ with a phosphorus concentration of 3.3×10^{19} (low concentration). 97 Ω/□ leads in our example to a specific resistance of 31×10^{-3} Ω mm (4.24) and the semiconductor can be contacted optimally with silver fingers. A higher sheet resistivity of e.g. 205 Ω/□ cannot be optimally contacted without losing efficiency. The reason for this is that the dark current J_o increases strongly at very low doping concentrations under the metal contacts [7]. This leads to low cell efficiencies. In this way the *pn*-junction can be evaluated by means of sheet resistivity.

4. ***Surface Recombination***

A large part of the recombination within solar cells can be attributed to surface recombination. The neighbouring lattice atoms are missing on the surface, so that foreign atoms, especially oxygen, can accumulate. Additionally, doping with foreign atoms, for example with phosphorous, also contaminates the surface and even intensify the recombination. As an example: In the case of a *p*-type silicon wafer, the phosphorus dopant enters from the surface. Finally, and most importantly, the metal contacts are lying on the surface—these contacts introduce additional contaminants and, act, thus as very active recombination centres. In the metal contacts (like in all metals), the Fermi level E_F lies within the Conduction Band. Therefore, there are very many free electrons in these contacts which are just ready and waiting to swallow all holes, which reach the metal layer. Metal contacts are, thus, zones of very high surface recombination. The theoretic treatment of surface recombination is similar to that of SRH recombination. For the production of solar cells, the reduction of surface recombination by passivation of the electrically active recombination centres has the highest priority. The surface recombination velocity S is a variable with the unit cm s^{-1}; it indicates how fast charge carriers recombine on the surface. Surface recombination can be calculated for electrons (e) or for holes (h). In the following it is calculated for holes. In *p*-type bulk material electrons are the minority carriers, but at the front-surface holes are the minority carriers because of the doping used for the formation of the *pn*-junction. For holes, surface recombination velocity S_h depends on the capture cross-section σ_h, on the density of electrons at the surface n_e and on the thermal velocity v_{th} according to the following equations:

$$S_h = \sigma_h \, n_e v_{th} \tag{4.25}$$

If we now multiply the surface recombination velocity S_h with the density of holes n_h, we get the surface recombination rate.

[19]The *pn*-junction is typically 350 nm thick.

$$R_{\text{surf-h}} = S_h n_h \tag{4.26}$$

As an example: the surface recombination velocities are about 200–600 cm s^{-1} for standard solar cells and about 60–80 cm s^{-1} for PERC cells (see Chap. 5).

In *n*-type bulk material holes are the minority carriers, but at the surface electrons are the minority carriers because of the formation of the *pn*-junction. The surface recombination velocity for electrons S_e can be calculated similarly as in (4.25) and (4.26).

References

1. P. Würfel, Physik der Solarzellen, in *Spektrum akademischer Verlag*, 2. Auflage (Spektrum Hochschultaschenbuch, 2000). ISBN 3-8274-0598-X
2. E. Hecht, A. Zajac, *Optics*, 4. Auflage (Addison-Wesley Longman, Amsterdam, 2003). ISBN 0-321-18878-0, S. 402
3. H.-P. Sperlich, *Meyer Burger Technologie AG, Analyse 2014, anlässlich einer Studie zusammen mit der Westsächsischen Hochschule Zwickau*
4. T. Marvart, L. Castaner, *Solar Cells: Materials, Manufacture and Operation* (Elsevier, 2005). ISBN 1856174573
5. W. Shockley, W.T. Read, Phys. Rev. **87**, 835 (1952); R.N. Hall, Phys. Rev. **83**, 228 (1951)
6. K. Lauer, *Untersuchungen zur Ladungsträgerlebensdauer in kristallinem Silizium für Solarzellen* (Universitätsverlag Ilmenau, 2005)
7. S. Werner, E. Lohmüller, S. Maier, S. Mourad, A. Wolf, Challenges for lowly-doped phosphorous emitters in silicon solar cells with screen-printed silver contacts, in *7th International Conference on Silicon Photovoltaics, Silicon PV 2017*. Energy Procedia **124**, 936–946 (2017)

Sylvère Leu was from 2008 to 2017 with the Meyer Burger Group, where he became a member of the executive board, as Chief Innovation Officer and Technology Officer (CIO/CTO). Since his retirement in 2017, he is active as technology consultant, for the Meyer Burger Group. Sylvère graduated from the Swiss Federal Institute of Technology in Zürich (ETHZ). He additionally obtained a Master's Degree in Business Administration at the University of St. Gallen (HSG). As a Swiss pioneer, Sylvère Leu started to work in photovoltaics 30 years ago. He constructed an industrially relevant laminator and sun simulator in his own company. Beside his job he worked as an associate lecturer at University of St. Gallen (HSG), for several years, in the field of industrial production. At the end of 2005 he was charged with building up an integrated 250 MW_p photovoltaic facility, including wafer, cell and module manufacturing, in Germany.

Detlef Sontag is Senior Technologist for Solar Cells at Meyer Burger (Germany) GmbH. since 2015—one of the world's leading PV equipment manufacturing companies. He graduated as a physicist at the University of Konstanz in 2000, and obtained his Ph.D. degree in 2004 working on novel wafer materials for Photovoltaics. In 2004 he joined Deutsche Cell GmbH, a subsidiary of Solar World AG, as quality assurance engineer with insights in the industrial crystalline silicon PV manufacturing processes along the whole value chain from block casting via cell processing to module assembling. Detlef Sontag joined the equipment manufacturer Roth & Rau AG as Senior Principal for cell technology in 2009. There he supported the planning and commissioning of the Technology centre and worked actively on the development of cell technologies like PERC and HJT including project coordination of corresponding R&D-projects.

Chapter 5
Crystalline Silicon Solar Cells: Homojunction Cells

Sylvère Leu and Detlef Sontag

Abstract In Chap. 3 we learned about the typical functioning of a semiconductor under the influence of light: Light can generate an electric current; this happens if an electron absorbs energy from the light and detaches itself from its atom—it becomes thereby a free charge carrier, which contributes to power generation. This is called the "photovoltaic effect". In the following, our aim will be to understand the use of the photovoltaic effect to a greater depth, so as to be able to design and fabricate solar cells with higher efficiencies. In order to better understand the cell concept, it is important to understand the preceding crystallization process and the wafering process. Thereafter we will discuss the development of the basic homojunction Al-BSF (Aluminium Back Surface Field) solar cell. The further development of the Al-BSF cell takes us to the PERC cell (Passivated Emitter Rear Cell). We will look at these cell designs in detail. Based on the potential for improvement of the PERC cell we will explain the cell concept of the PERT cell (Passivated Emitter and Rear Totally Diffused cell) and the different architectures of advanced cell designs such as TOPCon cell (Tunnel Oxide Passivation Contact). To conclude, we will briefly discuss the IBCs (Interdigitated Back Solar Cells), which achieve the highest efficiency so far for single-junction silicon solar cells.

5.1 Production of Silicon Wafers and Solar Cells

5.1.1 Production of Silicon Ingots

Crystalline solar cells used for large-scale terrestrial applications consist almost exclusively of silicon as base material. There are good reasons for this:

S. Leu · D. Sontag (✉)
Meyer Burger Technology A.G, Gwatt, Switzerland
e-mail: dsontag@web.de

S. Leu
e-mail: sylvere.leu@ciptec.ch

A. Shah (ed.), *Solar Cells and Modules*, Springer Series in Materials Science 301,
https://doi.org/10.1007/978-3-030-46487-5_5

- Silicon is the second most abundant element of our Earth's crust after oxygen. Weighted by atomic per cent, the earth's crust contains[1]:
  - 60.4% oxygen
  - 20.4% silicon
  - 6.3% aluminium
  - 2.9% hydrogen.
- The bandgap of silicon is 1.12 eV and is, thus, not too far from the optimal value for converting—in solar cells—sunlight into electricity.
- Silicon melts easily, has good mechanical properties and is easy to machine (sawing, polishing, etching).
- Silicon can be made into a highly pure form, as a single crystal. This means that silicon has a homogeneous crystal lattice, which is very regular and has few unwanted foreign atoms.
- With doping, the type and conductivity of silicon can be easily changed.
- Silicon is non-toxic.
- Silicon forms a native oxide layer, which serves as a high-quality insulator, whereupon different layers can be deposited.
- Silicon can be thoroughly cleaned. Ultrapure silicon for the semiconductor industry is manufactured in purity levels of up to 11N[2] in mass production. For low-efficiency solar cells, a purity of 7N is sufficient but, as the demand for higher efficiency solar cells increases, the specifications on silicon feedstock are also requiring lower concentrations of impurities.
- The vast amount of knowledge gained during the last 50 years by semiconductor industry in how to process silicon and how to make industrial tools for silicon-based devices. A lot of this knowledge has been used for Si PV as well.

Jöns Jacob Berzelius a Swedish chemist discovered silicon in 1823. Other semiconductors such as germanium and gallium arsenide have higher charge carrier mobilities and, thus, allow for higher switching speeds in integrated circuits. But Germanium has two disadvantages in connection with solar cells:

(a) Its bandgap is too small to form single-junction solar cells
(b) Germanium oxide is unstable and not suitable for surface passivation and insulation layers.

Germanium and Gallium Arsenide (GaAs) are both too expensive to be suitable for the production of solar cells for terrestrial applications

[1]The earth's crust forms the outer part of the earth and extends about 35–40 km into the earth's interior. Its composition is very diverse. If one arranges the elements of the earth's crust according to weight percent, a shift of the portions is noticeable in comparison to the distribution according to atomic percent: Oxygen (46.6%), silicon (27%) and aluminum (8%) are still the three most frequent elements.

[2]11N means: 11 nines. 11N is 99.999999999% pure.

The production of pure silicon ingots[3] takes place in four stages [1]:

(a) Production of raw silicon
(b) Preparation of trichlorosilane
(c) Production of high-purity silicon
(d) Production of ingots, either multicrystalline or monocrystalline.

(a) **Production of raw silicon**

Starting material is silica sand (quartz sand) SiO_2, which is mined above all in Brazil and China where the quartz sand is very pure. At ~1800 °C, this silica sand is reduced with coal (i.e. with carbon C) to metallurgical grade silicon with a purity of 98–99%:

$$SiO_2 + C \rightarrow SiO + CO \tag{5.1}$$

$$SiO + C \rightarrow SiC + CO \tag{5.2}$$

$$2SiC + SiO_2 \rightarrow 3Si + 2CO \tag{5.3}$$

We now have "technical silicon", e.g. raw or metallurgical silicon with a purity of 98–99%.

The next three steps (b) to (d) are made with the so-called Siemens process which was patented by the company Siemens A.G., in 1954.

(b) **Preparation of trichlorosilane**

Raw silicon is now finely ground; then it is converted in a fluidized bed reactor, with the aid of hydrochloric acid, to gaseous trichlorosilane. The process temperature for this step is approximately 300 °C:

$$Si + 3HCl \rightarrow H_2 + SiHCl_3 \tag{5.4}$$

(c) **Destillation of Trichlosilane**

Trichlorosilane boils at 31.8 °C; it is distilled in tall stainless steel columns. Thereby a lot of impurities are filtered out. With repeated distillation, the degree of purity can be increased.

(d) **Production of ultrapure silicon**

Trichlorosilane is broken down in a reducing atmosphere at around 1000 °C via the reaction

$$SiHCl_3 + H_2 \rightarrow 2Si + SiCl_4 + SiCl_2 + 6HCl \tag{5.5}$$

[3]This process step is very important, as it accounts for more than half of the total energy invested in the production of crystalline silicon solar modules.

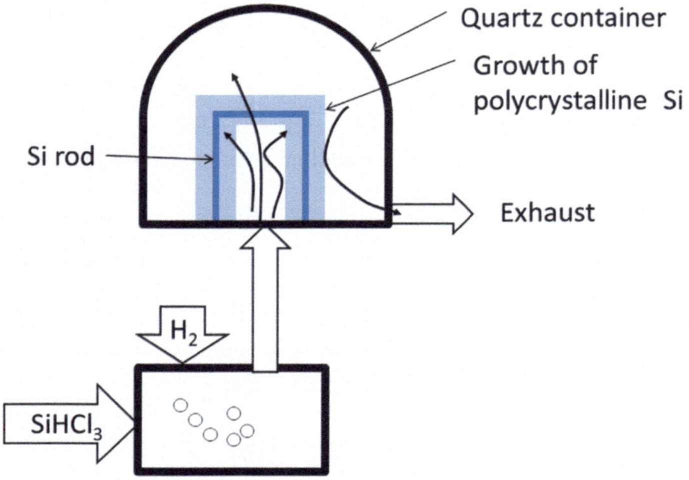

Fig. 5.1 The Siemens reactor process, which extracts pure silicon from trichlorosilane. The pure thin silicon rods, with a diameter of about 8 mm are electrically heated to about 1150 °C. Pure trichlorosilane flows around these hot silicon rods and polycrystalline silicon is deposited until the rod has a diameter of about 300 mm

so that Si atoms from the vapour are deposited on a silicon starting 'seed', usually consisting of a cylindrically arranged array of thin Si rods. It is the most widespread cleaning process for silicon—it results in pure silicon with a purity of 9 N–11 N, depending on how pure the trichlorosilane is. Figure 5.1 shows, in a schematic way, the process flow.

In the next step, the polycrystalline silicon rods grown in the Siemens reactor process are broken down in pieces of different sizes (chunks and chips) so that the various crucibles can be filled to produce either monocrystalline silicon or multicrystalline silicon. The aim of this process step is to produce, from the ultra-pure silicon, a silicon crystal (ingot) with, on the one hand few dislocations, and, on the other hand a material in which the desired concentration of the doping material is contained.

(e) **Production of monocrystalline ingots**

Czochralski Method

In the widely used **Czochralski[4] method (CZ)**, which is shown in Fig. 5.2, ultrafine silicon chunks are filled into a crucible, which consists of pure quartz glass SiO_2 and is coated with Si_3N_4. The quartz glass crucible is in its turn embedded in a graphite crucible, which supports it. The temperature resistance of these crucibles is over 1600 °C. The structure and coating of the crucible are crucial for the quality of the silicon [2]. After the quartz glass crucible has been filled with ultra-pure silicon

[4]Jan Czochralski was a Polish Chemist (1885–1953). He developed 1916 the Czochralski method for pulling single crystals from the melt.

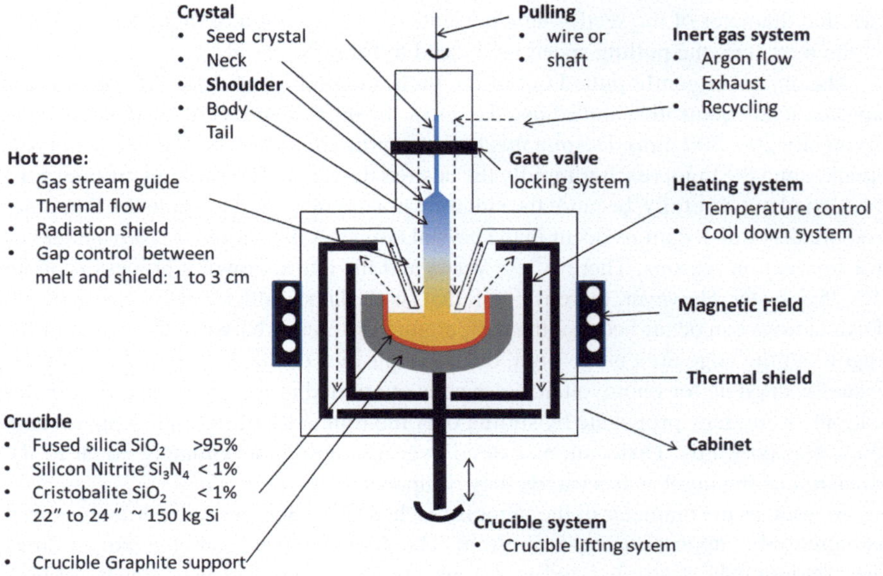

Fig. 5.2 Schematic representation of a puller for monocrystalline crystals according to the Czochralski method

(chunks and chips), it is electrically heated to approximately 1420 °C.[5] A shielding gas, usually argon, prevents impurities from entering the chamber; it also stops the oxygen from escaping out of the quartz glass crucible, by transport with the outward flow of argon.

Once the silicon has melted, a rotating and height-adjustable silicon seed crystal of 3–5 mm size is slowly approached to the melt without touching it; the goal is to bring the seed to the same temperature as the molten silicon. When the seed itself starts to melt, it is gently put into contact with the molten silicon in the crucible (**Dipping Step**). The crucible rotates in the opposite direction to that of the seed crystal. The rotation is important so that the heat distribution remains homogeneous and does not create thermal stress. Immediately before dipping, the melt is slightly cooled to a temperature just below the melting point.[6] The silicon atoms will now dock on the colder seed crystal, solidifying and adopting the orientation and structure of the seed crystal.

During the **Necking Step**, the seed is pulled out of the melt faster to reduce the diameter of the single crystal ingot to a minimum value (much smaller than the

[5]This is just above the melting point: silicon melts at 1412 °C.

[6]The Ostwald–Miers range (according to Wilhelm Ostwald and Henry Alexander Miers) is the temperature range in which the melting point is undershot during cooling in a liquid to a temperature, where crystallization does not take place as yet. Crystallization can thus take place on a seed crystal. This is just below the melting point: silicon melts at 1412 °C.

original diameter of the seed)—this prevents the propagation of dislocations. Then, in the next step, the pulling speed is reduced to form the shoulder.

The crystal is gently pulled out of the melt; the pulling speed (~0.5 mm/min) is automatically controlled and adjusted to grow the ingot diameter to the desired value (typically 200–300 mm). Despite this limitation the shoulder (see Fig. 5.2) can grow quickly and the ingot reaches rapidly the desired diameter. The necking process must be carried out carefully, because the situation must be avoided in which the thin inner rod breaks or dislocations occur in the crystal lattice. A flat shoulder is advantageous for production reasons. There is less waste and the entire drawing process is about 15–25% faster. However, there is a risk that, with flat shoulders, dislocations occur. Dislocations can occur because the temperature difference between the inside of the ingot and the edge area is too high, due to impurities or due to external vibrations. Usually, crystals for photovoltaic solar cells are pulled in the {100} plane. If dislocations occur, they propagate by sliding over the four sides of the {111} planes and, thus, slip outwards. Dislocations extend over a length approximately equal to the diameter of the ingot. **Afterwards** they disappear.

As soon as the diameter of the ingot is reached (200–300 mm), the drawing speed is increased to approximately 8 mm/min. The drawing speeds and the temperatures are continuously controlled and monitored, so that a constant diameter is formed. After the ingot has been pulled, it must be mechanically secured (using a locking system) and slowly cooled, so that no cracks are created due to stress. The end of the drawing process must not be abrupt—because this can trigger thermal shocks, which can lead to dislocations in the crystal lattice. Therefore, the end of the drawing process is initiated via a tapered tail. Incidentally, this is also where most of the impurities are found, so the act of pulling the crystal can also purify it. Using the Czochralski method, round ingots of up to 4 m in length, typically 2.5 m, and up to 300 mm (12 in.) in diameter are drawn. The filling of the quartz glass crucible is about 150 kg and can be increased with subsequent recharging to about 200 kg.

The disadvantages of the method are: (1) the wall of the quartz glass crucible can react with the silicon, which limits the resistivity by the penetration of impurities; (2) even if the crucible is coated with Si_3N_4, impurities from the coating layer can penetrate the silicon melt. This coating acts as a barrier and the impurities entering the silicon melt from the crucible are greatly reduced. Nevertheless the coating itself is a source of contamination. However, the total contamination (crucible and coating) is reduced; (3) finally oxygen from the quartz glass crucible wall penetrates into the silicon, forming SiO_2 according to the simplified reaction:

$$SiO_{2\text{quartz glass crucible}} + 2Si_{\text{melt}} \rightarrow Si + 2SiO \quad (5.6)$$

This is, in fact, inevitable. In a typical ingot, the concentration of interstitial oxygen is between 10^{17} and 10^{18} cm^{-3}. Because silicon has about 10^{23} atoms per cubic centimetre, oxygen contamination is typically between 0.1 and 1 ppm.[7]

[7] «ppm» means «**p**arts **p**er **m**illion».

The oxygen atoms are originally randomly distributed in the silicon; during crystal growth, various complicated morphological processes take place and as a consequence, the oxygen atoms can join together and form clusters, so-called "precipitates". Precipitates have various positive and negative effects, depending on how the process is conducted.

1. Precipitates lead to local disturbances in the crystal structure.
2. In a heat treatment process, this precipitation process can be partially controlled, in such a way that precipitates can be placed where the semiconductor is not active.
3. On the other hand, oxygen precipitates can serve as trap sites for metallic foreign atoms (gettering process)[8] [3].
4. Above a concentration of 10^{18} cm^{-3}, the solubility limit of oxygen in the silicon is reached and no further oxygen precipitates can be formed. This is also the reason that the oxygen concentration should be less than 10^{18} cm^{-3}.
5. If one has higher oxygen content, one will also have a more pronounced Light Induced Degradation (LID) in *p*-type material (see Chap. 10) because of the B–O (Boron–Oxygen) complexes.

The oxygen accumulates mainly at the top of the ingot while the impurities tend to be at the tail (bottom).

Typical lifetimes of passivated wafers are in the range of 1.5–10 ms. The specific resistance is typically between 0.5 and 7 Ω cm. To obtain high cell efficiencies, the rule of thumb is that the quality factor τ/ρ should be greater than 1 ms/Ω cm:

$$\tau/\rho > 1\,\mathrm{ms}/\Omega\,\mathrm{cm} \tag{5.7}$$

τ lifetime in ms
ρ specific resistivity in Ω cm.

The crucibles can be recharged two to three times in the hot state. After that, the impurities in the silicon become too large due to oxygen and carbon; the target value $\tau/\rho > 1$ ms/Ω cm (passivated wafer) can no longer be met. Once the Crucible has cooled to room temperature, it cannot be reused. Thanks to the use of a magnetic field (see Fig. 5.2), the oxygen content can be reduced and the ingress of impurities from the Crucible can be prevented. This is called the Magnetic Czochralski (MCz) process.

For *p*-type material boron is added, and for *n*-type material phosphorus is added to the silicon melt. Phosphorus has a segregation coefficient[9] of 0.35; boron has a

[8]Gettering process means a controlled modification of the silicon crystal by thermal processes to draw impurities far from the active part of the semiconductor, in order to reduce their potential degrading effects.

[9]The "**segregation coefficient**" is defined as the **ratio** ***K*** of the impurity concentration C_{solid} in the solid state (here: in silicon) **to** the impurity concentration C_{melt} in the melt: $K = C_{\mathrm{solid}}/C_{\mathrm{melt}}$. The segregation coefficient defines how well impurities are separated from the rest of the material. If the segregation coefficient is 0.1, this means that the impurity concentration C_{melt} in the melt is 10

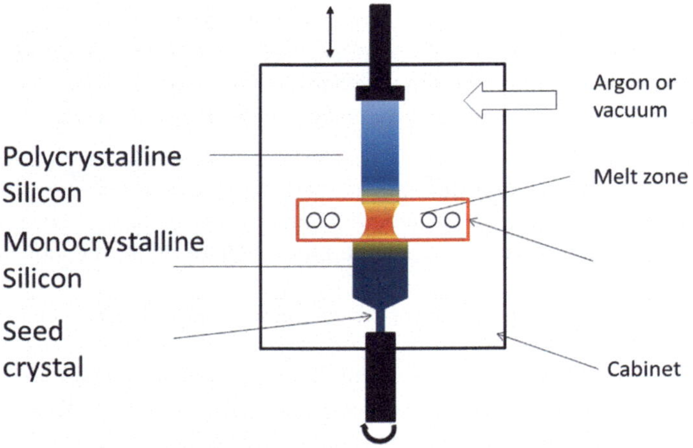

Fig. 5.3 Schematic representation of a monocrystalline float-zone puller

segregation coefficient of 0.7. Phosphorus is therefore less well absorbed by silicon and lingers longer in the liquid phase. It disperses less homogeneously than boron. Therefore, for *n*-type material the resistivity varies in a wider range (0.5–7 Ω cm) than for *p*-type material (1–3 Ω cm).

Float-Zone Method

If lower impurity concentrations are required, one employs the relatively expensive[10] **float-zone method (FZ)** according to Fig. 5.3—here, no crucibles are used. A polycrystalline silicon rod is clamped vertically and heated locally through an induction coil right up to the melting point. A seed crystal at the lower end of the rod initializes the drawing process.

The silicon rod now moves slowly down (or: the coil moves slowly upwards). The polysilicon rod melts in the melting zone and crystallization begins. Since the impurities have segregation coefficients of <1, they remain in the melt and migrate with the melt upward. Repeated zone melting (which is not a process used in the photovoltaic industry, due to cost) allows the production of extremely pure silicon rods. Actually, this technique is employed to produce the rods used in the Siemens process. Because no crucible is used in the float zone process, virtually no impurities

---

times higher than the impurity concentration C_{solid} in the solid state (here: in the solidified silicon). The closer the segregation coefficient is to 1, the more homogeneous will be the resulting "mixture" of the two materials. This is the reason that Phosphorous (P) ($K_P = 0.3$) and Boron (B) ($K_B = 0.7$) can be used for doping. Impurities like Oxygen and Iron have segregation coefficients of: $K_{Oxygen} > 1$ and $K_{iron} \ll 1$. Oxygen accumulates mainly at the top of the ingot while impurities like iron tend to be at the tail (bottom).

[10]Basically, the FZ process should be cheaper than the CZ process, because of faster rates of crystallization, higher throughput per puller and less energy consumption. However, the lower yield and the requirement of a machined polysilicon rod render this technique more expensive than the CZ process, when all cost factors are accounted for.

are present in the resulting crystal: resistivity up to 1000 Ω cm is obtained! That is a very high value. In the Cz process (see previous section) we obtain for *n*-type material resistivity from 0.5 to 7 Ω cm, and for *p*-type material resistivity from 1 Ω cm to 3 Ω. FZ silicon contains much less impurities, which is why the resistivity is so high. In photovoltaics, float-zone wafers are mainly used in research to compare solar cells, which are produced with different process parameters and to show the limits of these processes with regards to cell efficiencies. The advantage is that material-related differences can be ruled out in this comparison, because float-zone material is very pure.

A comparison between the Czochralski method and the float-zone method is given in Table 5.1.

Crystal growing is very energy intensive. Therefore, we will make here a rough estimate of the energy balance sheet. To produce 1 kg silicon, an equivalent energy investment of approximately 100 kWh has to be made. With 1 kg of silicon, 74 wafers of 180 μm thickness can be sawed and a solar module with 450 W_p (Watt peak) can be produced; this module can generate, in the Central European Climate, during a period of 25 years, 12,000 kWh Electricity. This rough calculation shows the positive energy balance of crystalline silicon solar cells.

Production of Multicrystalline Ingots

In the production of **conventional multicrystalline** ingots either the Bridgman process (or less widespread in PV the block casting process) is used. In Fig. 5.4, both methods are schematically shown.

A square quartz crucible ideally coated with Si_3N_4 is filled with polycrystalline silicon chunks, heated and melted. Thereafter, the melt is slowly cooled from bottom to top and the crystals begin to grow from bottom to top. The growth process is not directed as in monocrystalline crystal growth. Crystals grow very randomly and form larger areas with different microstructures. The goal is to grow large crystals with a vertical columnar structure, so that the number of grain boundaries within the wafer

Table 5.1 Comparison of Czochralski and float-zone methods

	Czochralski	Float-zone
Oxygen content (atoms/cm^{-3})	<1 × 10^{18}	<1 × 10^{16}
Carbon content (atoms/cm^{-3})	<1 × 10^{17}	<1 × 10^{16}
Metallic impurities	Some	Very few
Bulk lifetime of minority carriers (ms)	1–10	30–45
Relative production costs (%)	100	>200
Cell efficiency for PERC (%)	~22.5	25[a]
Cell efficiency *n*-type TopCon (ISE)	~23%	
Cell efficiency for HJT (M2) (%)	~24	26.6[b]

[a]World record for PERC on FZ by USNW, small size
[b]Kaneka, based on a IBC-HJT cell structure, M2

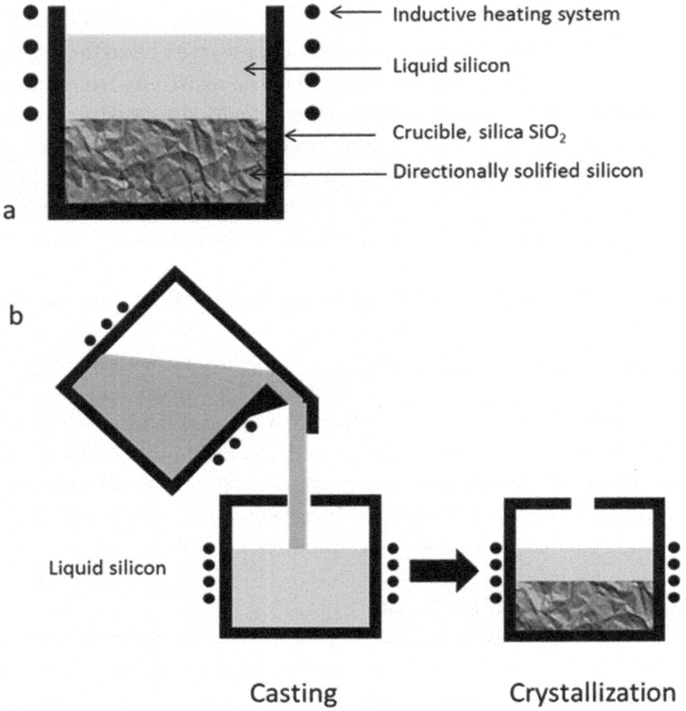

Fig. 5.4 **a** Bridgman Process: the melting process and the crystallization are carried out in the same Crucible; **b** block casting process: after the silicon has been melted in a first Crucible, it is crystallized in a second Crucible. In contrast to the Bridgman process, the crystallization time and the cooling time can be reduced. Also, different heating systems can be used, which allows for process optimization

remains low and that the grain boundaries are always perpendicular to the surface of the wafer. Typically, the crucible has a size of G6 or G8. G6 means that the ingot has a size of 6 × 6 bricks, with a brick having an area that corresponds to the area of a solar cell. The height of the crucible is about 30 cm. In this way, ingots can be produced relatively quickly and inexpensively. However, recombination centres are formed at the grain boundaries, dislocations and clusters of precipitates, so that the cell efficiency of multicrystalline solar cells is a bit lower than that of monocrystalline solar cells.

High Performance Multi Crystalline Technology (HP-mc-Si)

It was discovered later (~2011) that large grains also create large clusters of dislocations. Therefore, the HP mc-Si process was developed. With the introduction of **High Performance Multi Crystalline Technology (HP-mc-Si)** in 2011, the controlling of the grain size was possible. The smaller the grain boundaries, the less dislocations occur in the crystal—dislocations that appear when the grains are too large. Cell efficiencies were increased from 16.6% (2011) to >22% (2019)[11] in mass production, thanks to the introduction of HP-mc-Si. Furthermore the variation of cell efficiencies per ingot was also reduced—with this, the production yield was improved. Finally, thanks to HP-mc-Si, new cell concepts could be applied, such as PERC cell technology.

In the production of **HP-mc-Si** very fine silicon grains of the size 30–300 μm [4] are sprinkled on the bottom of the crucible as a **seed layer**. Thereafter, the silicon pieces (chunks and chips), which are obtained from the Siemens process, are filled in. The crystallization process starts from the bottom. It is important that the interspersed silicon grains do not melt completely, but can pass on their crystal structure to the melt. The small crystalline particles act as a seed for the crystal growth. Additionally the wall of the crucibles is roughened and coated with SiN, so that the crystal structure is predetermined in this way. Due to the fact that there are many interfaces, more than with the conventional multicrystalline production, the thermal stress remains small and dislocations in the crystal structure occur less frequently. Overall, the cell efficiency increases compared to the conventional multicrystalline process.

It is interesting that the grains become larger towards the top and so the higher the crucible is, the larger the grains become, which should actually be avoided because dislocations can appear [5]. Before the bricks are sawn into wafers, the bottom, top and sides of the brick have to be cut away because the contaminants are there. About 20 mm on each side must be separated and one can have non-melting particles at the bottom of the crucible to replace the silicon particles of the HP process. This reduces the yield of the process. On the other hand, the square solar cells are obtained directly from the bricks and the packing density in the module is larger than in the pseudo-square monocrystalline wafers.

Other Manufacturing Methods

Less common is the use of **mono-like multi-material**. This process is similar to that of HP-mc-Si. However, monocrystalline large seed wafers are used. Often, the centre of the mono-like multi-ingot is monocrystalline, but outside it is multicrystalline. Wafers with both crystal structures are difficult to texture and the process of optimizing the contact resistance in the presence of grains of different orientation on the surface represents another difficulty.

Kerf-free technologies are also the subject of research such as

- Controlled cleave with **energy beam**
- **Epitaxy**, in which a pure silicon layer is deposited by means of a gas phase, using e.g. silane (SiH_4), on a "dirty" silicon mother wafer lying on a graphite carrier). The mother wafer can be employed several times. Therefore, it is fabricated with many small silicon tips, so that the wafer grown on it can be easily removed. The advantage of such wafers is that full-square monocrystalline wafers can be produced, which have a larger area than pseudo-square wafers (see Sect. 5.1.2). The production also requires about 50% less energy than with Cz wafers.[12]

[11] Trina 21.25% (2015) and Jinko 22.04% (2017).

[12] Developed by Nexwave GmbH, Freiburg i.Br. Germany.

5.1.2 Wafering

Now that monocrystalline or multicrystalline ingots have been grown, they must be processed into wafers. Thereby, the monocrystalline ingots must first be marked to show how the crystal structure runs. The direction of the crystal structure {111}, {110} or {100} is marked along the entire ingot by a notch or flat. This type of marking is standardized in the semiconductor industry.

Figure 5.5 illustrates how the ingots are processed in bricks. The monocrystalline round ingots are cropped in a pre-wafering step and then made into a square form. During cropping both the top and the tail of the ingot are cut off. Squaring follows now: the round ingot is cut into a square form, over its entire length. Actually, it is not a fully square form, but a "**pseudo-square**" form, e.g. a square form with cut edges. In this way the solar cell surface becomes as large as possible and the packing density of the cell surface within the module can be optimized.

With increasing experience in crystal growing, the diameters of the ingots have been continuously increased a little, which is shown in Table 5.2. Of course, this also enhances the cell power because of the larger cell size.

The products cut-off during the squaring process are called "slabs". Slabs, as well as tops and tails are reused in the new melt. The now pseudo-squared ingots are cut

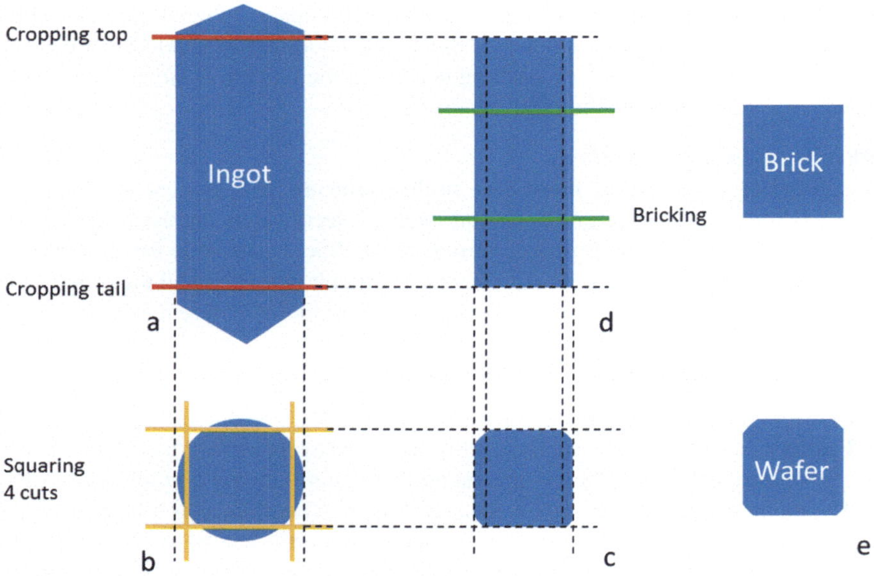

Fig. 5.5 Process steps from ingot to brick. **a** The round ingot is cropped first. Top and tail are separated and reused; **b** then the cropped ingot is squared. The four cut silicon parts are called slabs. The slabs are recycled; **c** the squaring process creates the pseudo-square surface, which is typical for monocrystalline cells; **d** the squared ingot is now sawn into bricks. Any areas with dislocations or too small diameters are sawn out and recycled; **e** a brick is about 15–40 cm long

Table 5.2 Ingot diameter, solar cell edge length and wafer size

Size	M0	M1	M2	M4	M6
Diameter (mm)	200	205	210	211	223
Flat length (mm × mm)	156 × 156	156 × 156	156.75 × 156.75	161.70 × 161.70	166 × 166
Wafer area (cm^2)	239	241	244	258	274
Enlargement (%)	100	100	101	107	114

M6 announced by LONGi Solar in 2019

into bricks, which are approximately 15–40 cm long. The bricks are finally glued to a 1 cm thick beam. The entire sandwich—Beam and Brick—is glued to a workpiece carrier. The workpiece carrier is inserted, by means of a tool holder system, into the wire saw. The loading length of the wire saw is approximately 70 cm. The typically 60 μm diamond wire cuts the bricks into 140–180 μm thick wafers. The cut is from bottom to top; one cuts up to the middle of the beam. The wafers hang on the beam, after cutting; and the beam hangs on the workpiece carrier. In a cleaning bath, the wafers are removed from the beam and the beam from the workpiece holder—by dissolving the adhesive in a vinegar bath. The wafers are then separated, cleaned, measured and sorted. The beam can no longer be used, but the workpiece carrier can be re-used.

In the case of square multicrystalline ingots, the entire ingot, which is about 30 cm high, is simultaneously cut into square bricks, of an area of e.g. $156 \times 156\ mm^2$. For multicrystalline ingots, the ingot size is indicated by “Gx”. For example, G6 means that $6 \times 6 = 36$ bricks can be cut out of the ingot. With an ingot size of G8 there are 64 bricks. After the ingot crystallization, a few centimeters (about 2–3 cm) from the top, bottom and each side are cut away because they contain an unacceptable level of impurities and contain many defects in their crystal structure. The bricks are finally cut into wafers using the same procedure as in the case of monocrystalline bricks and sorted after the cleaning process.

5.1.3 *Wafer Cleaning and Texturization*

After wafering, the wafers are cleaned and sorted and then given a texture.

Section 4.23, Chap. 4 (Optical and recombination losses) examines the different texture requirements for slurry-cut and diamond-cut wafers, both multicrystalline and monocrystalline, and describes their influence on light trapping. The exact description for wafer cleaning and texturing for monocrystalline cells is given in Sect. 7.5.2 (heterojunction). For HJT cells wafer cleaning and texturization is very important;

therefore these topics fit better into the chapter on Heterojunction (HJT) cells. For these reasons, this chapter does not deal with cleaning and texturization.

5.2 Cell Processing for the Al-BSF Cell

This section describes the basic steps for manufacturing a solar cell. We describe first of all the Al-BSF (Aluminium Back Surface Field) cell, which was the most widely used solar cell until 2015. It was then further developed and led to the PERC cell, which is described in Sect. 5.3.

5.2.1 Light Trapping by Texturization

The first step in cell production is the Texturization. To get a high efficiency, we increase the front side of the cleaned, microcrack free and saw damage free wafer surface. We increase the surface area by texturing the wafer. We have already described this in Chap. 4.

5.2.2 Formation of the pn*-Junction*

The presence of a charge-separating *pn*-junction is a prerequisite for a functioning traditional Al-BSF solar cell. In crystalline solar cells, one employs silicon as a semiconductor material—with boron and phosphorus as dopants. Silicon is tetravalent, so it has four valence electrons, through which a bond to neighbouring atoms can be implemented. Boron is trivalent and has, thus, one valence electron missing compared to silicon, while phosphorus is pentavalent and has one valence electron in excess compared to silicon. A boron-doped *p*-type silicon crystal has, therefore, more free holes (it is *p*-type, e.g. "***p***ositive"), in contrast with a phosphorous-doped silicon crystal, which provides more free electrons (it is *n*-type, e.g. "***n***egative"). A traditional solar cell consists of a *p*-region bordering on an *n*-region. Through charge exchange, one obtains, at the interface of the two regions, a *pn*-junction. which is associated with the following effects:

(a) ***Diffusion of phosphorous***

In the traditional process of crystal growing for solar cells, the silicon base material is doped with boron (to become *p*-type silicon) and ingots are grown. From these ingots thin wafers are cut out, which are processed into solar cells. Via a temperature-controlled diffusion process,[13] the top side of the wafer is doped with phosphorous.[14] At over 800 °C, one drives this phosphorous about 100–300 nm deep into the wafer.[15] By doing so, phosphorous atoms sit inside the crystal and occupy the place where normally a silicon atom should be. This is how the *pn*-junction is formed in mass production of standard solar cells. Due to the asymmetry between the negative and positive charge carriers, two effects now occur: charge carrier diffusion and drift through the creation of an electric field. These effects are illustrated in Fig. 5.6.

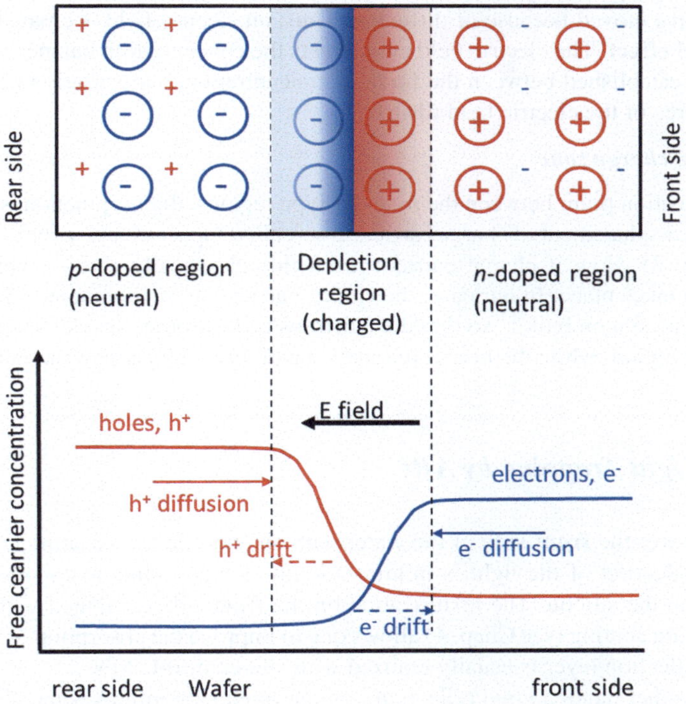

Fig. 5.6 Charge carrier diffusion, drift and space charge zone according to Fig. 3.9. The front side is doped with phosphorous (*n*-doped). The term “diffusion” in this figure means “charge carrier diffusion” (which take place in fully fabricated devices) and should not be confused with the term “phosphorous diffusion” which is used for the manufacturing process

[13]Some companies are also using ion implantation instead of diffusion.

[14]*n*-type wafer are doped with boron.

[15]The diffusion process deposits phosphorus around the surface of the solar cell, which will lead to a short circuit. Therefore the edges of the front side are etched away (Etch isolation) with a laser or a chemical bath and the front side is isolated from the back side. The phosphor-silicate glass (PSG) produced during phosphorous diffusion on the surface of the cell is also etched away (Phosphorous Glass Etching).

(b) ***Charge carrier diffusion***

First, both areas (*n*- and *p*-type) are electrically neutral. Due to the concentration gradient of electrons and holes, however, charge carrier compensation takes place: the electrons of the phosphorus-doped layer migrate into the boron-doped layer, because there are fewer electrons in that region, and conversely, the holes migrate into the phosphorus-doped layer. The **first effect** is thus a concentration-related diffusion. [16]

(c) ***Drift***

Now, holes are positively charged and electrons are negatively charged, and charge carriers have moved because of diffusion: Thus, an electric field is created. This is the **second effect**. The electric field counteracts the concentration balance and equilibrium is established between the force of concentration compensation (diffusion) and the force of the electric field (drift).

(d) ***Space charge zone***

At the transition point between the *p*- and *n*-type regions, the *pn*-junction is formed. Due to the displacement of charge carriers, a so-called "space charge zone" is formed (see Chap. 3). Now if charge carriers reach this electrically active zone, charge separation takes place. In our case, the charge carriers are predominantly generated in the *p*-type silicon bulk,[17] so that the electrons drift through the electric field into the *n*-type region, while the holes are repelled and thrown back into the bulk.

5.2.3 Light Trapping by ARC

Now we cover the **front side** of the wafer with an antireflection coating (ARC), so that the reflection of the light is minimized and as many photons as possible are absorbed in the silicon. The texturization on the front side combined with a good antireflection coating (see Chap. 4) allows one to improve the absorption of photons. The antireflection layer is usually realized with silicon nitride SiN_x.

On the other hand, a good reflectivity on the **back side** improves the absorption of low energy (long wavelength) photons, which are weakly absorbed by silicon and which pass through the entire solar cell and are reflected on the back side, back into the solar cell. This gives them a second chance to generate an electron-hole pair—in another passage, some more electron-hole pairs will be generated: the process can be repeated through multiple reflections. The better the light trapping is, the higher the electric current I_{sc}, which is generated, will be (see also Sect. 5.2.4d).

---

[16]This process can be clearly compared to a glass of water in which salt is sprinkled. Diffusion ensures that the concentration is equalized.

[17]The bulk is very much thicker than the *pn*-junction. Therefore, most of the charge carriers are generated in the bulk and not in the thin *n*-type layer.

5.2.4 *Passivation*

(a) **Bulk Passivation by Saturation of Impurities with Hydrogen**

In Chap. 4, we learned that impurities will be embedded into silicon—becoming, thus, recombination centres (defects), leading to Shockley-Read-Hall recombination. With the help of atomic hydrogen, it is possible to electrically deactivate (passivate) the impurities and to prevent electrons, which were generated as electron-hole pairs, from recombining there. Additionally, the antireflection layer composed of silicon nitride serves also as a source of hydrogen. On one hand, silicon nitride is a nearly ideal antireflection coating, and, on the other hand, it also contains a large proportion of hydrogen[18]—a proportion that can be adapted via the process control. The hydrogen atoms diffuse many micrometres into the silicon, even at temperatures around 200 °C and, thus, lead to a passivation of the defects. At temperatures > 800 °C, as used in the production of standard solar cells, the complete bulk is slowly flooded with hydrogen.

(b) **Bulk Passivation by Gettering**

In the production of the silicon crystal, foreign atoms can be incorporated in the silicon, defects can occur and dislocations generate stress in the crystal structure. All of these effects can cause recombination centres to be formed. In Fig. 5.7 some typical defects are illustrated.

The gettering process reduces contaminants in a wafer and increases the carrier lifetime. This process is not fully understood. One may just state the following: With the application of additional layers such as aluminium on the back and phosphorus on the front, regions are formed that act like a sink for foreign atoms [7]. Especially at high process temperatures, foreign atoms are easily released from the crystal compound and are pushed to the surface via substitution by P or Al. On the front side, so-called phosphorus glass is etched away after the phosphorus diffusion and, thus, the impurities are also removed. On the back, the foreign atoms are trapped in the aluminium. As a result, a part of the recombination centres can be eliminated.

(c) **Surface Passivation**

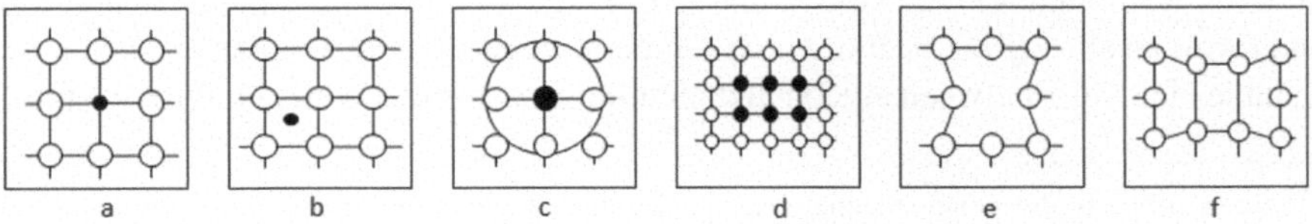

Fig. 5.7 Some typical defects in monocrystalline silicon. **a** a foreign atom (e.g. a dopant atom) occupies the place of a lattice atom (silicon atom); **b** a foreign atom lying in an interstitial space; **c** a substituted impurity generating stress in the crystal lattice structure; **d** precipitates (e.g. a group of oxygen atoms forming "lumps"; **e** vacancy (a position in the crystal lattice which is not occupied); **f** dislocation (crystallographic defect) in the lattice structure [6]

[18]The hydrogen enters into the silicon nitride layer during its fabrication process, which is based on a plasma (activated gas) containing silane, ammonia and hydrogen.

We have learned that recombination (see Chap. 4) can occur on the surface of the wafer because at that point the crystal structure is broken abruptly; creating so-called "dangling bonds" (see also Chap. 6). These recombination centres are also saturated by the hydrogen in the silicon nitride and can then no longer form recombination centres. Silicon nitride has, thus, two functions: (1) it acts as antireflection layer (leading to an increase in the short-circuit current I_{SC}) and (2) also as passivation[19] layer (leading to an increase in the open-circuit voltage V_{OC}). At the back of the solar cell we also have to ensure a passivation of the surface. There, we form a "Back Surface Field".

(d) **Back Surface Field**

Electron-hole pairs generated by long-wavelength light in the lower half of the wafer, where there is only weak diffusion and a negligible electrical field, should be collected, before they can recombine. Without any special measures they would be "sucked up" by the metal contact on the backside, where they would all recombine. To avoid this, one coats the backside with aluminium, which creates, thus, a small backside electric field. At higher process temperatures, the aluminium forms an electrically active alloy with the silicon, thereby establishing a heavily doped *p*-type region, generating a p^+/p junction. This reduces the effective surface recombination velocity of electrons at the back surface. As discussed in Chap. 4, surface recombination is one of the major loss mechanisms in solar cells. The so-called **Aluminium Back Surface Field (Al-BSF)** ensures that electrons, which are close to the back side, do not recombine there, but are directed back into the bulk of the silicon wafer. In addition to the passivation property, the aluminium layer simultaneously takes over the task of conducting the electric current on the back side—and, like the phosphorous in the case of phosphorus diffusion, also ensures that the silicon wafer is cleaned during the gettering process.

The backside metallization consists of an aluminium layer, which contains about 80% aluminium particles with a size of a few μm, and which is coated with a thin oxide layer. Another important component of the aluminium paste is 5% boron or borosilicate glass frits. The remainder consists of organic solvents to guarantee the rheology[20] and viscosity,[21] so that the paste can be printed using a Screen Printing process. At around 800 °C, silicon dissolves in the aluminium during the firing step and forms an alloy [8]. The entire high-temperature process takes about 40 s; thereby a stable eutectic[22] between silicon and aluminium is formed with about (by weight)

---

[19]Silicon nitride is also positively charged. The positive charge in the silicon nitride improves the accumulation of electrons at the front surface, creating an accumulation region in the *n*-type doped emitter and thereby reducing the effective surface recombination velocity.

[20]Rheology (from ancient Greek ῥεῖν "rhein") is the science that deals with the deformation and flow behaviour of materials.

[21]Viscosity is a property of liquids (fluids) and gases. The higher the viscosity, the thicker the fluid; the lower the viscosity, the thinner (more flowing) the fluid is.

[22]Eutectic alloys have a clearly definable melting point, the so-called eutectic point. Because in eutectics all components solidify simultaneously and this occurs at a much lower temperature than would be the case with the pure components, a fine and uniform microstructure is formed.

12.6% silicon in aluminium. In this process, aluminium atoms diffuse into the silicon crystal and displace the silicon atoms from the lattice sites. As a result, in addition to the boron doping of silicon, doping by the trivalent aluminium atoms takes place. This results in a small potential jump similar to the that of the *pn*-junction (see also Fig. 5.10 right): The potential jump ensures that electrons generated by sunlight are directed from the back to the front. The probability that electrons recombine at the back is, thus, significantly reduced. Thanks to the Aluminium Back Surface Field, an improvement in the effective surface recombination velocity is achieved. It is about 200–600 cm s^{-1} and is, thus, several orders of magnitude lower than in an aluminium-coated cell without the firing process.

5.2.5 *Metallization*

(a) **Silver Pastes**

On the front there is an antireflective layer consisting of SiN_x (silicon nitride) and on the back there is an aluminium layer. The aluminium layer conducts very well, so that we can easily take off the current on the back. On the front side, this is not possible because silicon nitride is an insulator. For this reason, metal layers are printed on the **front side** by employing a Screen Printing process. The pastes used for the Screen Printing process contain silver flakes and glass frits. At about 800 °C, glass frits are fired through the silicon nitride and make an Ohmic contact with the silicon. In this way, depending on cell size and layout, about 70–90 thin **fingers** (see also Chap. 9) in the range of 50 μm, and about 5 **busbars** are connected to the front side of the solar cell. During the construction of a solar module, one solders copper ribbons on the busbars, and the current can be taken away through these copper ribbons. To make the **back side** solderable, busbars containing silver are also printed on the backside of the cell before the rest of the backside is covered by an aluminium layer; aluminium itself is not easily solderable. Thus, the free charge carriers can leave the solar cell via the metal contacts on both sides and contribute to the flow of current in an external circuit.

(b) **Cell connections in the solar module** (see also Chap. 9)

With the *p*-type wafer material, the front side delivers negative and the back side positive charge carriers. For the construction of a whole module, the copper ribbons are used to connect the front of the first cell to the back of the second cell, etc. This corresponds to a series connection of the solar cells and leads to an addition of the individual voltages, whilst the current remains constant. Typically, 1–12 cells are processed into a so-called "string". Six strings are then connected in series to finally make up a whole module.

5.3 Functionality and Losses of the Al-BSF Cell

5.3.1 Architecture of the Al-BSF Cell

In the previous Sect. 5.1 we developed the fabrication processes and the physical concepts leading to the standard solar cell: to the so called Al-BSF solar cell. The Al-BSF cell (Aluminium Back Surface Field cell) consists of *p*-type silicon and has a cross-section, which is shown in Fig. 5.8.

A homogeneous emitter (*pn*-junction) represented as a n^{++} layer[23] heavily doped with phosphorous and having a thickness of ~300 nm is located on the textured silicon surface. In addition, there is an antireflection layer of silicon nitride SiN_x, with a thickness of about 80 nm, which greatly reduces the reflectance of the surface, so that as much light as possible is absorbed. During the deposition of the silicon nitride layer, it is enriched with hydrogen during the plasma process in order to passivate the surface of the cell. Hydrogen atoms are very small and diffuse easily through every material, especially at higher temperatures; and they do this also during the production of the standard solar cell. Hydrogen acts as a positive, negative or neutral atom and occupies open bonds of the crystal atoms. Electrons cannot recombine later

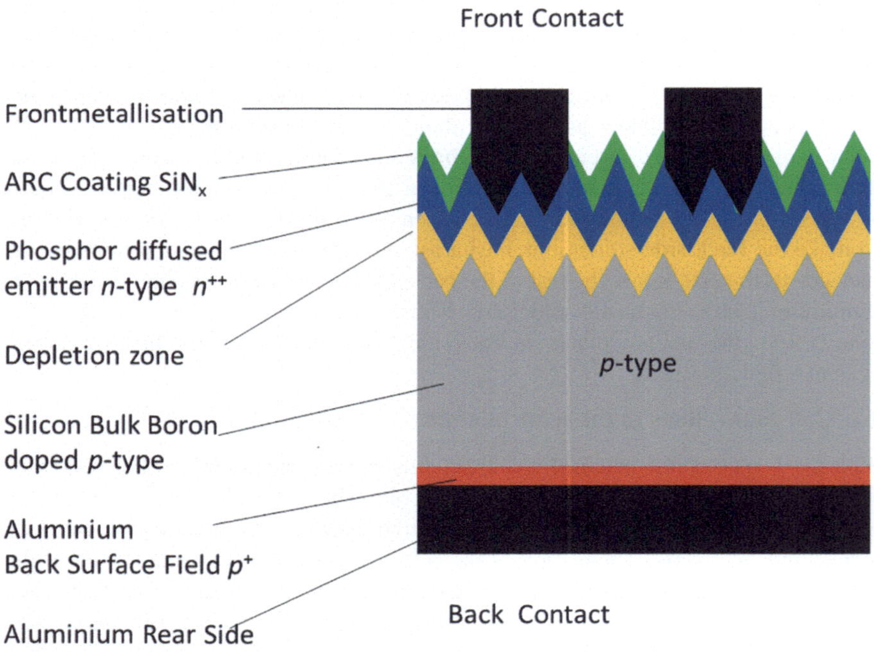

Fig. 5.8 Cross section of a Al-BSF homojunction solar cell with *p*-type wafer material

[23] n^{++} means higher doped with phosphorous than n^{+}. p^{++} means higher doped with boron than p^{+}.

at these defects—because they are already occupied by hydrogen—and their lifetime increases. On the reverse side, an aluminium layer with a thickness of ~30 μm is applied by screen-printing. A subsequent firing process leads to contact formation via an Al–Si alloy. This forms an electric field (Back Surface Field, BSF).

5.3.2 *Band Diagram*

We begin with a description of Fig. 5.9, so that we can apply it to the cell structure in Fig. 5.10. In Fig. 5.9, the band diagram is shown for an undoped and a doped semiconductor [9], as well as the Fermi level.[24] Under sunlight, the electrons flow in the direction of the *n*-side and the holes in the direction of the *p*-side.

The bandgap energy E_g is 1.12 eV for silicon. When irradiated by sunlight, carriers are generated. The field in the space charge zone causes the electrons in the conduction band to flow to the *n*-side, e.g. energetically downwards, and the holes in the valence band slide to flow to the *p*-side.

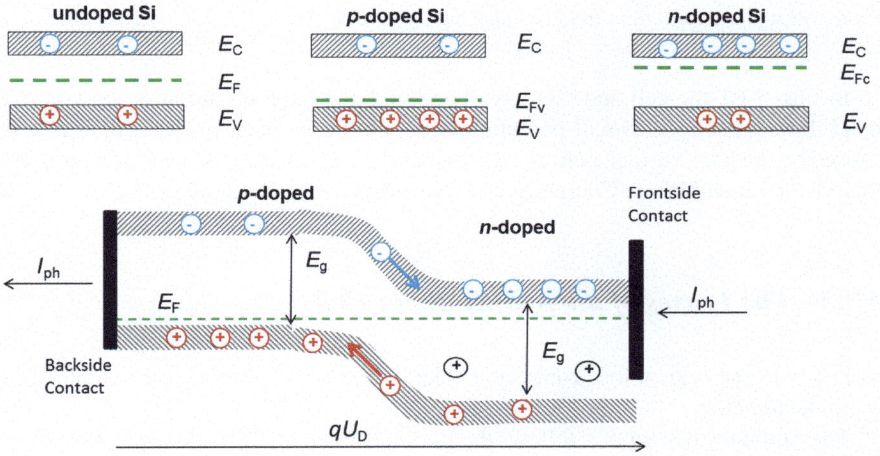

Fig. 5.9 The Fermi level for undoped silicon (top, left), for *p*-doped silicon (top, middle) and for *n*-doped silicon (top, right). In the undoped case the Fermi level is exactly in the centre of the bandgap. It is pushed down for *p*-doped silicon and it is pushed up for *n*-doped silicon. When one puts a *p*-doped silicon region and a *n*-doped silicon region together in the same device, the various Fermi levels have to align themselves, to form a single level for the whole device (lower part of the figure). This results in a potential step for conduction and valence band

[24]The Fermi level is the energy where the probability is just 50% to occupy this level. For undoped semiconductors, the Fermi level is exactly in the middle of the bandgap. For doped semiconductors, the Fermi level is near the edge of the conduction band for *n*-type material, since many electrons are present (*n*-doping) or close to the valence band edge in *p*-doped semiconductors.

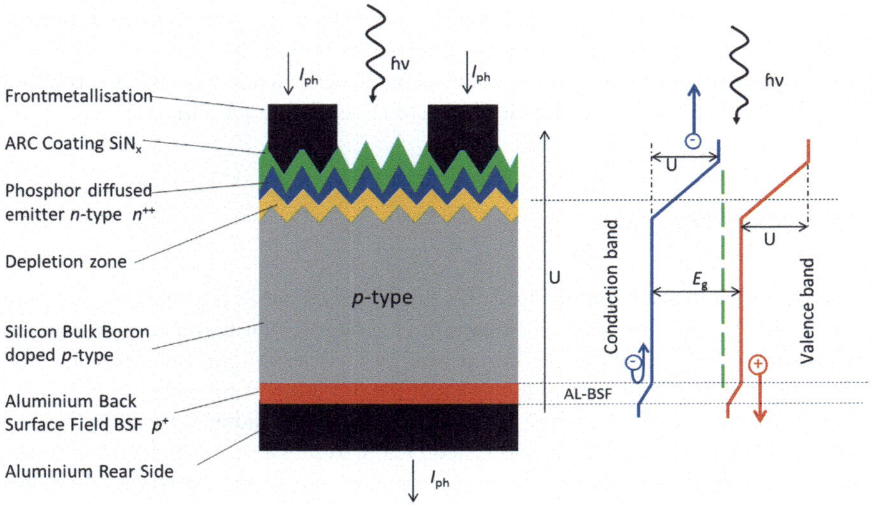

Fig. 5.10 The Al-BSF solar cell: cross section (left) with the corresponding bandgap diagram; the bandgap diagram (right) shows the increase in potential through the Aluminium Back Surface Field, which protects electrons from recombining at the back

In Fig. 5.10, the cell and the corresponding band diagram are now shown. The band diagram shows a small potential mountain on the back of the cell, which is caused by the back surface field. This repulses the electrons on the back and prevents them from recombining. Charge carrier transport is visualized in Fig. 5.11.

5.3.3 *The Losses of the Al-BSF Solar Cell*

In Fig. 5.11 the typical loss zones of a solar cell are marked; they are explained in more detail below.

The following description follows the 7 points in Fig. 5.12. The losses listed below are based on their geometric origins within the cell.

① (a) The photon has too little energy or the light strikes the solar cell at an incorrect angle. It will be reflected at the surface or it may enter silicon but will not be absorbed, it may be reflected by the back surface or escape silicon
   (b) Parasitic absorption by free carriers (FCA)

② (a) Recombination losses underneath the metal contact
   (b) Ohmic losses on the metal contact
   (c) Shading losses due to the metal contacts
   (d) Recombination at the surface

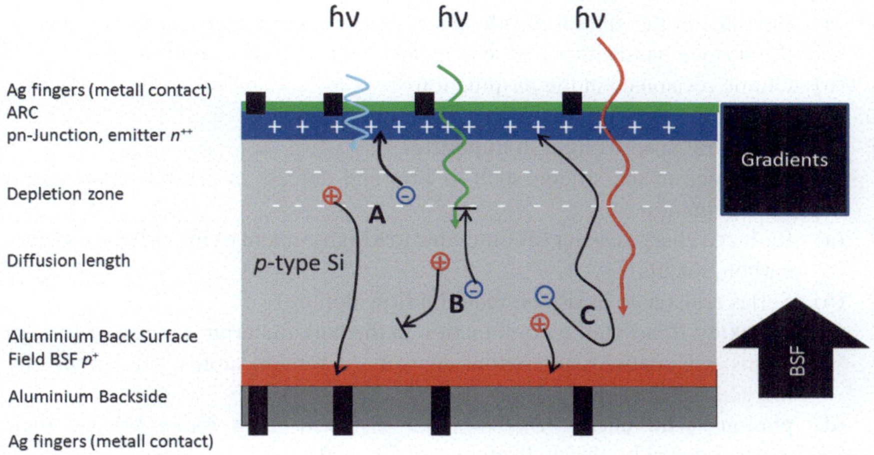

Fig. 5.11 Simplified charge carrier transport. Blue light with a short wavelength does not penetrate deep into the solar cell. Red light penetrates far into the solar cell. The electron-hole-pairs are now forced through different gradients within the space charge zone. The gradients are generated by chemical potentials or electrical potentials resulting in Drift and Diffusion. Situation A: The electrons are driven upwards and the holes downwards and reach the contacts. Situation B: Some electrons and holes recombine at impurities in the crystal (or at the surface). Situation C: The Al-BSF pushes electrons back and prevents them from recombining at the back

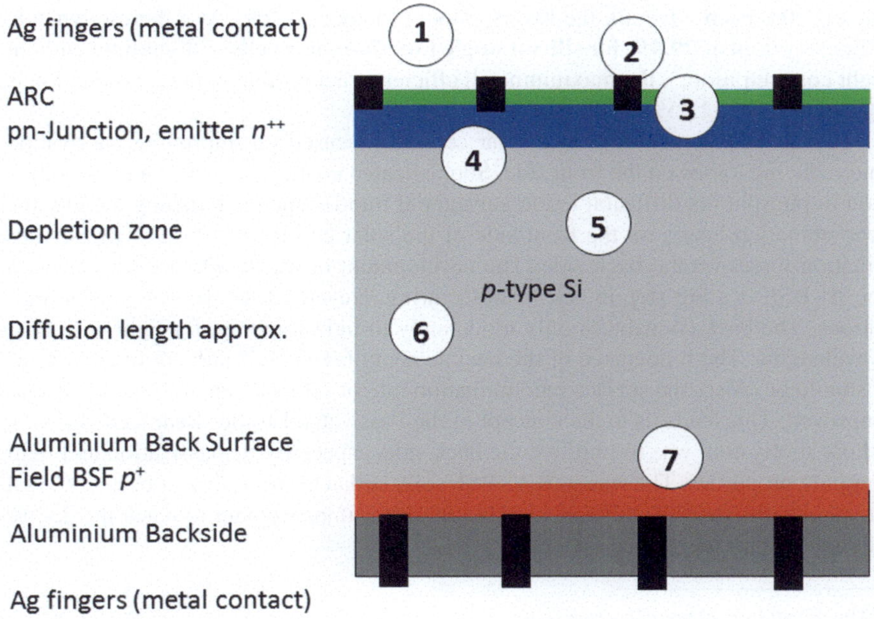

Fig. 5.12 Typical loss zones (1–7) of a standard solar cell

③ (a) Pinholes in the silicon nitride layer, which affect negatively the quality of the surface passivation and lead locally to increased recombination
(b) Ohmic resistance in the *pn*-junction

④ Auger recombination

⑤ (a) Scattering absorbs light in impurities.
(b) Impurities in the silicon such as iron and defects in crystal lattice cause recombination.

⑥ (a) Reduced charge carrier lifetimes due to a high content of impurities (oxygen, carbon, metals)
(b) Series resistance in silicon material (low doping)
(c) Shockley-Read-Hall recombination in the bulk material because of foreign atoms, dislocations, interstitial deposits of foreign atoms, precipitates by oxygen, vacancies
(d) photons with energy $E_{ph} = h\nu > E_{g.}$ will most likely release their excess energy by thermalisation (see Chap. 3)

⑦ Recombination on the back side in spite of passivation and back surface field.

The losses can be also classified according to their nature:

(a) Optical losses: ①, ②c, ⑤a, ⑥d
(b) Recombination losses: ②a, ②d, ③a, ④, ⑤b, ⑥a, ⑥c
(c) Ohmic losses: ②b, ③b, ⑥b.

They can be split up as follows: recombination losses: 2.5%; Ohmic and optical losses 2.0% each. Overall, the losses amount to about 6–7%. At a thermodynamic efficiency limit of **29.4 % for silicon single junction solar cells** with sunlight without light concentration,[25] the maximum cell efficiency achievable in mass production is approximately ~23.5% (=29.4–6%).

The development of standard solar cell was focused on improving passivation and reducing losses on the front side. Sophisticated coating processes and optimization in phosphorus diffusion led to substantial improvements, reducing thereby the recombination losses on the front side of the solar cell to one third of the recombination losses on the back side. The development of the standard solar cell with the Al-BSF is a big step in cell manufacturing. Nevertheless, the cell shows limitations. The back contributes only moderately to light trapping, for light with long wavelengths. The reflectance of the back is about 60–70%,[26] and not over 90%, as it should be. Also, the surface recombination rate of 200–600 cm s^{-1} can be further improved. This leads us to the concept of the Passivated Emitter Rear Cell (PERC), whose motivation was to improve the back side, especially its passivation and light trapping properties. The standard Al-BSF solar cell achieves today, at best, 20% cell efficiency. So, there is still considerable room for improvement to reach the 23.5% efficiency limit mentioned above.

[25]This is valid for all homojunction cells.

[26]Long-wave light penetrates the solar cell and can escape from the cell on the back. To prevent this, the light-trapping on the back can be improved so that the light on the back is reflected and directed back into the cell.

5.4 Motivation for the Development of the PERC Cell (Passivated Emitter Rear Cell)

In the following, some important physical relationships are explained, in order to improve the Al-BSF cell. This shows where exactly we can improve the Al-BSF cell. And this leads us to the concept of the PERC cell (Passivated Emitter Rear Cell).

5.4.1 Lifetime and Diffusion Length

Lifetime and diffusion length are described in Chap. 3. Here we consider the absolute magnitude of lifetime and diffusion length in a solar cell. In very good quality Cz mono-crystalline silicon wafers, the lifetime can be several milliseconds (ms). For example, assuming a lifetime of 1 ms, the diffusion length L of electrons in p-type silicon is approximately 1900 µm, with a boron doping of about 10^{16} cm^{-3}. This is the statistical distance that electrons can migrate on an average before recombining. In comparison to the thickness of the solar cell of typically 160 to 180 µm, the average transmission distance of the electron is thus far enough to reach the front-side solar cell surface and reach the n-region via the space charge zone. If it does not reach the front of the solar cell within its lifetime, the electron recombines with a hole, loses its energy mainly in the form of heat and falls back into the valence band. The lifetime is, thus, a measure of the quality of the wafer material at a given doping concentration. At a first glance it seems that lifetime is not the limiting factor.

5.4.2 Doping Versus Recombination

The more the silicon is doped, the higher the recombination rate will be, because there will be more impurities. Particularly in p-type silicon, electrically active complexes of boron and oxygen can be formed, which act as strong recombination centres. On the other hand, if the material is not so strongly doped, the dark current (J_0, see Chap. 3) increases. This can be imagined in such a way that with low doping less excess carriers are generated and the few recombine at the metal contacts. However, if the lifetime of the silicon is increased, high cell efficiencies can also be achieved with lowly doped solar cells. This requires better silicon material. If the pn-junction is lowly doped the sheet resistivity (see Chap. 4) is increased and can reach more than 150 Ω/square. Today it is assumed that the optimal sheet resistance is approx. 100 Ω/square with regards to the most commonly used metallization paste materials.

The contact resistance between the phosphorus-doped n-region and the contacting pastes on the front side also depend on the doping concentration in the silicon and on the quality of the silicon (lifetime). Lifetime and Doping must be balanced. Conclusion: The higher the doping, the better one can contact the silicon. Conversely, the higher the doping, the lower is the lifetime of the minority carriers. Here there is a conflict between doping and lifetime.

5.4.3 Surface Recombination Velocity on Front and Back Sides

However, some of the **most active recombination centres** are at the surfaces of the silicon wafer with the partially unsaturated bonds of the surface atoms. Indeed, at these surfaces, the crystal structure is abruptly interrupted in its periodicity. Defects occur which provide energy levels within the bandgap, thereby making it easier for the carriers to recombine. This also changes the profile of the bands and it creates band distortions. As discussed in Chap. 4, the surface recombination velocity S is a measure of how fast a charge carrier recombines on the surface. For open, unpassivated silicon surfaces, it is in the range of 10^5–10^6 cm s^{-1} and drastically limits the open-circuit voltage V_{OC} of the solar cell. This can be visualized in such a way that the minority carriers (in *p*-type wafers the electrons) flow with the speed S to the surface and recombine there, emptying, thus, the space charge zone. For an increase in V_{OC} and an associated increase in the efficiency of the solar cell, it is therefore imperative to minimize the surface recombination velocity by passivation of the surface, e.g. by deactivation of the recombination centres.[27] We can now discuss how that can be done at the front and back surfaces.

Surface Recombination on the Front Side

As the standard solar cell has demonstrated, excellent surface passivation has been achieved thanks to silicon nitride. The recombination velocity is about 200–600 cm s^{-1} for standard solar cells (Al-BSF-cells), so that an improvement over unpassivated silicon surfaces by several orders of magnitude was obtained.

Surface Recombination on the Back Side

The surface recombination velocity of the solar cell on the back side is 500–1000 cm s^{-1} and is about three times higher than that on the front side. In standard solar cells (Al-BSF cells), the passivation of the back side usually takes place via the metallization with an aluminium layer. Aluminium has the advantage that it is, like boron, trivalent and results in *p*-doping of silicon. If aluminium diffuses into a *p*-doped (boron-doped) silicon crystal, the doping concentration at this point increases, so that even more holes are created on the back side and the Fermi level

[27] A measure for **assessing the quality of the passivation** is the open circuit voltage V_{OC}. The better the passivation, the less electrons recombine and the higher open circuit voltage V_{OC}. On average, V_{oc} is 650 mV for standard solar cells. The recombination velocity depends strongly on the surface doping concentration.

continues to approach the edge of the valence band. For the solar cell, this means that a near-surface electric field is formed on the back side, which pushes the minority charge carriers (in the case of *p*-doping, the electrons) back into the crystal bulk and, thus, prevents recombination at the back. Expressed numerically, this reduces the back surface recombination velocity S_{rear} from 10^5 to 10^6 cm s^{-1} to the range of 500–1000 cm s^{-1}. Such values of S_{rear} were sufficient for a long time, since recombination in the highly doped emitter (*pn*-junction) and at the interface gave rise to even higher recombination velocities on the front side, and, thus, at that time, the back side did not represent the limiting zone for the efficiency of the solar cell. As a result of a worldwide focus on the optimization of the front side of the solar cell, based on the development of new silver pastes for contacting high-impedance *pn*-junctions (emitters) the back side of the solar cells became (around 2010) the focus of attention as a limiting factor in cell efficiency.

5.4.4 *The Structure of the PERC Cell*

In particular, the PERC solar cell concept (Passivated Emitter Rear Cell) has shown a pronounced potential for increasing performance with only moderate changes in the cell production process. The advantage of this cell concept is that it has an "evolutionary" character, whereby the existing process plants for standard solar cells are preserved, and only small expenses for additional equipment are necessary. In the following, therefore, the concept of the PERC cell will be introduced; the differences between the standard cell (Al-BSF cell) and the PERC cell will be discussed in more detail. Figure 5.13 illustrates the fabrication process sequences for the two cell concepts. It should be noted that the three coating processes in the PERC cell can be realized by a single machine.

Figure 5.14 shows the PERC cell in cross-section. In order to reduce the surface recombination velocity, the entire back side is first treated with a 1–2 nm thick, high-quality silicon oxide layer.[28] This oxide layer (SiO_2) has a similar crystal structure as silicon, thus, very effectively saturating the free surface bonds of the silicon and forming an excellent interface with the silicon crystal. Thereafter, the cell is passivated with a non-conductive dielectric layer. Aluminium oxide is particularly suitable for this, because aluminium oxide is an excellent insulator. In contrast to SiO_2, Al_2O_3 has a rhombohedral crystal structure. This leads to a mismatch with SiO_2, whereby the Al atoms of the Al_2O_3 compound have, at the boundary to SiO_2, a free bond that tends to trap electrons, forming thereby static, negative charges. The result is a so-called "field effect"[29] that additionally protects the minority charge carriers (the electrons) in silicon from recombination at the SiO_2 interface [10]. In this way, surface recombination velocities of 10–20 cm s^{-1} can be achieved industrially.

[28]This is optional and can be done (as usual) by oxidation (not deposition).

[29]By "field effect" we wish to indicate that there is an electric field, which is built up like in the Al-BSF cell, so that the electrons on the back side are reflected back into the cell, before recombining.

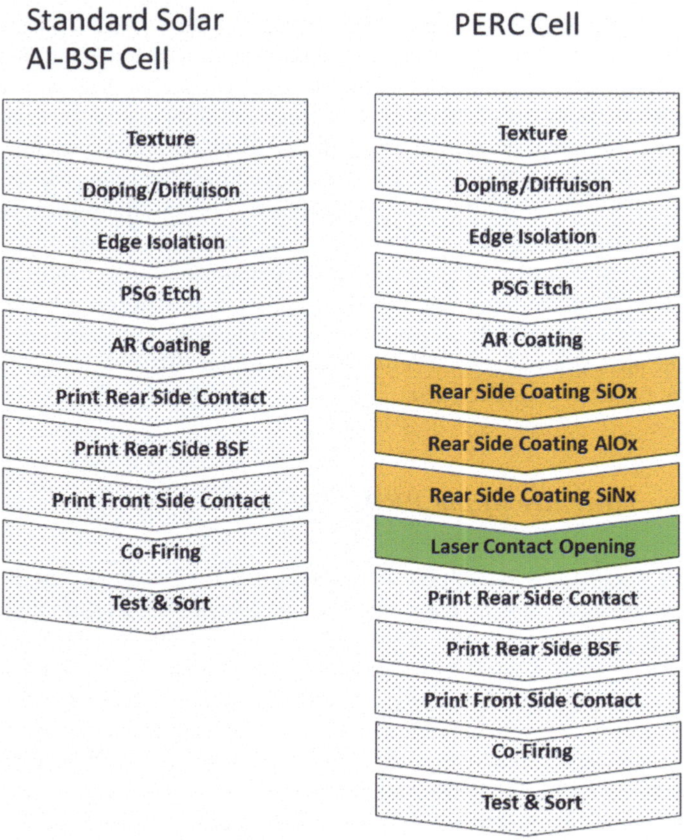

Fig. 5.13 Schematic comparison of the two basic cell concepts: standard Al-BSF cell (left) and PERC cell (right). It is easy to see that the PERC concept focuses exclusively on the back of the solar cell. The Back Side Coating consists of three layers: SiO_x, AlO_x, SiN_x. The sensitive aluminium oxide (AlO_x) layer is protected by the silicon nitride layer (SiN_x). For PERC cells, only two additional machine processes are needed: the coating process and the laser contact-opening process. In the PERC process, the back side SiO_x coating is optional and can also be done by oxidation. PSG means: Phosphor-silicate Glass and is described in footnote 15. Edge isolation is described as well in footnote 15

On special samples a reduction of the surface recombination velocity to less than 5 cm s^{-1} is even possible.

For various reasons such as cost, throughput and excellence of layer quality, the PECVD process has become established in the industrial environment, for the deposition of the $Al_2\,O_3$ layer. Here, trimethylaluminium (TMA) and N_2O react in a plasma-assisted process according to the reaction equation

$$24N_2O + 2Al(CH_3)_3 \rightarrow 24N_2 + Al_2O_3 + 6CO_2 + 9H_2O \tag{5.8}$$

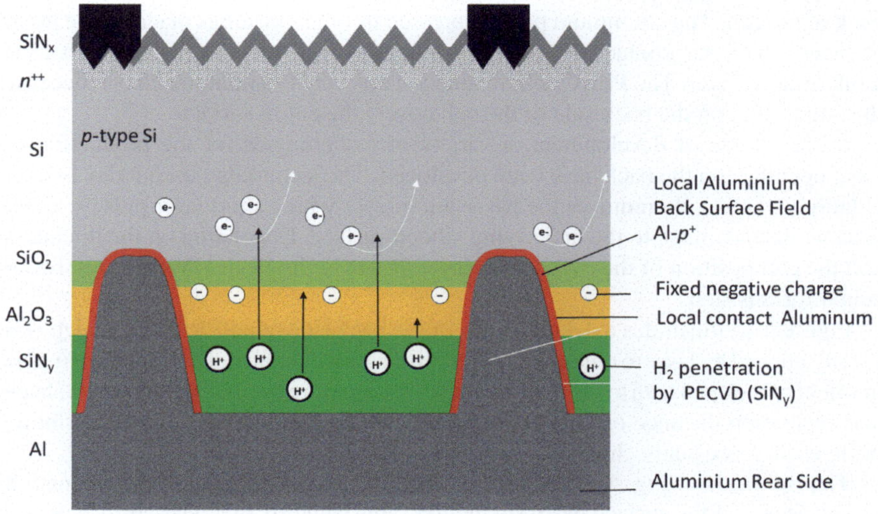

Fig. 5.14 A typical PERC cell. The drawing is not to scale. See the text for a complete description of the processes and interactions shown in the figure

The Al_2O_3 layer is highly hygroscopic and binds water vapour very easily. The layer is, thus, sensitive to moisture. This means that one needs a protective layer, a SiN_y layer, which is usually implanted in the industrial environment using the same PECVD method. In addition to the protective effect, an optical adaptation of the back side of the solar cell is thereby achieved (see Chap. 4, Light Trapping), provided the layer thickness is correctly chosen. Low-energy photons with large wavelengths, which have penetrated through the entire solar cell can be reflected back into the solar cell very effectively through such an optimized layer stack, consisting of Al_2O_3 and SiN_y: the stack functions as an optical mirror. In this way, the photons get another chance to create an electron-hole pair in the bulk silicon. Compared to the Al-BSF standard cell with an average back-reflection of 65%, the dielectric layers of the PERC technology can increase the average back-reflection to 89%.

In contrast to the standard Al-BSF solar cell, the silicon nitride passivation layer (SiN_y) hinders the actual contacting of the cell with the metallization paste, because SiN_y is an insulator and the generated charge carriers therefore cannot contribute to the current flow. To counteract this problem, we must enable a local contact between the metal paste and the underlying silicon wafer. This is done by locally opening the SiN_y layer as well as the Al_2O_3 layer with a laser, so that the backside aluminium paste can be connected to the silicon and forms a contact. If these local contacts, that is to say the laser openings, are located at a certain distance from each other, the corresponding charge carriers can reach them. The distance between the contact points has an influence on the series resistance. The design and optimization of the openings, their size and their distance from each other has been the subject of many studies. Figure 5.13 shows the cross-section of a PERC cell and a "top view" of the back of the cell. The aluminium paste is pressed through openings made by the laser, to ensure an Ohmic contact. The laser openings can be connected to one another via a full-area Al layer. The PERC cell obtained in this way is a monofacial cell because the aluminium on the back side of the cell covers the entire surface.

In the course of development, a variety of technologies for the production of local openings on the back have been developed. The technology used today is laser ablation: One locally removes the passivation layer with a short laser pulse without causing deep damage to the underlying silicon surface. Depending on the thickness and the composition of the passivation layer, lasers in the IR or UV range are better suited for this step.

Figure 5.15 illustrates a detailed view of the back side of a PERC cell, which was locally opened by laser to obtain a circular form with a diameter of about 160 μm. The quotient of the open area to the total area gives the aperture ratio *f*. Depending on shape and separation distance of the openings and also on the type of Al screen-printing paste used, *f* is usually chosen to be between 1 and 10%.

The right side of Fig. 5.15 shows the close-up of a point that has been opened. It can be seen that the surface has been melted. This suggests that near-surface cracks have formed in the silicon due to the thermal stress. If the damage is too high, there will be significant losses in the cell parameters such as open circuit voltage and fill factor. A short, wet-chemical over-etching of the back side can be used to remove laser damage and avoid associated performance losses.

The metallization of the back is done by screen printing according to the procedure for standard Al-BSF solar cells. The subsequent firing of the Al paste, however, requires greater care. Micrographs for prepared cross-sections of metallized PERC back sides fired with different parameters are shown in Fig. 5.16. Below is the silicon crystal on which a layer of Al paste lies. One can clearly recognize the granular structure of the paste. Due to its small thickness of less than 20 nm, the Al_2O_3 passivation layer can hardly been seen in Fig. 5.16. Due the inevitable presence of cavities there is a risk of void formation. Depending on the firing temperature of

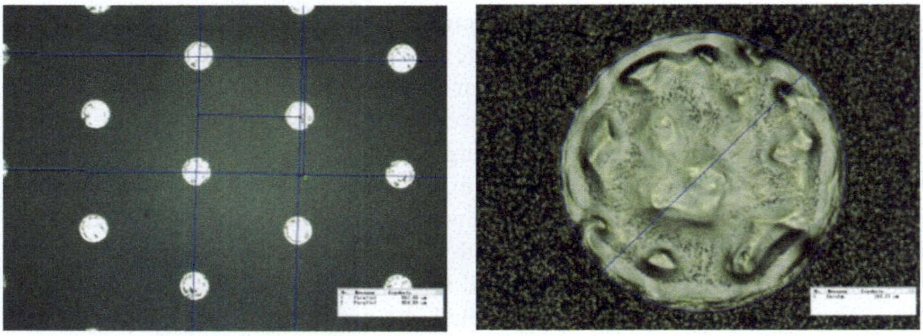

Fig. 5.15 Detailed view of the back side of a PERC cell, which was locally opened by laser, with circular openings of about 160 μm diameter. The distance between the openings is 740 μm, so that in this example, an aperture ratio $f \approx 4.2\%$ is obtained. On the right the surface of a contact opening, produced by laser is represented. *Courtesy* Meyer Burger Technology AG

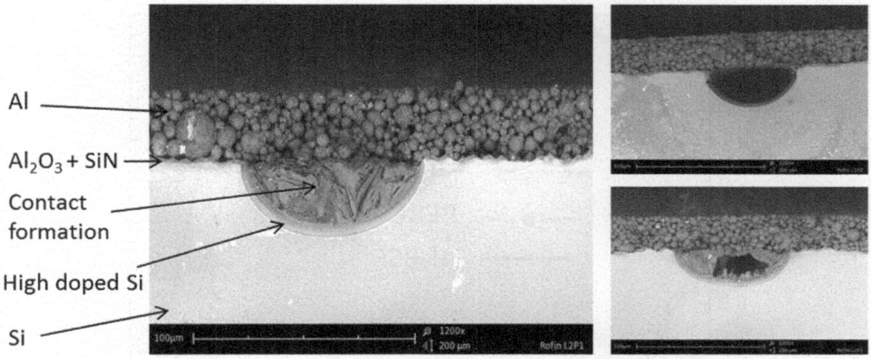

Fig. 5.16 Prepared cross-sections (grind cuts) of local openings of the passivation layer after metallization and firing. Depending on the firing parameters, an Al–Si alloy forms in the laser aperture (left) or voids are formed in the silicon (right). *Courtesy* Meyer Burger Technology AG

about 800 °C, the heating and cooling ramps, the paste composition, but also the aperture ratio and the thickness of the Al paste, the areas below the local openings remain filled or so-called "voids" arise. The underlying mechanism for the formation of voids is described in more detail in [11]. Obviously, in the worst case, the high firing temperature between 750 and 800 °C can generate voids. As a result, charge carriers that have to pass from the crystal via the contact into the Al layer can pass only at the edge of the cavity; thereby, the electrical resistance is increased.

If one chooses the process conditions favourably, the contact area remains filled and a silicon-aluminium alloy is formed. At the edge of the contact area, an approximately 2–5 μm wide strip is observed. This strip consists of highly doped silicon and is responsible for the creation of a Back Surface Field. One obtains a back side passivation, which is characterized by the average surface recombination velocity, averaged over the entire surface of the back side. The surface recombination velocity, which was introduced in Chap. 4, can be divided into two parts: Into a first part, where there are no laser openings and where an excellent passivation is obtained with a surface recombination velocity $S_{pass} \approx 10$–$20\ \text{cms}^{-1}$, and into a second area with the laser openings, where the passivation is interrupted, and the surface recombination velocity S_{met} is 500–1000 cm^{-1}, e.g. here S is in the range of what is obtained for standard Al-BSF solar cells. The average surface recombination velocity was derived in the literature by using an experimental approach to loss analysis (see [8, 12]). The average surface recombination velocity S_{back} for the back side is thereby found to be:

$$S_{back} = \left(\frac{R_{bulk}\,(p) - \rho W}{\rho D} + \frac{1}{f' S_{met}} \right)^{-1} + \frac{S_{pass}}{1 - f'} \tag{5.14}$$

$R_{bulk}(p)$ [$\Omega\ \text{cm}^2$] series resistance, which depends on the contact geometry,
ρ [Ω cm] resistivity of the wafer,

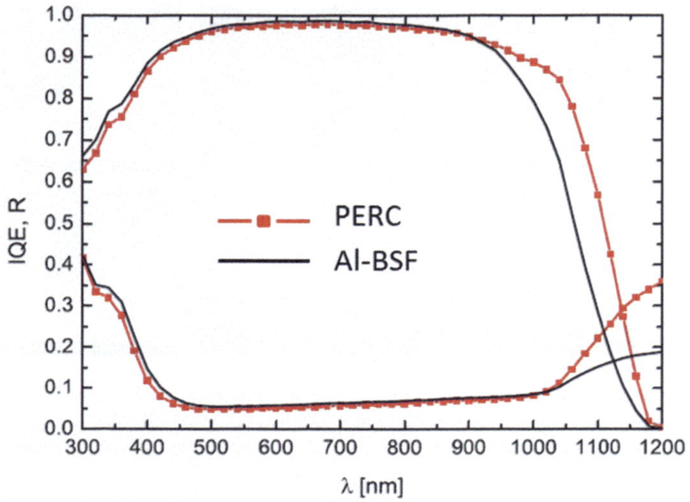

Fig. 5.17 Internal quantum efficiency, IQE (upper curves) of a PERC and of a standard Al-BSF solar cell. Also shown are the associated reflection curves (lower curves). In the longer wavelength range, from 900 nm onwards, the advantages of a PERC backside compared to the standard Al-BSF backside are clearly visible [7]

W [cm]	thickness of the wafer,
D [$cm^2\ s^{-1}$]	diffusion coefficient of the minority carriers (here electrons in a p-type substrate, so that $D \sim 36\ cm^2\ s^{-1}$),
f' [%]	aperture ratio,
S_{met} [$cm\ s^{-1}$]	surface recombination velocity of the local contacts,
S_{pass} [$cm\ s^{-1}$]	surface recombination of the passivated intermediate regions.

After laser processing, the original aperture ratio f changes by up to 50% due to the aluminium/silicon alloy formation, becoming $f'\left(\frac{f'}{f} \approx 1.5\right)$. Today, values for S_{back} between 60 and 80 cm s^{-1} are achieved.

What is the benefit of improved backside passivation for the solar cell parameters?

For this, we first consider the internal quantum efficiency[30] (IQE) of a PERC cell compare it with that of a standard Al-BSF solar cell. Figure 5.17 compares the internal quantum efficiency curves and the associated reflection curves of a PERC cell with those of a standard Al-BSF solar cell. While hardly any differences can be observed in the short-wave spectral range (for blue light i.e. for light that is absorbed near the front), there is a clear difference in the long-wave spectral range (red light) from about 900 nm onwards. The energy yield[31] of the PERC cell has significant

[30]Definitions of internal quantum efficiency (IQE) and external quantum efficiency see Chap. 3, Sect. 3.6.

[31]By **"energy yield"** we mean here the electrical energy delivered by the cell at MPP (maximum power point) divided by the energy of light reaching the cell.

advantages and is in places 30% higher than the energy yield of a standard cell. In addition, it can be observed that the reflection for the PERC cell increases from about 1000 nm onwards. The reason for this is the improved optical mirror formed by the passivation layer on the back side of the cell: this improved optical mirror allows a second passage of unused photons through the solar cell, thereby giving them a second chance to generate an electron-hole pair.

In addition to the improved optical mirror on the back and the associated higher short-circuit current density J_{SC}, higher open-circuit voltages V_{OC} are found in PERC solar cells. V_{OC} is a function of the saturation current density J_0 and, thus, depends on the quality of the surface passivation—which is excellent in PERC cells. The fill factor FF increases with the increase in V_{oc}, but, on the other hand, FF also depends on the series resistance R_s. The contacts on the PERC back side are limited to very small laser openings. Thus, the charge carriers have to make a longer path to reach the contact. The series resistance R_s is thereby increased. As a consequence, the fill factor FF tends to be lower for PERC solar cells, but this loss is minimised in well-designed PERC cells and is compensated by the increase in V_{oc} and the increase in short-circuit density J_{sc}.

A further development is the bifacial design of the back side. This is illustrated in Fig. 5.18. In PERC bifacial solar cells the back side has to be transparent. This is achieved by using aluminium tracks, joined together in one direction, instead of a full aluminium layer. Indeed, the aluminium layer is interrupted in the bifacial design. Between the aluminium tracks the back SiN_y is lying

As a result, stray light that hits the back of the solar cell due to the albedo[32] effect (see Chaps. 2 and 10) can be captured. However, less light coming in on the front side is captured by the solar cell, as a part of the this light exits from the back of the solar cell, without having been used; this occurs, because the aluminium mirror on the back is now partially transparent. Furthermore, the resistance losses also increase due to the lower conductance of the narrow Al tracks.

On the whole, the positive effects of the rear side of the PERC predominate and lead compared to the standard solar cell to a significant increase in efficiency—of more than $1\%_{abs}$. Today's PERC cells achieve efficiencies of 22–23%. Aspirations are aimed at further improving the passivation, for example via a passivation of the contact area as promoted by the TOPCon[33] cell concept. The TOPCon (Tunnel Oxide Polysilicon Contact) cell concept is described in Sect. 5.5.2. Cell concepts with passivation on both sides, like the concept of the HJT cell will be discussed in the next chapter (Chap. 7).

---

[32] Albedo is here the ratio between the light striking the back of the cell and the light striking the front. (See also Chaps. 2 and 10)

[33] The TOPCon cell concept was first presented by Fraunhofer (ISE) in 2013.

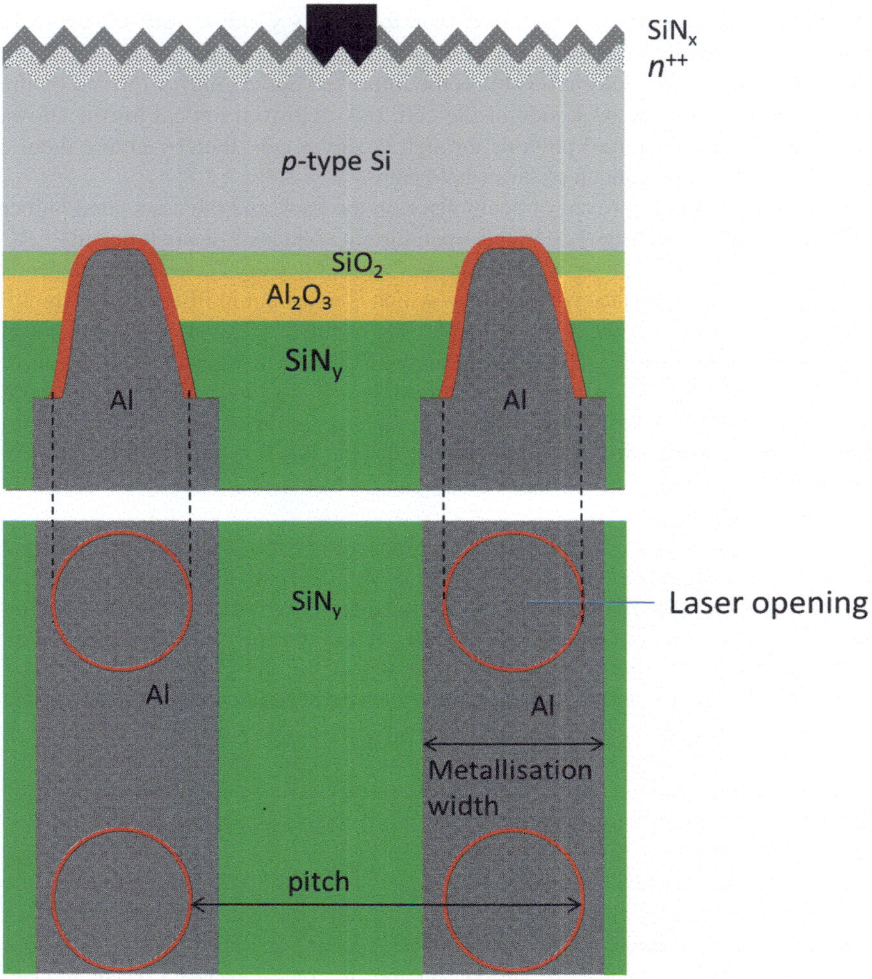

Fig. 5.18 Basic structure of the PERC back side in bifacial design. Above is the cross-section of the PERC cell and below is the arrangement of the laser openings on the back side

5.5 Other Homojunction Cell Concepts

5.5.1 PERT Solar Cells (Passivated Emitter and Rear Totally Diffused)

We have learnt that in the Al-BSF cell a back-surface field could be built up internally thanks to the aluminium paste. As a result, the electrons at the back side of the cell are repelled and sent back into the cell; they are then collected at the front side, where they finally leave the solar cell and contribute to the production of electricity. In the passivated emitter rear contact (PERC) cell, the back surface has been improved with respect to the Al-BSF cell, by building up an Al_2O_3 layer, covered by a protective layer of SiN. In addition to an improved passivation, an additional back surface field is also created here. In the Al-BSF cell the entire back side is coated with aluminium and constitutes an Ohmic contact with high recombination rates. In the PERC cell, contacting the back side is only done via small openings. This improves the passivation of the back side; the surface recombination decreases. Such contacts are also applied to the PERT cell. The difference between a PERC cell and a PERT cell is that the back side of the PERT cell is further improved. This will be described now.

To further improve the back side, one can dope the latter even more to further reduce recombination of electrons at that location. This leads us to the PERT cell (Passivated Emitter and Rear Totally Diffused). In the PERT cell a dopant diffusion is also carried out at the back side, in the same way as on the front side, but with the opposite "polarity" of the dopant: with boron on the back side, instead of phosphorous as used on the front side. Now the back side of the PERT cell is very strongly doped with boron, so that a p^{++} layer is created. Figure 5.19 shows the cross-sections of the three types of homojunction cells: the basic homojunction Al-BSF, PERC and PERT cells. For the sake of simplicity, all three cells are drawn here with *p*-type wafer material so that it becomes easier to compare the different layers. The PERT

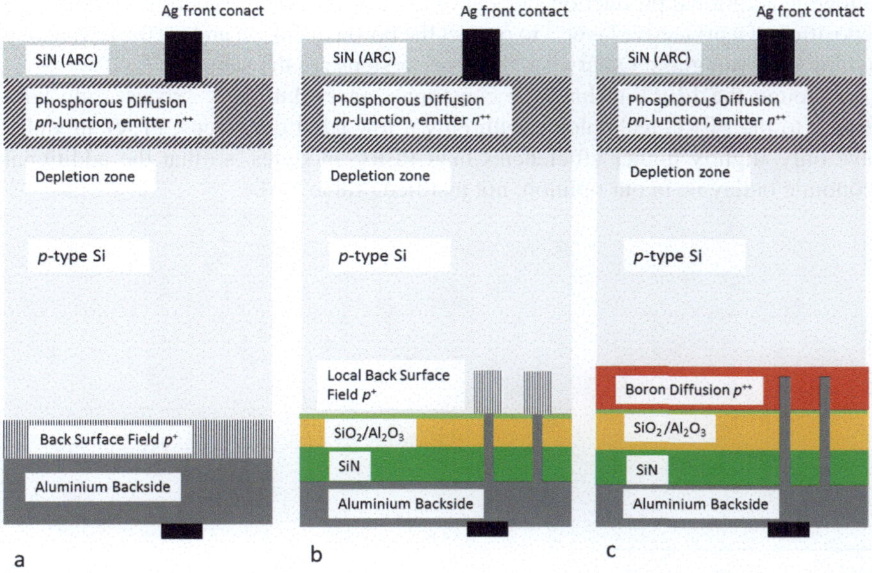

Fig. 5.19 Cross-sections of the three main types of homojunction cells: **a** Al-BSF (Al-doped-BSF) cell; **b** PERC cell; **c** PERT cell. The drawings are not to scale. The charge carrier transport corresponds to that in Fig. 5.11. For the sake of simplicity, all three cells are drawn here with *p*-type wafer material

cell has a better passivation than the PERC cell due to the additional p^{++} layer, as shown in Fig. 5.19 with the boron layer. Thanks to these improvements, the PERT cell has a slightly higher efficiency than a PERC cell: about $0.5\%_{abs}$ higher.

In practice, hardly any *p*-type PERT cells are produced—for the production of PERT cells one chooses monocrystalline *n*-type wafer material. The reason is: A solar cell with a very high degree of passivation (like the PERT cell) should only be used together with silicon material of a very high quality, such as *n*-type monocrystalline wafers. In the PERT cell aluminium is only used for contacting the cell and not for forming the BSF, which is formed by the second diffusion (Fig. 5.19c). PERT cells have the following economic disadvantages compared to PERC cells:

1. **Formation of the back surface field (BSF)**: In a *p*-type PERC cell, the aluminium generates the local BSF with a built-in field created by the negative charges in the aluminium oxide layer. In a type PERT cell a second diffusion step is needed for generating the BSF.
2. **Co-Firing**[34]: In a *p*-type PERC cell, there is co-firing of the phosphorous diffusion and metallisation. No additional protection layer is needed. In a *n*-type PERT cell, there is co-firing of the boron diffusion and metallisation. An additional protection layer (SiO_x) is needed, to protect the boron-diffused p^{++} layer from interference with the Al_2O_3 layer (Fig. 5.20).

In the case of the *n*-type rear-emitter PERT, there is no BSF (there is one rear-emitter and one FSF (Front Surface Field). For the fabrication of a PERT cell the following additional production steps have to be introduced: (a) boron diffusion, (b) deposition of a protective layer[35] to protect the boron diffusion and (c) the retroactive etching of boron glass. Otherwise, the process steps are the same:

Although the PERT technology constitutes an evolutionary process step with respect to the PERC technology, with only a few additional steps, PERT modules have only slightly higher efficiencies than PERC modules, so that the additional economic outlay is, in our opinion, not justified (Table 5.3).

---

[34] During the Co-Firing process, two steps take place simultaneously: (1), the metallization paste applied by the screen printing step is fired through the SiN and (2) the dopands are driven into the bulk to form the *pn*-junction.

[35] After the boron diffusion, a protective layer is applied so that subsequent phosphorus diffusion and chemical texturing do not "destroy" the effects of the boron diffusion.

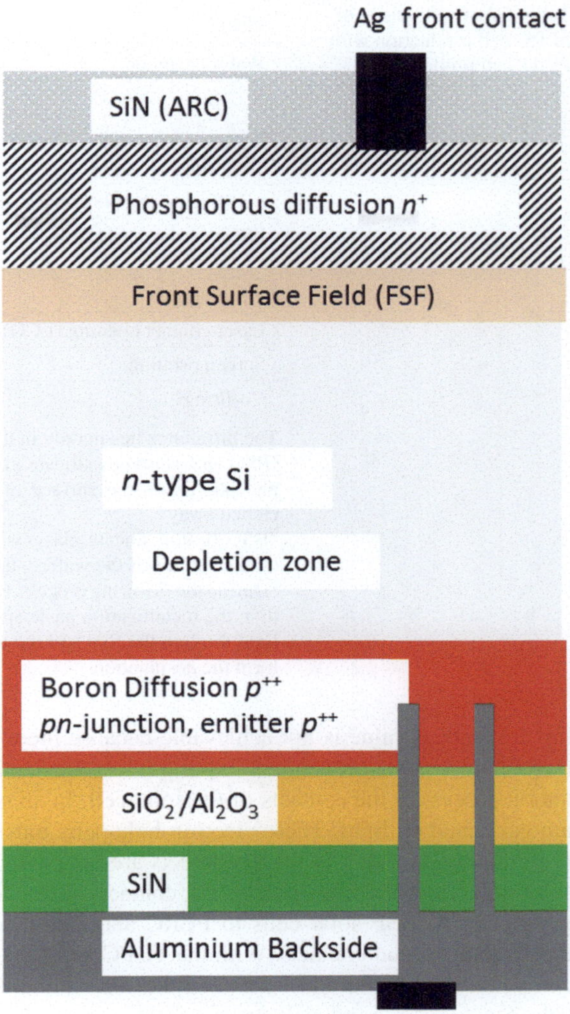

Fig. 5.20 *n*-type PERT cell with back-side emitter. Here the front side is equipped with the so called Front Side Field (FSF), in analogy to the Back Surface Field (BSF); in the case of the FSF, the holes are reflected back, and not the electrons (as with the BSF). It can be seen that an additional diffusion steps is required, here a Phosphorous diffusion on the front side

5.5.2 Homojunction Cells with Fully Passivated Contact[36]: The TOPCon Cell (Tunnel Oxide Passivation Contact)

From a thermodynamic point of view, a maximum efficiency of 29.4% can be achieved with silicon solar cells (consisting of a single junction) under non-concentrated sunlight (AM1.5) (Shockley-Queisser-Limit).[37] In practice, however,

[36]Passivated contact only on one side

[37]With Tandem cells one can achieve theoretically more than 29.4%, employing thereby materials with different bandgaps. There are, of course no tandem cells, which consist only of crystalline

Table 5.3 Comparison of PERC cell production with PERT cell production

PERC	PERT
Wafer cleaning	Wafer cleaning
	Boron-diffusion
	Rear protection layer
Texturing	Texturing
n-Diffusion (by phosphorous)	*n*-Diffusion (by boron)
PSG[a] + dielectric etch	PSG + BSG[b] + dielectric etch
Passivation	Passivation
Laser contact opening (LCO)	LCO
Screen printing	Screen printing
Co-firing[c]	Co-firing

The difference lies mainly in the boron diffusion
[a]PSG (phosphorous silicate glass) is formed after diffusion with phosphorus on the surface of the wafer—this surface must be etched away
[b]BSG (boron silicate glass) is formed after diffusion with boron on the surface of the wafer—this surface must be etched away
[c]During the Co-firing process, two steps take place simultaneously: first, the metallization paste applied by the screen printing step is fired through the SiN and the dopands are driven into the bulk to form the *pn*-junction

this theoretical value is not achievable because there are various loss mechanisms (see Chap. 4). One of the most important loss mechanisms is given by the recombination process at the contacts of the solar cell. In all solar cells, electron-hole pairs are generated by light. These electron-hole pairs must leave the solar cell, in order to produce electricity. For this, contacts are necessary. In the homojunction silicon solar cells described up to now, the contacts were silicon-metal contacts. As one goes from Al-BSF solar cells to PERC and PERT cells, one reduces the area of these metal-silicon contacts. With the TOPCon[38] [13] solar cell concept one avoids completely a direct contact between the silicon and the metal contact on the back side: One uses a "sandwich structure" consisting of a SiO_2 layer of approximately 2 nm thickness and a strongly *n*-doped polycrystalline layer (n^+-poly-Si) to form the contact, as shown in Fig. 5.21. The charge carriers can tunnel through the SiO_2 layer.[39] Thus, the recombination rate at the back side of the solar cell is drastically reduced, because there are no Ohmic contacts anymore. The additional tunnel oxide,

silicon. See Chap. 6 for the example of a microcrystalline silicon/amorphous silicon tandem (so-called "micromorph" tandem). In space applications, one frequently makes use of tandems, based on various materials within the Al-Ga-As material system. As these tandems are not of importance for terrestrial applications, they will not be treated in this Book.

[38]In 2013, Fraunhofer ISE published the TOPCon cell architecture [13]. **TOPCon** means Tunnel Oxide Passivation Contact.

[39]See Chap. 7, for a detailed description of the tunnel effect

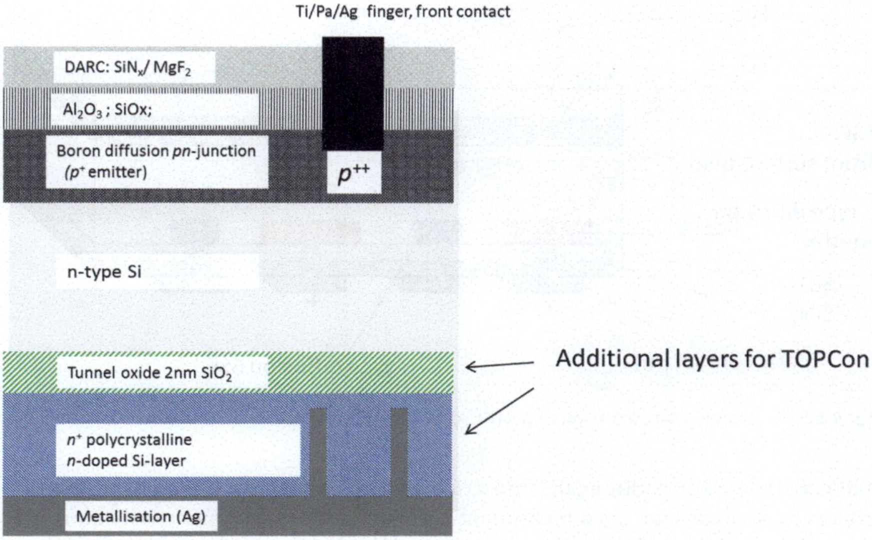

Fig. 5.21 Cross-Section of a TOPCon cell as presented by Fraunhofer ISE 2013 [13]. Two layers (tunnel oxide and n^+ polycrystalline layer) form the passivated contact of the TOPCon cell

with a thickness of approximately 2 nm, takes over the high charge carrier-selective current conduction via the tunnel effect (Fig. 5.21). The thin polycrystalline layer enhances passivation and is conductive. This makes it possible to take over the high passivation properties of a-Si layers (see Chap. 6), which is reflected in high V_{oc} values and at the same time obtain high fill factors, since the layers are not amorphous but polycrystalline and conduct the current in a better way. On a wafer substrate of 20×20 mm^2, Fraunhofer ISE achieved the world record for double-sided contacted solar cells. A cell efficiency η of 25.3%, a fill factor $FF > 82\%$, a short-circuit current density $J_{sc} > 42$ mA/cm^2 and an open-circuit voltage V_{oc} of 714 mV were achieved. The change in the base resistance of the *n*-type material due to the unfavourable segregation factor has no adverse effect on cell efficiency and the entire *n*-type ingot can be used (see Sect. 5.1.1). The reason is the selective current conduction: because of the tunnel effect, the current flows only in one direction. Another advantage of such a solution is to maintain the high temperature gettering process and still obtain the passivation properties of the amorphous layers [14]. This cell concept therefore has the potential to play a leading role in photovoltaics.

Since 2013, research has been carried out by various institutes in order to lead the cell into mass production. Various challenges have to be solved, which are briefly explained below.

The oxide process is produced either by thermal oxidation, dry oxidation or by wet chemistry with HNO_3. Depending on the process temperature it can be a self-limiting process that automatically stops at a thickness of approximately 1.2 nm. Achieving thin layer thicknesses is not a problem, but the homogeneity of the layer presents

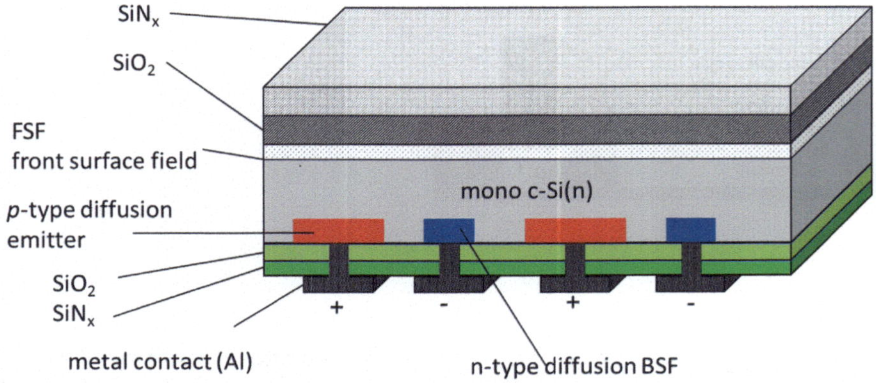

Fig. 5.22 Schematic representation of an IBC cell

major challenges. In addition, the thin oxide layer does not have sufficient passivation properties, so hydrogen must be brought to the interface for proper passivation. The *p*-doped amorphous silicon which is deposited, in the next step, with a thickness of about 30 μm contains hydrogen. However, this hydrogen is not delivered optimally to the interface. The reasons for this are not properly understood. Hydrogenation is therefore a challenge. As a workaround one can imagine depositing SiN on the amorphous layer in order to bring the hydrogen into the interface. In fact, it is known that SiN promotes hydrogenation and H^0 is generated.[40] Today (2020) average cell efficiencies of 23.5% are achieved in production.

5.5.3 *Back-Contacted Cells: IBC Cells (Interdigitated Back Contact)*

In the previous chapter, we saw that recombination losses can be reduced by using passivated contacts. There still remain optical losses on the front side due to shadowing by the contact layers. These optical losses can be avoided if the contacts are moved to the back. Such cells are called **Interdigitated Back Contact cells** (**IBC**).

Figure 5.22 shows schematically the structure of an IBC cell. The *n*-area alternates with the *p*-area. In order to avoid short circuits between the *n*-region and the *p*-region, an undoped area between the two regions must be maintained: this area does not contribute to the production of electricity. Thus, the emitter does not cover the entire back surface, resulting in some additional losses.

Some minority carriers have to "travel" a longer path to reach the emitter, resulting in a small additional loss called "electronic shading". However, the fabrication of

[40] Another solution is to apply the formal gas annealing process (FGA) using a furnace. However, this process produces H^+ and not H^0. H^0 is the form of hydrogen most suited for passivation. H^0 is electrically neutral, H^+ has a positive charge and H^- has a negative charge.

such a cell structure is quite complex. An alternative is to apply the IBC concept to heterojunction cells (see Chap. 7). Here, the *p*-region is next to the *n*-region and the emitter can be enlarged compared to the cell concept described in the present Section. Since the thin *p*- and *n*-doped amorphous layers have very low conductivity, they can be placed immediately adjacent to each other. HJT-IBC cells have excellent passivation properties due to the amorphous layer. With such a structure, Kaneka Inc. has achieved 26.7% cell designated-area[41] efficiency on 79 cm^2 and 26.6% designated-area efficiency on a larger cell of 180 cm^2 [15].

With passivated "**pol**ycrystalline silicon on **o**xides" (**POLO**) contacts [16], the ISFH institute (**I**nstitut für **S**olarenergie**f**orschung in **H**ameln) achieved in 2018 a designated-area cell efficiency of 26.1%, with an open circuit voltage V_{oc} of 726.6 mV, a short-circuit current density I_{sc} of 42.6 mA/cm^2, with a fill factor *FF* of 84% on a 4 cm^2 *p*-type Wafer and an IBC structure. The polycrystalline *p*-regions and the polycrystalline *n*-regions lie side by side and are mutually isolated by an intrinsic polycrystalline Si(*i*)-layer.

References

1. H. Jakob, Reinstsilicium – Werkstoff des Computerzeitalters (Praxis der Naturwissenschaften Chemie, Aulius Verlag Köln Deubner & CO KG, 1988)
2. P. Capper, *Bulk Crystal Growth in Electronic, Optical and Optoelectronics Materials* (Wiley, The Atrium, Southern Gate, 2005)
3. J.D. Murphy, M. Al-Amin, K. Bothe, M. Olmo, V.V. Voronkov, R.J. Falster; The effect of oxide precipitates on minority carrier lifetime in n-type silicon, published online 7 December 2015. J. Appl. Phys. **118**, 215706 (2015)
4. A. Hess, P. Krenckel, T. Trötschler, T. Feherenbach, S. Riepe, Development of high performance multicrystalline silicon with controlled seeding (Fraunhiofer Institut for Solar Energy Systems ISE; publica.fraunhofer.de/dokumente/N-521450.html, 2018)
5. G. Stokkan, A. Song, B. Rynigen, *Investigation of the Grain Boundary Character and Dislocation Density of Different Types of High Performance Multicrystalline Silicon* (Crystal Journal Published by MDPI, 2018)
6. A. Hein, Einflüsse von Verunreinigungen im Silicium und in der Ätzlösung auf das anisotrope Ätzen von Silicium in wäßrigen KOH-Ätzlösungen. Dissertation, TU-Berlin 2000, Berlin 2000
7. A. Luque, S. Hegedus, *Handbook of Photovoltaic Science and Engineering*. ISBN 13:978-0-471-49196-5(H/B) (Wiley, 2005)
8. S. Gatz, et.al. Analysis of local Al-doped back surface fields for high efficiency screen-printed solar cells. Energy Procedia S. 318–323 (2011). https://doi.org/10.1016/j.egypro.2011.06.143
9. D.A. Neamen, *Semiconductor Physics and Devices, Basis Principles* (McGraw Hill, 2003)
10. P. Saint-Cast et al., A review of PECVD aluminium oxide for surface, in *27th PVSEC, 2012, European Photovoltaic Solar Energy Conference and Exhibition* (Frankfurt, 2012)
11. E. Urrejola Davanzo, Aluminium-silicon contact formation through narrow dielectric openings, Ph.D. thesis, at University Konstanz, Konstanz, 2012

[41] The "designated area" is the area of the wafer on which the active solar cell is located. Often a large wafer is used for research cells, but the solar cell is only located on a small area of the entire wafer—the "designated area"—The term "designated area" is only used in Research and Development.

12. B. Fischer, Loss analysis of crystalline silicon solar cells using photoconductance and quantum efficiency measurements, Ph.D. thesis at University Konstanz, Konstanz, 2003
13. J. Benick; TOPCon – „Überwindung der fundamentalen Hindernisse für eine neue Weltrekord-Siliciumsolarzelle", Fraunhofer Institut für solare Energiesysteme ISE, Oktober,2013 https://www.ise.fraunhofer.de/de/forschungsprojekte/topcon.html
14. F. Feldmann, M. Bivour, C. Reichel, M. Hermle; S.W. Glunz, A passivated rear contact for high-efficiency n-type silicon solar cells enabling high Voc and FF > 82%, in *28th European Photovoltaic Solare Energy Conference and Exhibition*, 30 Sept. 2013
15. K. Yoshikawa, H. Kawasaki, W. Yoshida, T. Irie, K. Konishi, K. Nakano, T. Uto, D. Adachi, M. Kanematsu, H. Uzu, K. Yamamoto, Silicon heterojunction sola cell with interdigitated back contacts for a photoconversion efficiency over 26%, nature energy, published 20 March 2017, volume 2, article number: 17032
16. C. Klamt, M. Rienäcker, F. Haase, N. Folchert, R. Brendel, R. Peibst, V. Krausse, and J. Krügener, Intrinsic poly-crystalline silicon region in between the p^+ and n^+ POLO contacts of an 26.1%-efficient IBC solar cell, Brussels, Belgium, 26.09.2018, in *35th European Photovoltaic Solar Energy Conference and Exhibition, Brussels*, 2018

Sylvère Leu was from 2008 to 2017 with the Meyer Burger Group, where he became a member of the executive board, as Chief Innovation Officer and Technology Officer (CIO/CTO). Since his retirement in 2017, he is active as technology consultant, for the Meyer Burger Group. Sylvère graduated from the Swiss Federal Institute of Technology in Zürich (ETHZ). He additionally obtained a Master's Degree in Business Administration at the University of St. Gallen (HSG). As a Swiss pioneer, Sylvère Leu started to work in photovoltaics 30 years ago. He constructed an industrially relevant laminator and sun simulator in his own company. Beside his job he worked as an associate lecturer at University of St. Gallen (HSG), for several years, in the field of industrial production. At the end of 2005 he was charged with building up an integrated 250 MWp photovoltaic facility, including wafer, cell and module manufacturing, in Germany.

Detlef Sontag is Senior Technologist for Solar Cells at Meyer Burger (Germany) GmbH. since 2015—one of the world's leading PV equipment manufacturing companies. He graduated as a physicist at the University of Konstanz in 2000, and obtained his Ph.D. degree in 2004 working on novel wafer materials for Photovoltaics. In 2004 he joined Deutsche Cell GmbH, a subsidiary of Solar World AG, as quality assurance engineer with insights in the industrial crystalline silicon PV manufacturing processes along the whole value chain from block casting via cell processing to module assembling. Detlef Sontag joined the equipment manufacturer Roth & Rau AG as Senior Principal for cell technology in 2009. There he supported the planning and commissioning of the Technology centre and worked actively on the development of cell technologies like PERC and HJT including project coordination of corresponding R&D-projects.

Chapter 6
Amorphous Silicon Solar Cells

Arvind Shah

Abstract This chapter will first describe, in Sect. 6.1, the deposition method, the physical properties and the main use of hydrogenated amorphous silicon (a-Si:H) layers. The deposition technique commonly used for a-Si:H layers and cells is plasma-enhanced chemical vapour deposition (PE-CVD). The role of broken bonds (or "dangling bonds") within amorphous layers will be highlighted. Special mention will be made of the Staebler-Wronski effect: the increase of dangling bonds, which act as recombination centres, under the influence of light-exposure. The present main use of a-Si:H layers is in the form of passivation layers within the Heterojunction (HJT) solar cell, which will be described in the next Chap. 7. Section 6.2 will deal with amorphous silicon solar cells. First, the ***p-i-n*** structure necessary for amorphous silicon solar cells will be introduced; thereafter, typical characteristics of amorphous silicon solar cells will be given and the advantages and disadvantages of such solar cells listed. It will, thus, become evident, why the amorphous silicon solar cell is the ideal candidate for the generation of electric power in the indoor situation. In Sect. 6.3, our focus will shift to microcrystalline silicon (μc-Si:H) layers. Again, the plasma-assisted deposition technique used for these layers will be discussed. μc-Si:H has been, in the past, mainly propagated for use together with a-Si:H, in a tandem cell configuration; however, with the strong decrease in the price of wafer-based crystalline silicon solar cells, these tandem cells are, at the moment, no longer economically attractive.

A. Shah (✉)
EPFL (PV-Lab), Neuchâtel, Switzerland
e-mail: arvind.shah@epfl.ch

A. Shah (ed.), *Solar Cells and Modules*, Springer Series in Materials Science 301,
https://doi.org/10.1007/978-3-030-46487-5_6

6.1 Amorphous Silicon: Deposition Method and Layer Properties

6.1.1 Deposition of Amorphous Silicon with Plasma-Enhanced Chemical Vapour Deposition (PE-CVD)

Amorphous silicon thin films for solar cells are at present almost exclusively deposited by plasma-enhanced chemical vapour deposition (PE–CVD) either from silane (SiH_4) or, preferably, from a mixture of silane and hydrogen (Fig. 6.1).

The amorphous silicon thin films produced by PE–CVD contain about 5–15% of hydrogen atoms. The hydrogen atoms are essential, as they passivate a large part of the inherent defects in these semiconductor films.

Amorphous silicon thin films were first deposited by PE–CVD by Chittick et al. [2]; Walter Spear and Peter Le Comber and their research group at the University of Dundee continued this work in a systematic manner in the 1970s. In a landmark paper published in 1975 [3] (see also [4]) they showed that a-Si:H layers could be doped by adding to the plasma discharge—either phosphine (PH_3) to form *n*-type layers or diborane (B_2H_6) to form *p*-type layers: They demonstrated that the conductivity of these thin amorphous silicon layers could thereby be increased by several orders of magnitude. This is illustrated in Fig. 6.2.

The pioneering work of Walter Spear and his research group at the University of Dundee made it possible to use *hydrogenated amorphous silicon* (a-Si:H) to fabricate diodes and thin-film transistors; the latter are used for the active addressing matrix in liquid crystal displays.

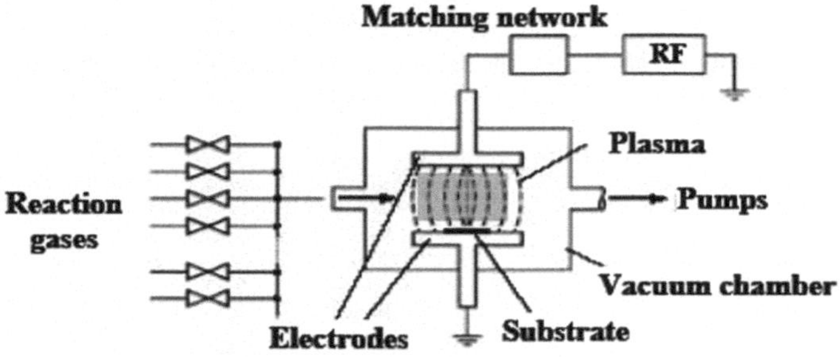

Fig. 6.1 Schematic representation of a PE-CVD deposition system. To deposit amorphous silicon layers one uses the following Reaction gases: Silane (SiH_4), Hydrogen (H_2) and the doping gases—either phosphine (PH_3) for *n*-type layers—or diborane (B_2H_6), for *p*-type layers. Reproduced from [1], with the kind permission of the EPFL Press

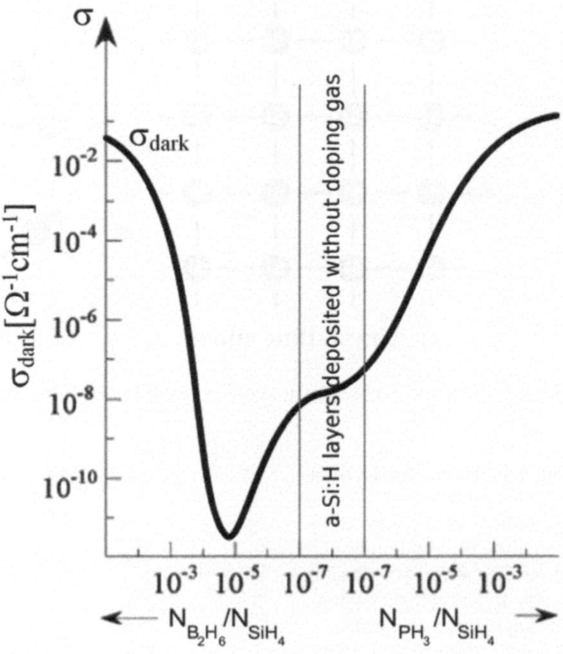

Fig. 6.2 *Measured* values of dark conductivity σ_{dark} for amorphous silicon layers, deposited by PE-CVD on glass. On the *x*-axis is indicated the gas-phase doping ratio N_{PH3}/N_{SiH4} (for *n*-type layers) and N_{B2H6}/N_{SiH4} (for *p*-type layers). Measurements are from [3]. "a-Si:H layers deposited without doping gas" are layers deposited without the addition of phosphine or diborane to the PE-CVD deposition chamber. Because of the presence of oxygen in the latter, these layers have a slightly *n*-type character

6.1.2 Physical Properties of Amorphous Silicon Layers

1. **General structure**

Crystalline solids, such as monocrystalline silicon wafers, have a fully regular and periodic structure; they possess what is called both *short-* and *long-range order*.

In such a crystalline silicon network (or crystalline *matrix* as it is also called), each silicon atom is bonded to four neighbouring silicon atoms. Figure 6.3a schematically shows this situation. Amorphous solids, such as hydrogenated amorphous silicon (a-Si:H) do not have a fully regular and periodic structure, but rather a random or "chaotic" structure: They only have short-range order (see Fig. 6.3b).

In crystalline silicon, the *bond angle*—i.e., the angle between two adjacent bonds—is fixed at a value of 109°28′ and remains the same throughout the whole crystalline network (Fig. 6.4). The *bond length*, or distance between two neighbouring silicon atoms within such a network, is also fixed and remains constant throughout the whole network at a value of approximately 0.235 nm.

In amorphous silicon, both the bond angles and the bond lengths vary in a random fashion: there is a whole distribution of values. As an example: The bond angles in a-Si:H have a random distribution centred on 109°28′ and a standard deviation

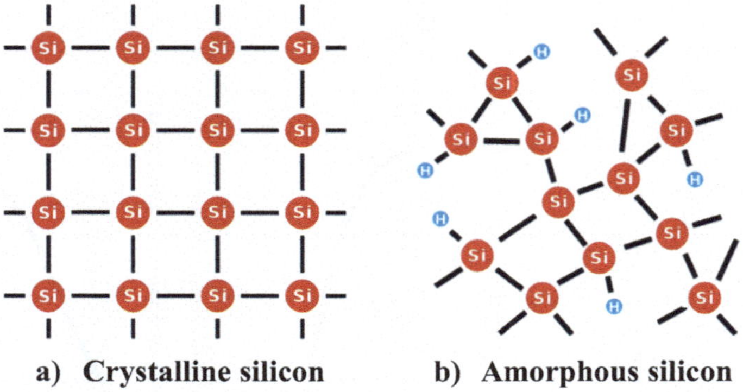

Fig. 6.3 Schematic representation of **a** Crystalline silicon matrix and **b** amorphous silicon matrix

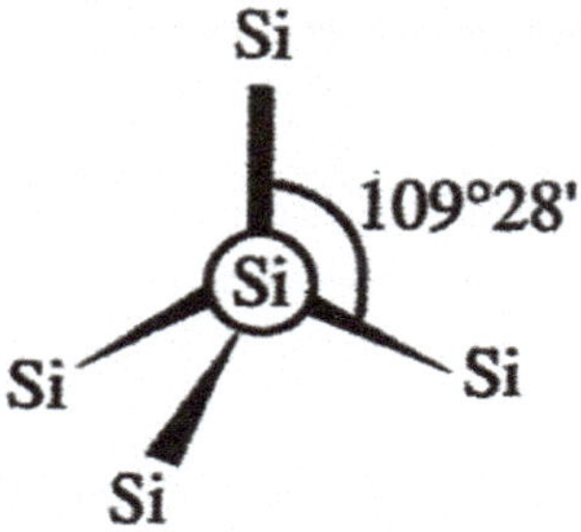

Fig. 6.4 Atomic model for a silicon atom within a crystalline silicon network, indicating the bond angle formed between two adjacent bonds. In amorphous silicon, this angle has a distribution of values. Reproduced from Shah [1] with permission of the EPFL Press

of 6°–9°. The "best"[1] a-Si:H has bond angles with a narrow distribution (standard deviation of 6°–7°).

Because of the random or "chaotic" structure of amorphous silicon, not all silicon atoms within the amorphous layer, find four other silicon atoms as "next neighbours". From time to time, there is a silicon atom, which only has three other silicon atoms as "next neighbours". This specific silicon atom has, therefore, a "broken bond" or "dangling bond" as it also called (see Fig. 6.5a). In amorphous silicon layers, as deposited by Plasma-Enhanced Chemical Vapour Deposition (PE-CVD) from silane (SiH_4), a large proportion (over 99%) of the original dangling bonds are "passivated" by hydrogen, during the deposition process: The "passivated" dangling bonds have a hydrogen atom sitting on them, as represented in Fig. 6.5b. The "passivated" dangling bonds do not act as recombination centres and do not constitute gap states; they will therefore not be counted as dangling bonds in the following discussion. In some very rare cases, a silicon atom has only two other silicon atoms as "next neighbours"—they constitute a "SiH_2-configuration" as represented in Fig. 6.5c.

[1] «best» meaning «usable for the production of devices, such as solar cells and thin-film transistors».

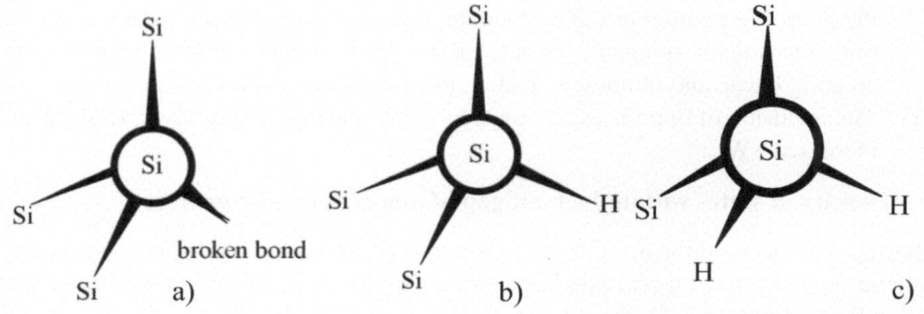

Fig. 6.5 Model for silicon atom with **a** unpassivated dangling bond (acting as recombination centre or mid-gap state) **b** dangling bond "passivated" by a hydrogen atom (and no longer acting as a dangling bond) **c** two hydrogen atoms connected to it (SiH_2-configuration). Reproduced from [1], with the kind permission of the EPFL Press

2. **Light-induced degradation or Staebler-Wronski Effect (SWE)**

Because of the random or "chaotic" structure of amorphous silicon layers, these layers change their properties, when exposed to light: Some of the dangling bonds passivated by a hydrogen atom (Fig. 6.5b), lose their hydrogen atom, under the influence of light—this was discovered as early as 1977 by Staebler and Wronski [5]. It was designated as the "Staebler-Wronski Effect (SWE)" Since then, there has been, right up to 2005, a huge research effort to find ways of suppressing the SWE, albeit without any real success [6, 7]. Well, although, one does not know how to fully suppress the SWE, many things are known about the SWE:

(a) If the amorphous silicon layer contains many SiH_2-configurations (Fig. 6.5c), the SWE will be more pronounced. Such SiH_2-configurations are created if the deposition is done too rapidly by increasing the "RF Power" in the PE-CVD deposition system (Fig. 6.1).
(b) SWE is a reversible effect: By heating the amorphous layer, during a few hours at about 200 °C (so-called "annealing" process), the original state is restored.
(c) SWE is an "asymptotic effect". If one continues exposing the layers to light over a very long period (typically many thousands of hours), one reaches what is called a "stabilized final state"—with roughly ten times more dangling bonds than in the beginning.
(d) SWE can be influenced by the deposition parameters (Fig. 6.1): an increase in deposition temperature [8] or an addition of atomic hydrogen in the Reaction gases [9] will both lead to layers with a less pronounced SWE.
(e) SWE is closely linked to the presence and behaviour of hydrogen[2] within the amorphous layer. In this context there is an optimal value for the hydrogen content of the amorphous silicon layer: This value is around 10 atomic % of hydrogen, meaning that there is 1 hydrogen atom for 10 silicon atoms. At this value, hydrogen is "helpful", because it mainly passivates dangling bonds. If

[2]Interestingly, hydrogen also plays a major role in the instability of perovskite solar cells.

the hydrogen content is higher, then the "excess" hydrogen atoms are no more linked to silicon atoms, but wander about freely within the amorphous layer, creating havoc and ultimately leading to an increase in SWE.

(f) The inclusion of impurities like oxygen within the amorphous layer, leads to an increased SWE.[3]

3. **Density of states within the bandgap of amorphous silicon layers**

Because of the random or "chaotic" structure of amorphous silicon, amorphous silicon does not have a real bandgap, like crystalline silicon. In comparison with crystalline silicon, a-Si:H layers have:

(a) An "equivalent bandgap", which is filled with gap states, acting (partly) as recombination centres. This is shown schematically in Fig. 6.6.
(b) A higher "equivalent bandgap", higher than the "real bandgap" of crystalline silicon, i.e. approximately 1.75 eV instead of 1.12 eV

Figure 6.6 indicates the density of states $N(E)$ within the "equivalent bandgap" (called in the Figure "mobility gap") of typical amorphous silicon layers. The "midgap states" act as recombination centres. Their density is increased by the SWE. (Fig. 6.7).

4. **Optical properties of amorphous silicon layers**

The optical properties of a-Si:H layers are very different from those of c-Si. From a practical point of view, the main differences are:

(1) Absorption starts at a higher value of photon energy for a-Si:H → i.e. at a shorter wavelength λ of light: at $\lambda \approx 700$ nm for a-Si:H as compared to $\lambda \approx 1100$ nm for c-Si

---

[3]There is a long story behind this simple statement—when work started at IMT Neuchâtel, after 1985, on amorphous silicon layers and solar cells, there was the "justified hope" that by suppressing (reducing) the oxygen content within the layers we were depositing, we would be able to totally remove the SWE effect. We therefore invested heavily in so-called "gas purifiers", so as to be able to produce a-Si:H layers with very low oxygen content. To our misfortune it turned out that these layers had just the same intensity of SWE as layers containing more oxygen. It was only above a certain threshold that oxygen led to an enhancement of SWE. However, when we started to deposit microcrystalline silicon layers, with our equipment, the gas purifiers were absolutely decisive: We at IMT Neuchâtel became the first laboratory able to produce microcrystalline silicon solar cells with conversion efficiencies η well above 5%. Thanks to this discovery, the Author of this chapter received in 2007 the Becquerel Prize. **More importantly the Author of this Chapter would wish to communicate to the Readers the following lessons he learnt**:

1. **Never underestimate instability effects**. They generally turn out to be unavoidable. no matter what tricks one tries to do. This should be remembered in the context of the present hype regarding **perovskite** solar cells. These cells are at present unstable, and may remain so, no matter what tricks researchers try to do.
2. **Research often leads to results you are totally unable to predict**—our layers with low oxygen content were of no use for amorphous silicon solar cells, but they were a decisive asset, when we started our work on microcrystalline solar cells. The present heavy research investment in perovskite solar cells will probably also lead to some quite unexpected results.

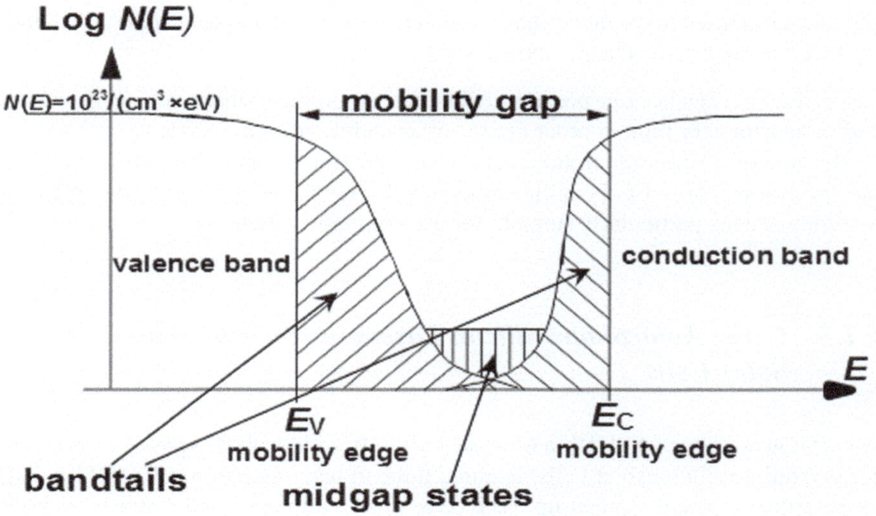

Fig. 6.6 Diagram showing schematically the density of electronic "gap states" $N(E)$, which are a result of the amorphous nature of the semiconductor. The density of states $N(E)$ is represented logarithmically and as ordinate (on the vertical axis); the energy E of the corresponding electronic state is represented as abscissa (on the horizontal axis). The "gap states" are situated in the mobility gap, i.e. in the energy range between the valence band edge E_V and the conduction band edge E_C. The "midgap states" result from the dangling bonds and constitute recombination centres. The bandtails are also characteristic of the amorphous nature of the layers. Adapted (with modifications) from [1]

Fig. 6.7 Sketch indicating the difficulty of doing research on amorphous silicon. *Courtesy* Dji-Illustrations, Neuchâtel

(2) Amorphous silicon layers have a much stronger absorption than crystalline silicon layers (see Chap. 3, Sect. 3.2.2)

Because of (2) it became possible to fabricate amorphous silicon solar cells, which had to be kept very thin, in order not to suffer unduly from the SWE.

Because of (1) these amorphous cells have a spectral response that is different from the spectral response of crystalline silicon cells (see Chap. 3, Sect. 3.6), rendering amorphous cells particularly suitable for use with indoor lighting.

6.1.3 Using Amorphous Silicon Layers in Heterojunction Solar Cells

Amorphous silicon (a-Si:H) thin films are currently widely used as passivation layers for crystalline silicon solar cells, leading, thus, to heterojunction cells (HJT cells), as described in Chap. 7, next-up. HJT cells work with passivated contacts on both sides. These contacts, consist of an approximately 5 nm thick layer of *intrinsic* amorphous silicon (which is directly touching the crystalline silicon wafer), followed by an appropriately 15 nm thick *doped* (*p*-type or *n*-type) a-Si:H layer. A transparent conductive oxide (TCO), usually indium tin oxide (ITO) caps the doped layers, to provide lateral charge transport to the screen-printed Ag fingers. The typical structure of a HJT cell is shown in Fig. 6.8, for the example of a Bifacial HJT solar cell, as developed by IMT Neuchâtel for Meyer Burger [10].

The critical interface is between the c-Si wafer and the intrinsic a-Si:H layer. Figure 6.9 shows a micrograph of this interface, presented by Sara Olibet in her Ph.D. thesis [11]. As noted there "Best passivation necessitates an abrupt and flat interface of the a-Si:H layer to c-Si". One remarks in Fig. 6.9, the crystallographic abrupt interface, which is indeed essential for passivation properties.

It is to be expected that intensive research will continue in this field, with the goals of (a) improving further the already excellent passivation properties; (b) studying the long-term behaviour of these structures—in order to completely avoid any danger of encountering degradation effects; (c) increasing throughputs and simplifying production steps—in order to reduce the fabrication cost of HJT solar cells.

6.2 Amorphous Silicon Solar Cells

For a recapitulation on this type of solar cells, see also [12].

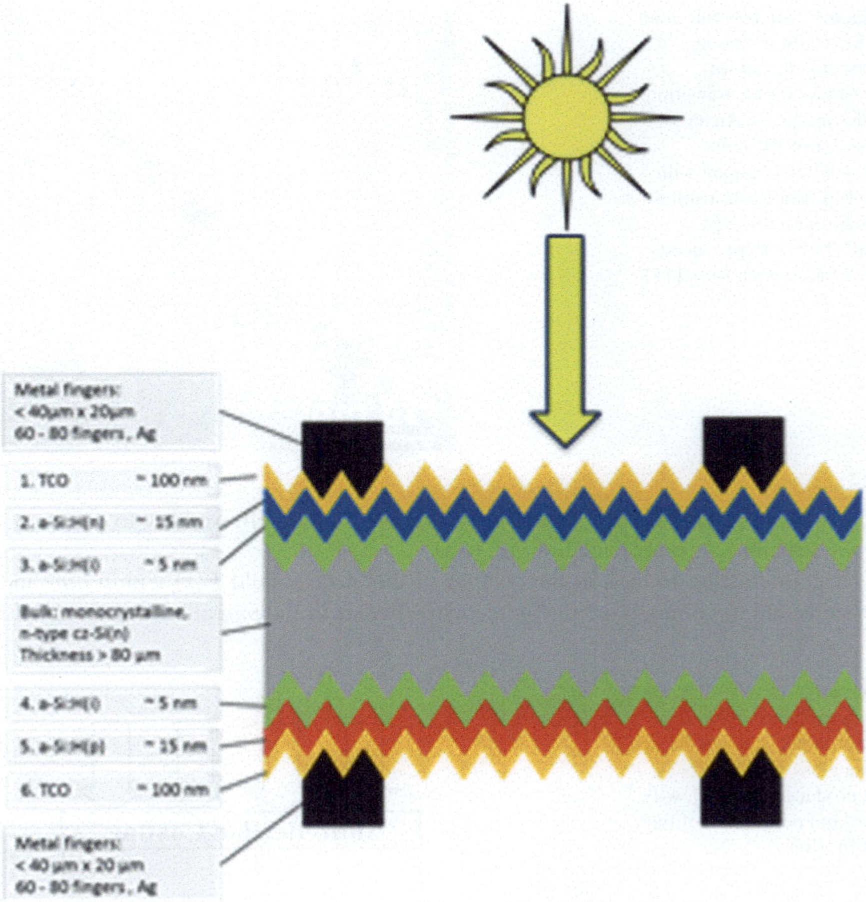

Fig. 6.8 Cross-section of a texturized bifacial HJT solar cell with *pn*-junction at the back and double-sided passivation

6.2.1 *The* p-i-n *Structure Used for Amorphous Silicon Solar Cells*

Because of the very poor electronic quality of intrinsic[4] amorphous silicon (a-Si:H) layers, it is not possible to use *pn*-diodes, to form solar cells. Indeed even the very best intrinsic a-Si:H layers have diffusion lengths, which are well below 1 µm. Therefore, diffusion is not a suitable transport mechanism for collecting the photo-generated carriers within an a-Si:H solar cell. Sufficient collection only becomes possible with the help of an electric field E, i.e. by using drift as transport mechanism. To obtain a

[4]Doped (a-Si:H) layers have even poorer quality.

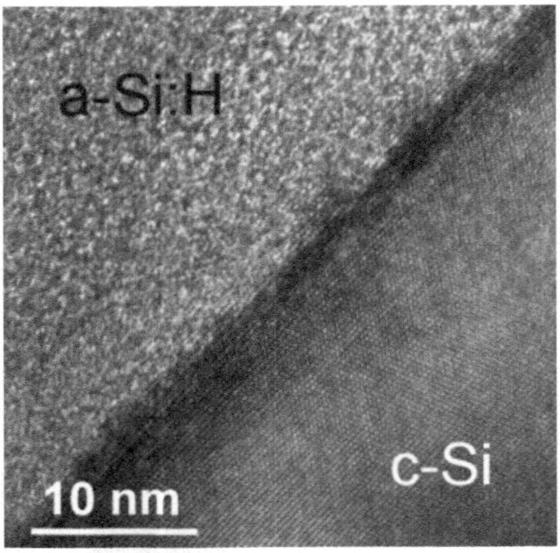

Fig. 6.9 Micrograph of an a-Si:H/c-Si interface, showing the abrupt crystallographic transition. Micrograph by Aïcha Hessler of PV-Lab Neuchâtel, obtained with high-resolution transmission electron microscopy (HR-TEM). Reproduced with permission from [11]

substantial electric field *E*, in the major part of the solar cell, we have to go over to a *pin*-structure, as represented in Fig. 6.10.

It is interesting to look at the profiles of the electric field $E(x)$ within *pin*- and *pn*-cells and to compare them (Fig. 6.11): Whereas in the *pn*-cell, the electric field

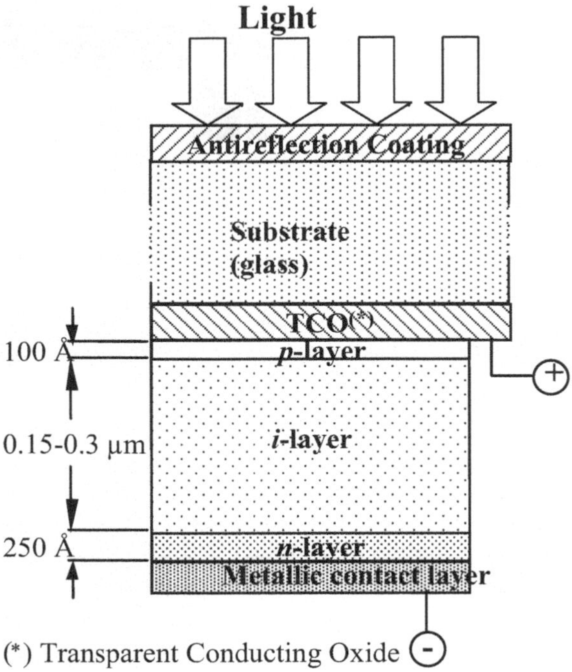

Fig. 6.10 Sketch showing the structure of a *pin*-type amorphous silicon solar cell on a glass substrate. Reproduced from [1], with the kind permission of the EPFL Press

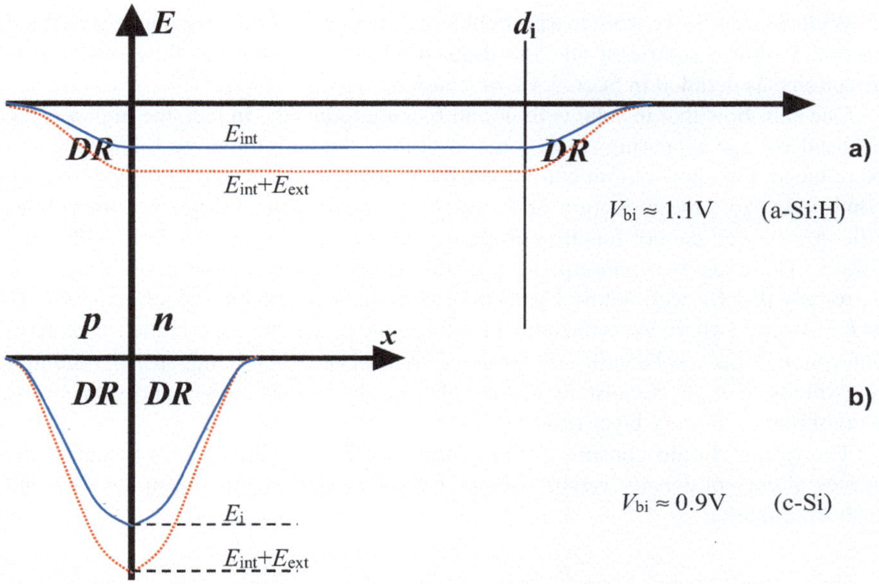

Fig. 6.11 Internal electric field $E_{int}(x)$, for **a** *pin*- and **b** *pn*-type diodes; note that for zero external (applied) voltage, the integral $\int E_{int}(x)dx$ is, for both types of diodes, equal to the built-in voltage V_{bi}, a parameter, which is approximately equal to 1 V, for all forms of silicon. E_{int} is the internal electric field for zero applied voltage; E_{ext} is the additional electric field due to an externally applied negative voltage *V*; $E = E_{int} + E_{ext}$ is the total electric field; $\boldsymbol{d_i}$ is the thickness of the ***i***-layer; ***DR*** is the depletion region. Reproduced, with modifications, from [1], with the kind permission of the EPFL Press

extends only within the relatively thin depletion region (**DR**), in *pin*-cells, it extends over the whole ***i***-layer. This is the reason, why we use *pin*-cells for amorphous silicon.[5]

For the following considerations, we shall talk of "diodes" and not (yet) of "solar cells"—this will make things simpler to understand. In Fig. 6.11 we have a blue line: this is the profile $E(x)$ of the electric field, when no external voltage is applied to the diode.

In Fig. 6.11 we also have a red line: this is the profile $E(x)$ of the electric field, when a negative external voltage $V = \int -E_{ext}(x)dx$ is applied to the diode: the electric field is increased, for a negative value of *V*: this is the situation with reverse-biased diodes, i.e. diodes, which hardly conduct any electric current, at all.

On the other hand, the electric field is decreased, for a positive value of *V*: this is the situation with forward-biased diodes, i.e. diodes, which freely conduct electric current. The corresponding profile $E(x)$ of the electric field is not shown in the figure, because it is no more flat, but has a curved form, which is not easy to calculate.

[5]There is, on the other hand, a disadvantage in using *pin*-cells: *FF* and V_{oc} will be basically lower than in *pn*-cells. This fact is experimentally verified; the theoretical justification remains to be done.

With solar cells, we work in a particular regime of the diode—the photo-generated current within a solar cell and the usual diode forward current flow in opposite directions, as detailed in Sect. 3.4.2 of Chap. 3.

One can show that in solar cells V and E_{ext} are reduced.[6] In fact, the higher is the external voltage appearing at the contacts of the solar cell, the more V and E_{ext} will be reduced. For short-circuit conditions, the reduction will be zero. For open-circuit conditions (V_{oc}) the reduction of V and E_{ext} will be a maximum. An amorphous silicon solar cell cannot function properly without a strong electric field within the ***i***-layer. There are two consequences of this (a) amorphous solar cells have to be extremely thin (d_i well below 1 μm) in order to have a high enough electric field E, as $E = V_{bi}/d_i$, (where V_{bi} is the built-in voltage and d_i the thickness of the i-layer); (b) amorphous silicon solar cells will have—with respect to their bandgap—particularly low values of V_{oc},[7] because at V_{oc} the electric field breaks down and very strong recombination in the ***i***-layer results.[8]

The reader should consult, at this point, Fig. 3.12 in Chap. 3, to visualize the curves of current density versus voltage, for solar cells, in general, in the dark and with illumination.

6.2.2 Fabrication of Amorphous Silicon Solar Cells and Modules

The fabrication of amorphous silicon (a-Si:H) solar cells and modules—described first for the example of an a-Si:H cell deposited on a non-transparent substrate (metal or polymer[9] foil), involves the following steps:

1. Cleaning of the Substrate
2. Deposition of the back reflector (BR) acting as back contact[10] (e.g. silver)
3. Deposition of the a-Si:H layers, in a PE-CVD system
4. Deposition of the front contact (here zinc oxide)
5. Performing cell interconnection by laser scribing and deposition of further metallic layers

---

[6]To understand this, consider Fig. 3.12 in Chap. 3.

[7]The reader should consult Fig. 3.17 of Chap. 3 in this context.

[8]There are many reasons for the low value of V_{oc} in a-Si:H solar cells. The "breakdown" of the internal electric field described here is just one of them. The presence of bandtail states (see Fig. 6.6) is another reason.

[9]Polymer foils are mostly transparent to light, but one prefers to avoid letting the sunlight pass through them, because they, in general, suffer from "yellowing" under the effect of UV light.

[10]The back contact is usually textured, in order to increase the light-trapping properties of the cell (see Chap. 4). Amorphous silicon (a-Si:H) solar cells have to be kept extremely thin (thickness below 0.2 μm), so as to maximize the internal electric field E_{int}, and, thus, allow for satisfactory collection of the photo-generated electrons and holes. Therefore, light-trapping is absolutely essential for a-Si:H cells.

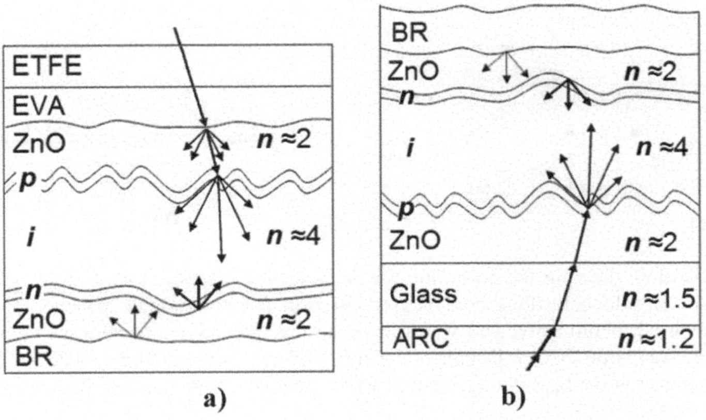

Fig. 6.12 Sketches showing the basic structures and the light management schemes (See Footnote 10) of amorphous silicon solar cells: **a** deposited on a metal or polymer substrate, **b** deposited on a glass substrate (In Fig. 6.12b the glass substrate is called «superstrate», just to indicate that during the future use of the solar cell, the substrate will be situated above with the light coming through it.). The sketches are not drawn to scale—the goal being here merely to illustrate the succession of different layers and the light-trapping schemes used. Note that the metal or polymer substrate itself is not shown; it is in (**a**) just below the BR. Abbreviations used: BR: Back Reflector; ZnO: textured zinc oxide layer, produced e.g. by LP-CVD (Low-Pressure Chemical Vapour Deposition); ARC: Antireflective Coating; ETFE: Ethylene Tetra-Fluor-Ethylene polymer foil; EVA: Ethylene-Vinyl Acetate foil. The symbol *n* is the refractive index of the layer. Reproduced from [1], with the kind permission of the EPFL Press

6. Attaching the covering layers (here EVA (Ethylene-Vinyl Acetate) and ETFE—Ethylene Tetra-Fluor-Ethylene)

The resulting structure can be seen in Fig. 6.12a. Figure 6.12b shows, on the other hand, the example of an a-Si:H cell deposited on a transparent substrate (glass).

One of the main advantages of all thin-film solar cell technologies is that the interconnection of individual cells to form entire modules, can be done during the cell production itself, and does not require separate wiring and soldering operations, by another machine (see also Chap. 8). This is illustrated very schematically in Fig. 6.13. This is termed as a "monolithic" interconnection scheme—so as to indicate that the interconnection steps are carried out on the same bloc of material (on the same "monolith").

Other characteristics of the manufacturing processes for a.Si:H solar cells and modules are:

1. Can be fully automated
2. Is capital intensive, but not labour-intensive
3. Many different deposition substrates can be used
4. Cells and Modules of all sizes and forms can be manufactured
5. Input Materials are generally cheaper and contain less "grey energy" than the materials used for most other cells and modules

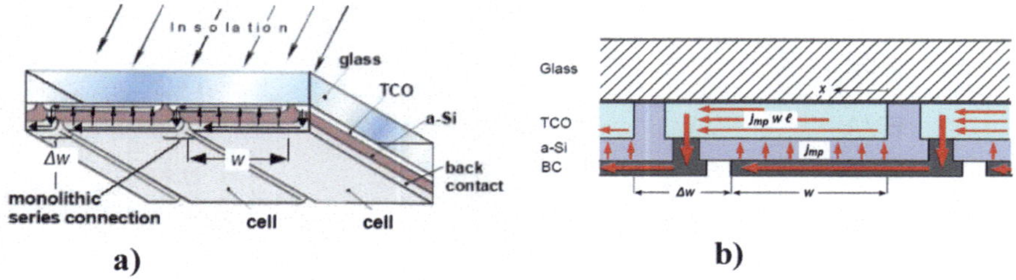

Fig. 6.13 Sketches showing the monolithic interconnection scheme, for connecting individual amorphous silicon solar cells (deposited on glass) to form an entire module: **a** General principle—Abbreviations used: photo-active cell width w; photo-inactive interconnection width Δw; Transparent Conductive Oxide TCO. **b** Details of the current flow—Abbreviations used: current density at maximum power point J_{mp}; Back Contact BC. Reproduced from [1], with the kind permission of the EPFL Press

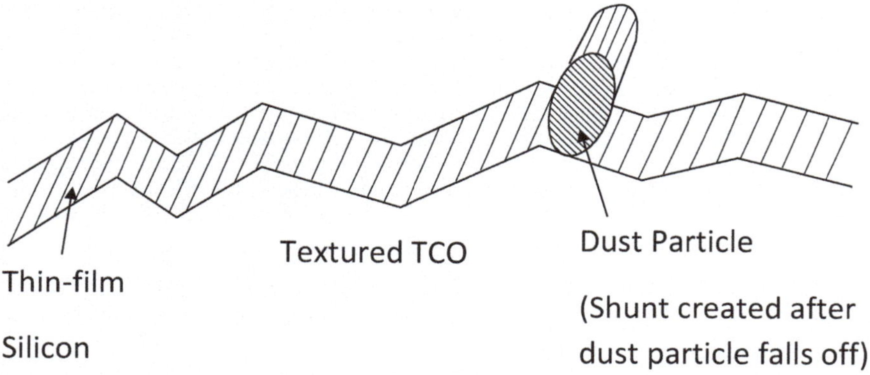

Fig. 6.14 Sketch showing the formation of a shunt within the solar cell, by a dust particle present during deposition. Reproduced from [1], with the kind permission of the EPFL Press

6. Is very sensitive to the presence of dust and other particles

Figure 6.14 shows what happens, if dust particles are present → Shunts will be created in the cells.

6.2.3 *Properties of Amorphous Silicon Solar Cells*

The main properties of a-Si:H solar cells and modules can be summarized as follows:

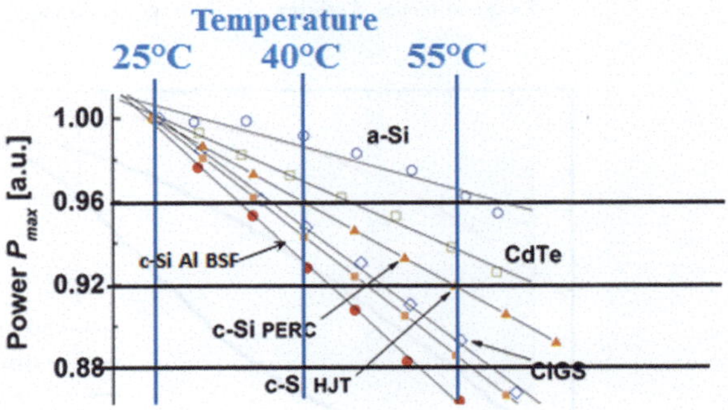

Fig. 6.15 Effect of operating temperature T on the normalized value of the output power P_{max}, at MPP, of typical solar modules, for various cell technologies. Adapted from [13]

(a) **Very low Magnitude of Temperature Coefficient**

Amorphous silicon (a-Si:H) solar cells and modules have excellent temperature behaviour. We can see that in Fig. 6.15: Their conversion efficiency drops (in relative terms) only by about 4%, for a 30 °C increase in temperature, whereas the "best" c-Si modules have a corresponding drop of 8%. We would therefore expect a-Si:H solar modules to be particularly suited for use in tropical countries. However, the very low conversion efficiencies obtained by these cells (see point c) hereunder) precludes such a usage.

(b) **Excellent low-light behaviour**

For lower light intensities, a-Si:H cells are, in general, far superior to c-Si cells. This is shown in Fig. 6.16. For details, see Chap. 3, Sect. 3.5.3.

(c) **Lower efficiency and narrower spectral response (than c-Si cells)**

a-Si:H solar cells have, in general, a stabilized efficiency of roughly 6% [14], whereas crystalline silicon solar cells can reach 20%. Their External Quantum Efficiency (EQE) curve, and consequently also their spectral response curve, are also much narrower, as can be seen in Fig. 6.17. a-Si:H solar cells are therefore, at present, limited to "niche" applications.

(d) **High Values of Power to weight ratio possible**

Amorphous silicon (a-Si:H) solar cells, when deposited on polyimide (PI) foils, are very light (in weight). This basically opens up specific applications in aerospace technology—wherever the weight of the power supply and not its surface area counts. This would be the case for equipment mounted on Zeppelin-type airships, [11] used for low-cost civil surveillance purposes (Table 6.1) [15].

[11] See [15] for further information. See: https://www.researchgate.net/publication/309424622.

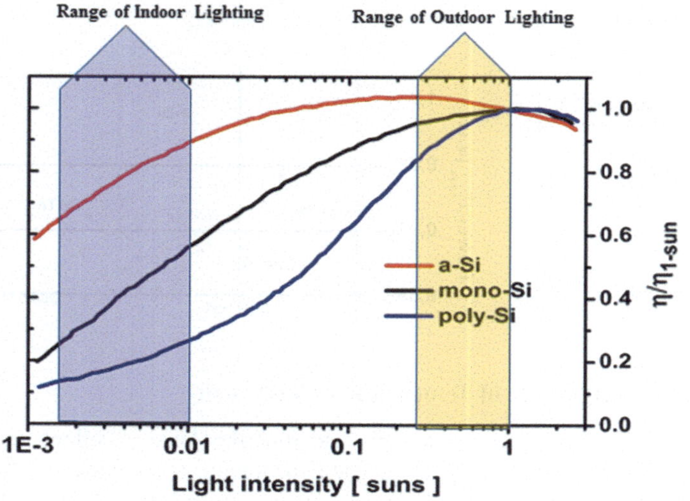

Fig. 6.16 Efficiencies of three types of solar cells as a function of light intensity: The superior behaviour of amorphous silicon (a-Si) cells in the range of Indoor Lighting is clearly visible. Note that the Figure shows ***relative*** efficiencies only: efficiencies at various light intensities divided by the efficiency at one sun; Figure is based on data in [1] (Fig. 4.28)

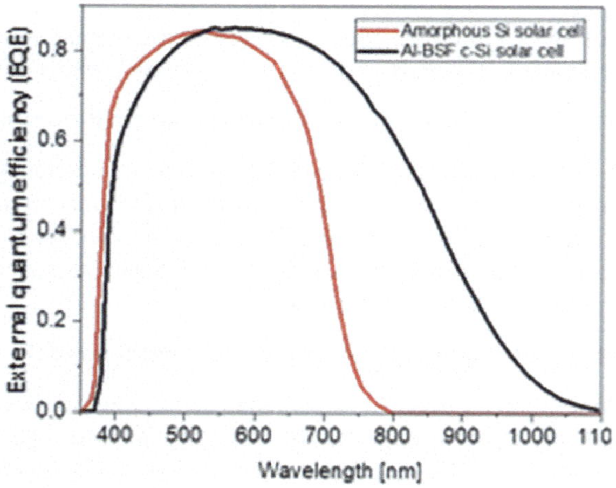

Fig. 6.17 External quantum efficiency (EQE) curves for an amorphous silicon solar cell and a crystalline silicon solar cell (of the Al-BSF type, see Chap. 5). Figure is based on data in [1] (Fig. 4.56). Note that modern c-Si solar cells have much broader EQE curves, so the cell uses more of the solar spectrum

6.2.4 Applications of Amorphous Silicon Solar Cells

There are a multitude of small objects, fitted with solar cells as power supplies. These range from calculators and other objects of general utility (garden lamps, remote controls for shutters), backpacks (rucksacks), over solar watches, to solar gadgets and toys.

Table 6.1 [a]Specific power density at the cell level of commercially available space cells and for a-Si:H on polyimide (PI) as a function of the substrate thickness, as well as the achieved (or expected) specific power at the solar array level

Technology	Efficiency (%)	Substrate/ wafer thickness	Cell specific power density (W/kg)	Array specific power density (W/kg)
c-Si	16	c-Si, 130 μm	676	<100
DJ ($GaInP_2$/GaAs)	22	Ge, 140 μm	354	<100
TJ ($GaInP_2$/GaAs/Ge)	30	Ge, 140 μm	483	<100
a-Si:H	6	PI, 50 μm	940	>500
a-Si:H	6	PI, 5 μm	3680	>1000

Abbreviations used: Double Junction *DJ*; Triple Junction *TJ*. Reproduced from [1], with the kind permission of the EPFL Press

[a]Efficiency Values given in this Table are from the Year 2010 (when [1] was published). Today (2020) much higher values for space cells are reached. However, the scope of this Table is not to present commercially available space cells, but merely to illustrate the potential of a-Si:H to increase the power density in (W/kg) of space cells, in view of specific applications as detailed in Sect. 6.2.4 under the heading "power supplies for dirigibles"

In this category, the **solar calculator** is ***the*** example of a solar-powered object of general utility: When the first pocket calculators came on the market, after the introduction of low-cost transistors and integrated circuits in the '70s, they were powered by batteries, which had to be replaced very frequently. It was Kuwano [16] and his team at Sanyo Inc., who in the early 80s replaced the batteries with small a-Si:H solar cells, developed also by Sanyo. Since then solar calculators (Fig. 6.18) have completely replaced battery-driven calculators. To power a calculator with a solar cell, is technically a well-adapted solution, as no electricity storage is needed here: One only uses one's calculator, when light is available.

Fig. 6.18 Example of a solar-powered calculator. *Photograph* Arvind Shah

Fig. 6.19 Example of a solar-powered watch. *Photograph* Arvind Shah

Solar watches are a case for themselves. Basically, it is easily possible to power a watch with a solar cell plus an accumulator or a super-capacitor, for storing the electric current during the night, or other periods, when sufficient light is not available. This is very convenient for the proprietor of the watch, as it saves the cost and botheration of regular battery replacement. Amorphous silicon (a-Si:H) solar cells are particularly suited for watches, because of the ease of integration of the very thin a-Si:H cells into watches, their flexibility (which renders them unbreakable) and their excellent low light performance. So far, only a small proportion of all electronic watches are powered by solar cells, but this should change in the future—as solar watches become more and more popular, their production volume will increase and their price will come down. Figure 6.19 shows the example of a solar watch.

For the whole remaining multitude of **solar-powered objects**, little is known about the type of solar cells they have. We can guess that most of the objects for outdoor use have c-Si solar cells, whereas objects for indoor use are probably mainly powered by a-Si:H solar cells. Figure 6.20 shows a solar power supply, intended for outdoor use and powered with an a-Si:H solar module.

In a completely different realm, one has repeatedly proposed to use a-Si:H solar cells as **power supplies for dirigibles** [15] intended for civil (and military) surveillance. The very high power to weight ratio of a-Si:H solar cells make them the ideal candidate for this application (see Sect. 6.2.3d). However, these solutions have, to date, never come "off the ground".

For actual larger-scale energy applications (e.g. for supplying current to isolated houses, and for providing electricity in grid-connected applications) a-Si:H solar cells do not, at all, play a role anymore. This is because c-Si solar cells and modules can do the same job, at lower costs and with a much smaller solar module surface Thus, amorphous silicon solar cells are today mainly used for indoor applications (Fig. 6.21).

Fig. 6.20 Example of a solar power supply (1.5 W, 12 V, 100 mA) used for maintaining the batteries of recreational and all-terrain vehicles, snow mobiles, and cars. The thin-film amorphous silicon solar technology (on glass) offers benefits, because of its low cost and since it operates well, even on cloudy days. Photograph courtesy of Greg P. Smestad/Sol Ideas Technology Development

Fig. 6.21 A well-functioning Indoor gadget powered by an amorphous silicon solar cell can be a true delight and bring music and joy into many Dance Studios *Courtesy* Dji-Illustrations, Neuchâtel

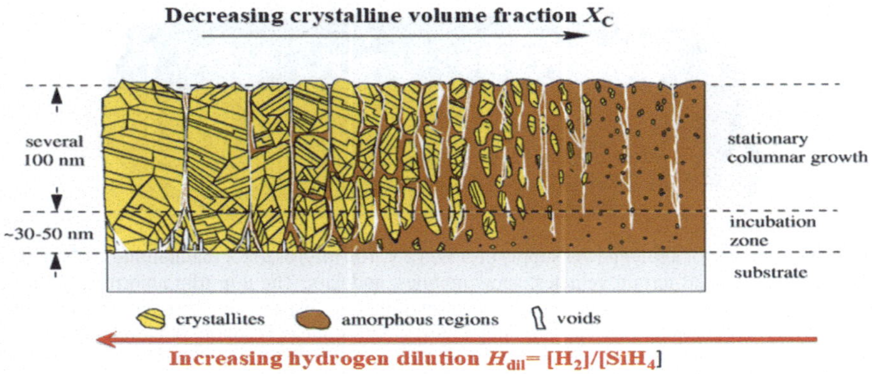

Fig. 6.22 Schematic illustration of the deposition of microcrystalline silicon by PE-CVD, from a mixture of Silane (SiH_4) and Hydrogen (H_2), in function of the hydrogen dilution ratio $H_{dil} = [H_2]/[SiH_4]$, where $[H_2]$ stands for the hydrogen gas flow and $[SiH_4]$ for the silane gas flow. Reproduced from [1], with the kind permission of the EPFL Press

6.3 Microcrystalline Silicon

6.3.1 Deposition of Microcrystalline Silicon Layers

If one increases the hydrogen (H_2) gas flow in the PE-CVD deposition system shown in Fig. 6.1, one obtains—instead of amorphous silicon, so-called "microcrystalline silicon (μc-Si:H)"—i.e. a material containing tiny crystallites (silicon nanocrystals with a diameter of 10–20 nm) embedded into an amorphous silicon matrix. Microcrystalline silicon is, thus, always a mixture of a crystalline phase and of an amorphous phase. The "crystalline volume fraction X_C" is the parameter used for characterizing the "degree of crystallinity" of μc-Si:H layers: $X_C = 0$ means the layer is completely amorphous; $X_C = 1$ means the layer is completely crystalline. Figure 6.22 illustrates how, with increasing hydrogen gas flow, the crystalline volume fraction X_C of the layers also increases. Figure 6.22 also indicates that μc-Si:H layers possess a highly complex microstructure: voids are inevitably formed; the crystallites tend to group themselves into columns.

6.3.2 Microcrystalline Silicon Solar Cells

For a recapitulation on this type of solar cells, see also [12].

Microcrystalline silicon solar cells can only be obtained for X_C values between 0.4 and 0.6. Lower X_C values lead to amorphous cells; higher X_C values lead to unstable microcrystalline cells. In fact, a certain amount of amorphous filling material is needed, in order to passivate the crystallites and to "cement" them together, to form a structure without voids. On the other hand, it is important to deposit the intrinsic (***i***) microcrystalline layer (which serves as the absorber layer and plays the decisive role of absorbing photons and generating holes and electrons) on a highly crystalline ***p***-type layer. Otherwise, the ***i***-layer will start its growth, during the first 50 nm, in the amorphous phase—leading, thereby, to solar cells, which are completely unusable.

6.3.3 The Microcrystalline/Amorphous or "Micromorph" Tandem Solar Cell

Microcrystalline silicon solar cells have been used as bottom cells, in a tandem structure, together with amorphous silicon top cells, to form what was called the "micromorph" tandem solar cell. The corresponding cell structure is shown in Fig. 6.23. These tandem cells were, in the period 2005–2009, considered to be one of the most promising options for future Photovoltaics. The reasons being the following:

(a) These cells use a very low quantity of silicon base material
(b) They have potentially a very low manufacturing cost
(c) No toxic materials are involved in the cell structure
(d) Very low-cost mass production seemed imminent
(e) The combination of microcrystalline silicon and amorphous silicon constitute theoretically the ideal combination of two different bandgaps for a tandem cell [17].

Indeed in 2016, relatively high module efficiencies (of over 10%) were reported by Industrial R and D laboratories [18].

In the years 2007–2009, many Industries invested heavily in the purchase of equipment for the large-scale production of modules based on micromorph tandems.

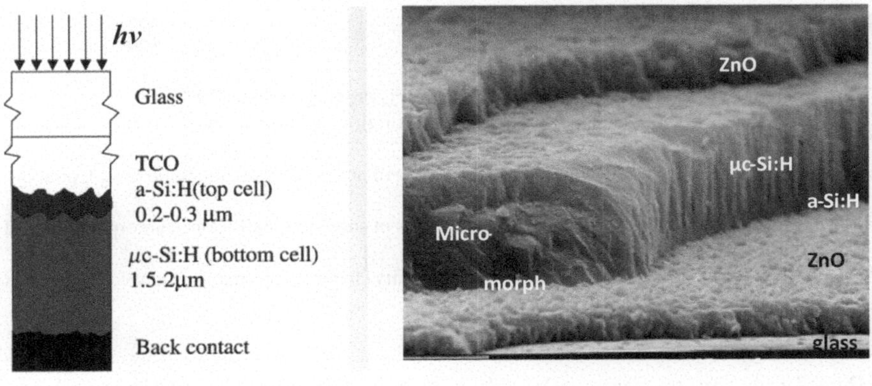

Fig. 6.23 a-Si:H/μc-Si:H or "micromorph" tandem solar cell: **a** basic structure; **b** electron micrograph. Reproduced from [1], with the kind permission of the EPFL Press

However, none of these Industries were able to make a profit from their investment. Indeed, the period after 2009, led to the complete abandon of the micromorph tandem. In retrospect we can analyse the reasons for this fiasco:

A. The equipment for the large-scale production of modules based on micromorph tandems was sold at an exaggeratedly high price, leading to commercial prices for micromorph tandem modules, which were far too high and could not compete with the prices of other modules, especially c-Si modules.
B. We all, who were then working in this field, failed to notice what was happening at the same moment in the sector of c-Si modules:

    (a) Unprecedented rapid technological progress in the c-Si sector, with a remarkable improvement, both in the increase of efficiencies and in the reduction of the material used.
    (b) The entry of Chinese manufacturers into the field. These manufacturers, brought, with massive government support, c-Si modules onto the market[12], with prices well below those of all European and U.S. module producers.

C. We also failed to notice the shift of PV market segments during the same period—from relatively small PV units to very large installations, where area was the dominant factor.

In conclusion, amorphous silicon solar cell development taught us a great deal about thin film solar cells in general and what is necessary to produce a useful, large-scale commercial solar module technology. At present, the only use of these types of solar cells and modules by themselves is in niche markets. The R&D work on a-Si:H also taught us a great deal about the use of disordered materials in electronic devices, and it led to their use as passivation layers in crystalline silicon solar cells, such as HJT.

References

1. A. Shah (ed.), *Thin-Film Silicon Solar Cells* (EPFL Press, Lausanne, 2010)
2. R.C. Chittick, J.H. Alexander, H.F. Sterling, Preparation and properties of amorphous silicon. J. Electrochem. Soc. **116**, 77–81 (1969)
3. W.E. Spear, P.G. Le Comber, Substitutional doping of amorphous silicon. Solid State Commun. **17**, 1193–1196 (1975)
4. W.E. Spear, P.G. Le Comber, Electronic properties of substitutionally doped amorphous Si and Ge. Phil. Mag. B. **33**, 935–949 (1976)
5. D.L. Staebler, C.R. Wronski, Reversible conductivity change in discharge produced amorphous silicon. Appl. Phys. Lett. **31**, 292–294 (1977)

[12]This led, in the years to come, to the breakdown of almost all European PV module manufacturers, both those for thin-film silicon modules and those producing c-Si modules. Thus, the Chinese Government attained their goal, which was to dominate the PV market—a market considered by them, to be of strategic importance.

6. A. Kolodziej, Staebler-Wronski effect in amorphous silicon and its alloys. Opto-Electron. Rev. **12**(1), 21–32 (2004)
7. T. Shimizu, Staebler Wronski effect in hydrogenated amorphous silicon and related alloy films. Jpn. J. Appl. Phys. **43**, 3257–3268 (2004)
8. R. Platz, S. Wagner, C. Hof, A. Shah. S. Wieder, B. Rech, Influence of excitation frequency, temperature, and hydrogen dilution on the stability of plasma enhanced chemical vapor deposited a-Si: H. J. Appl. Phys. **84**, 3949–3953 (1998)
9. S. Guha, J. Yang, A. Banerjee, B. Yan, K. Lord, High-quality amorphous silicon materials and cells grown with hydrogen dilution. Sol. Energy Mater. Sol. Cells **78**, 329–347 (2003)
10. C. Ballif, S. De Wolf, A. Descoeudres, Z. C.Holman, Amorphous Silicon/Crystalline Silicon Heterojunction Solar Cells, in *Advances in Photovoltaics: Part 3* vol. 90; ed. by G. P. Willeke, E.R. Weber, Semiconductors and Semimetals (Elsevier, Amsterdam 2014), pp. 73–120
11. S. Olibet, Properties of interfaces in amorphous/crystalline silicon heterojunctions, Ph.D. Thesis at the "Faculté des Sciences" of the University of Neuchâtel (2008), Section 4.6
12. A. Shah, *Thin-Film Silicon Solar Cells*", Chapter IC-1 in *Practical Handbook of Photovoltaics,* 2nd edn. ed. by A. Mc Evoy et al. (Elsevier, Amsterdam 2012), pp. 209–281
13. A. Virtuani, D. Pavanello, G. Friesen, Overview of temperature coefficients of different thin film photovoltaic technologies, in *Proceedings of the 22nd EU PVSEC Conference* (2007, Milano)
14. M. Zeman, R.E.I. Schropp, Amorphous silicon based solar cells and modules. Appl. Solar Energy **31**, 22–29 (1995)
15. K. Ghosh, A. Guha, S.P. Duttagupta, Power generation on a solar photovoltaic array integrated with lighter-than-air platform at low altitudes. Energy Convers. Manag. **154**, 286–298 (2017)
16. Y. Kuwano, K. Yoshida, Sanyo's flat diamond: Integrating R & D into the business. Long Range Plan. **30**, 473–492 (1997)
17. A. Shah et al., Basic efficiency limits, recent experimental results and novel light trapping schemes in a-Si:H, μc-Si:H and "micromorph tandem" solar cells. J. Non-Crystall. Solids, **338–340**, 639–645 (2004)
18. J.S. Cashmore et al., Improved conversion efficiencies of thin-film silicon tandem (MICROMORPH™) photovoltaic modules. Sol. Energy Mater. Sol. Cells **144**, 84–95 (2016)

Arvind Shah is the Founder of the Photovoltaics Research Laboratory (PV-Lab), at the Institute of Microtechnology (IMT), in Neuchâtel, Switzerland. PV Lab Neuchâtel has done pioneering work in the establishment of low-cost production methods for solar cells based on silicon: It introduced a novel plasma-assisted deposition method called "VHF deposition" permitting a significant increase in the deposition rate for thin-film silicon layers. It also introduced microcrystalline silicon, deposited by VHF plasma, and with very low oxygen content, as novel absorber layer, within thin-film solar cells. From 1979 to 2005, Arvind was Professor at the University of Neuchâtel. From 1987 to 2005, he was additionally part-time professor at the EPFL Lausanne. In 1975 he founded and co-directed the Centre for Electronics Design and Technology (CEDT) at the Indian Institute of Science in Bangalore. CEDT is now one of India's leading University Centres in the field of Electronics. Since 2006, Arvind has been active as scientific consultant to the PV Lab and to various Industries, in Europe, India and the USA. Arvind received the Swiss Solar Prize, together with Johannes Meier in 2005. He received the Becquerel Award in 2007.

Chapter 7
Crystalline Silicon Solar Cells: Heterojunction Cells

Sylvère Leu and Detlef Sontag

Abstract In contrast to conventional crystalline homojunction cells, heterojunction cells (HJT cells) work with passivated contacts on both sides. This chapter explains the functioning of such passivated contacts; it discusses the tunnel effect: an effect, which is important for these contacts. The role of the various layers within HJT cells is described. The advantages and disadvantages of the various cell architectures for HJT cells are explained. Since high-efficiency HJT cells usually consist of *n*-type material, the difference between *n*-type and *p*-type material is described in more detail: the term "capture cross-section" is introduced. Capture cross-sections play a decisive role in the recombination mechanisms studied with the goal of differentiating *n*-type and *p*-type silicon. Thereafter, the fabrication procedures for HJT cells are discussed; in particular, the advantages of texturing in the crystal plane (100) for monocrystalline cells. The authors also describe how the depth of microcracks can be measured with the Bevel method—and how much wafer material has to be etched off, so as to obtain a surface free of microcracks. The last section shows the favourably low TC-values of HJT cells.

7.1 Introduction

In Chap. 5 we have already learnt that passivation of the reverse side can be improved if a permanent electrical back surface field (BSF) is established by suitable process steps. In the discussion of PERC cells, we also found that aluminium oxide has better passivation properties than a silicon-aluminium alloy fabricated with an aluminium paste on the back.

In order to further increase the passivation quality, an additional adaptation of the cell structure is necessary. Increasing the passivation quality increases both the open circuit voltage and the fill factor FF and, thus, also the efficiency η of the solar cell.

S. Leu · D. Sontag (✉)
Meyer Burger Technology A.G., Gwatt, Switzerland
e-mail: dsontag@web.de

S. Leu
e-mail: sylvere.leu@ciptec.ch

A. Shah (ed.), *Solar Cells and Modules*, Springer Series in Materials Science 301,
https://doi.org/10.1007/978-3-030-46487-5_7

This can be obtained with the heterojunction cell. Heterojunction technology (HJT) combines the positive characteristics of amorphous thin-film technology with those of crystalline technology, as shown schematically in Fig. 7.1

It shows how heterojunction cells are constructed by combining the architecture of an amorphous cell and a crystalline cell. The efficient amorphous surface passivation layers *p-i and i-n* are used to passivate the crystalline silicon bulk. Amorphous cells are very thin (<1 μm), whereas conventional crystalline cells have typically a thickness of 140–160 μm. Heterojunction cells combine a high photon absorbance of a thick silicon bulk material with the extraordinary passivation properties of amorphous silicon [1]. Without losses in efficiency the thickness of Heterojunction solar cells can be reduced down to 80–100 μm. In Fig. 7.2 some typical examples for applications are presented.

The Role of the Amorphous Silicon thin-film Layer is

- To provide an excellent passivation effect.

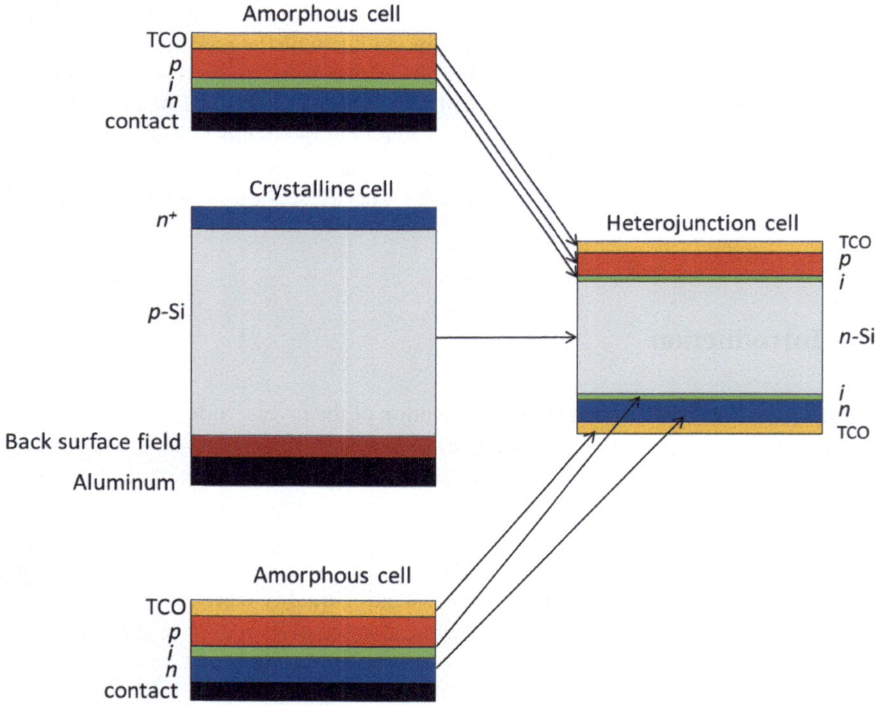

Fig. 7.1 Schematic illustration of the Heterojunction (HJT) cell. The HJT cell is a combination between an amorphous cell and a crystalline cell. Figure is not to scale

Fig. 7.2 Application of three different cell technologies. **a** Amorphous silicon thin -film technology with bended modules to supply a bus stop, **b** crystalline silicon technology on a flat roof and **c** HJT technology in bifacial design on a curtain-type façade

Thereby, the holes (minority charge carriers), are hindered by the band offsets to reach the cell contacts and, thus, recombination at the contacts is effectively suppressed [2].
(Note that in Chap. 6, amorphous silicon layers and solar cells have been discussed in more detail).

The Role of the Crystalline Silicon Bulk is

- To convert very efficiently a large part of the sunlight's spectrum into electrical energy, thanks to a favourable bandgap.
- To enable textured surfaces, reducing, thus, reflection—and allowing more light to enter into the solar cell.
- To provide high mobility of the charge carriers leading to high conductivity, and, thus, to higher electric currents.
- To provide absorber layers with negligible degradation.

Heterojunction cells combine the positive aspects of amorphous thin-film technology with those of crystalline technology. In a nutshell, it can be said that the amorphous layer serves as a passivation layer, leading to the reduction of recombination centres at the surface, and that the crystalline bulk acts as an «energy conversion machine».

Due to the excellent passivation quality of amorphous silicon, the recombination of charge carriers on the surface of the crystalline silicon wafer is no longer the main limiting factor for the efficiency η of the solar cell (see also Chap. 3). Rather, the quality of the wafer material comes to the fore. The easiest way to improve wafer quality is to use *n*-doped wafer material instead of the commonly used *p*-doped material. *n*-doped wafers are doped with phosphorus. In these *n*-doped wafers, the lifetime of the minority charge carriers (holes) is roughly 5 ms (ms). As a comparison: in *p*-doped wafers the lifetime of the minority charge carriers (electrons) is approximately 1 ms.

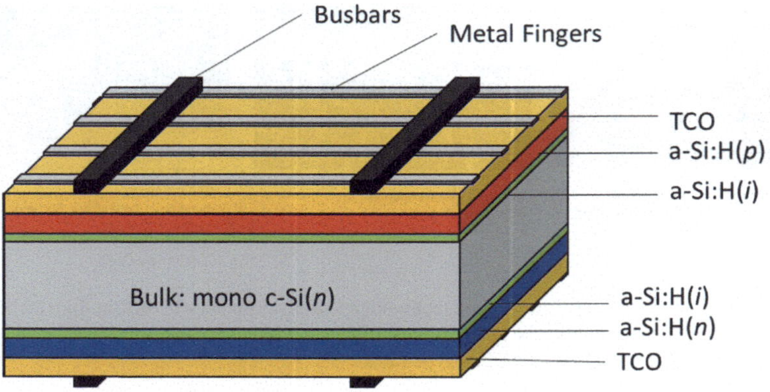

Fig. 7.3 Cross-section of a bifacial heterojunction cell. The *p-n* junction is in this case on the front side

As a result, the diffusion length[1] is about double as high in *n*-doped material than in *p*-doped wafers.

The structure of a typical heterojunction cell can be seen in Fig. 7.3. We can identify a total of **six layers** at the top and bottom. On each side there are two amorphous layers: an intrinsic one a-Si: H(*i*) and a doped layer a-Si:H(*n* or *p*); the latter are covered by TCO layers. The following section describes this simple and symmetrical structure and presents its outstanding advantages over other cell concepts.

7.2 Cell Structure

7.2.1 The Hetero-Contact

(a) ***The Ohmic Contact***

Different coatings of silicon surfaces show different passivation qualities.

For example, **aluminum oxide** passivates the cell surface in a better way than the aluminium-silicon alloy used in «standard Al-BSF solar cells». With aluminium oxide passivation layers (see Chap. 5, PERC solar cells), open-circuit voltages $V_{oc} = 660$ mV can be achieved. However, since the aluminium oxide, which is applied, as a passivation layer on the back side of the PERC cell, is a very good insulator with a resistivity of 10^{12} Ωm @ 20 °C, a direct contact between the solar cell and the metal layer—through the aluminium oxide layer—must be created, so that the charge carriers find their way. Such a direct contact is usually made in the PERC cell

[1]Diffusion length $L = (\tau * D)^{1/2}$. D is the diffusion constant and τ is the lifetime in seconds. Diffusion length describes the average length a carrier moves between generation and recombination.

with openings in the aluminium oxide through which a conductive aluminium paste is used to contact the crystalline silicon (see Chap. 5). A high-temperature step in the range of 850 °C forms, as in the «standard Al-BSF solar cell», an aluminium-silicon «eutectic[2]» resulting in an Ohmic contact between aluminum and silicon, but this time only at the openings of the aluminum oxide layer. At these contact points not only Ohmic resistance losses appear but, like in standard Al-BSF solar cells, an inferior passivation occurs, which results in reduced efficiency η.

The question now arises as to what an optimal contact might look like—a contact that has no or only very little recombination and no Ohmic losses. A reduction of the recombination losses can be achieved, for example, by making the openings even smaller; but this increases resistive losses. Very high doping of the silicon at the openings reduces resistivity losses but increases recombination. (Auger recombination, see Chap. 4.) Laser-Fired Contacts [3] constitute another way to reduce Ohmic losses. However, all these processes have additional process steps—a fact that renders cell production more expensive.

(b) ***Passivated Contacts***

Another, very elegant possibility to avoid direct metallic contact to silicon is to use thin passivation layers with a thickness of only a few nanometres. Charge carriers can «tunnel[3]» through the passivation layer and recombination effects cannot occur since direct Ohmic contacts are not present. In addition, such passivated contacts can form selective contacts, e.g. they conduct either only electrons or only holes. This possibility is implemented in the heterojunction cell.

(c) ***The Amorphous Double-Layer Construction of the HJT Cell Leads to the Highest Passivation***

The first silicon-based heterojunction cells used doped amorphous material **directly applied on the silicon** and achieved satisfactory passivation properties. However, it has been shown that in this structure the charge carrier density is very high, so that increased surface recombination takes place [4]. Finally, in 1992, **an intrinsic, undoped amorphous interlayer** sandwiched between the silicon crystal and the doped amorphous layer was found to provide even better passivation properties [5]. The recombination rate could be reduced and the fill factor *FF* increased. The cell was named «HIT», which stands for «**H**eterojunction with **I**ntrinsic **T**hin Layer©». Sanyo achieved 23% cell efficiency with this structure in 2009. In 2014, Panasonic obtained another breakthrough by applying passivated contacts on both sides, e.g. on the front and on the back side. The efficiency η increased to 25.6%. In 2017,

---

[2]The term «eutectic» designates a homogeneous mixture of substances that melts or solidifies at a single temperature that is lower than the melting point of either of the constituents.

[3]The term «tunneling» comes from quantum physics: If an electron encounters a potential barrier, it can pass through the barrier with a certain probability, even if its energy is lower than that of the barrier. That would not be possible according to classical physics. This probability is called the «residence probability»—it is given by the wave function representing the electron. The electron is described in quantum mechanics by a wave: as a wave it can also extend to the other side of the barrier. OR, in other words, it can just «tunnel» through the barrier.

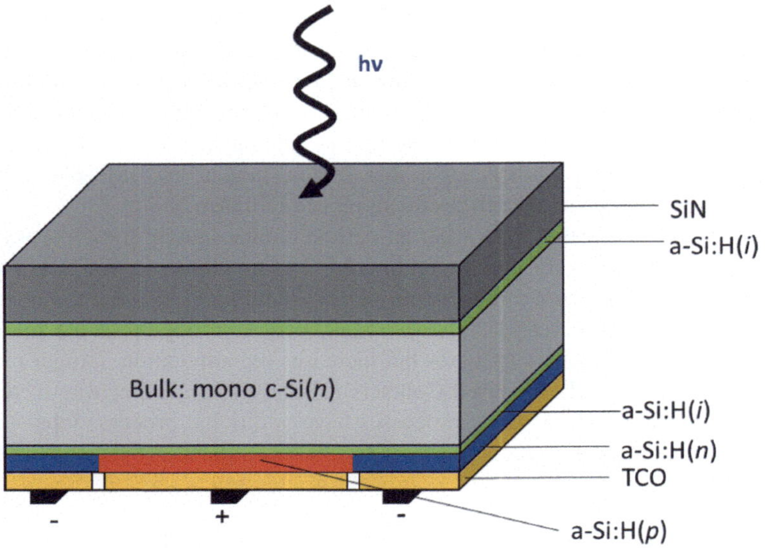

Fig. 7.4 Simplified cross-section of a heterojunction cell in the «back contact implementation». On the front, **no** fingers shadow the cell. The electricity is taken off on the back. The *pn*-junction is on the back; the *n*-region alternates with the *p*-region

Kaneka even achieved $\eta = 26.6\%$ cell efficiency on a 6×6 inch HIT cell, in which the contacts are only on the back side (back-contact cell). «HJT-**IBC** (**I**nterdigital **B**ack **C**ontact)» have the highest efficiencies. This is because the losses due to the shading of the metal fingers on the front side are completely eliminated and because there is no parasitic absorption on the front side; in fact, there is no TCO layer and no doped a-Si on the front side (see Fig. 7.4). Today, HJT-IBC cells are not in mass production due to the high manufacturing costs. However, with a clever design, the costs per peak-Watt (W_p) can be kept at the same level as with «normal» HJT cells. Figure 7.4 gives a schematic illustration of such an HJT-IBC structure.

The best measure for the quality of the passivation is the open-circuit voltage V_{oc}, which, is in a HJT cell, in general, higher than 740 mV, e.g. V_{oc} is about 15% higher for an HJT cell, than for conventional PERC cell, where we only obtain ~660 mV today.

7.2.2 *The Basic Structure of a Heterojunction Cell*

(a) *Two Different Semiconductors*

Heterojunction technology uses two materials with different energy bandgaps. «Hetero»[4] means "different" in Greek. This is the big difference with the so-called "homo[5]-junction cells" discussed in Chap. 5 (PERC cells). In the case of the heterojunction silicon solar cell, the following two materials are used: crystalline silicon (c-Si) and amorphous silicon (a-Si: H). Amorphous silicon has a bandgap of 1.7 eV, in contrast to crystalline silicon with 1.12 eV. In this configuration, the bandgap of the c-Si (*n*) lies **within** the bandgap of the a-Si:H silicon (see Fig. 7.5. drawings A and B). Such a configuration is called a «heterojunction type 1[6]» contact. Other heterojunction contacts are:

1. those relating to amorphous and microcrystalline silicon tandem cells,
2. those between Cadmium telluride and Cadmium Sulfide (CdTe/ CdS),

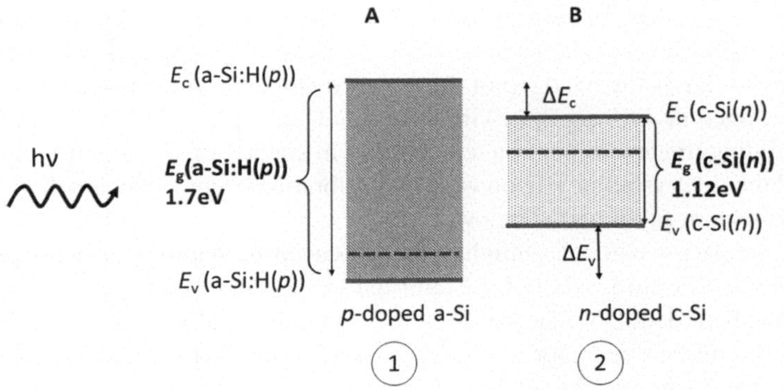

Fig. 7.5 The bandgap energy of crystalline silicon is 1.12 eV and the bandgap energy of amorphous silicon is ~1.7 eV. The dashed violet line shows the Fermi level. The blue line shows the conduction band edge and the red line represents the valence band edge. Drawings A and B show the two semiconductors. "A" represents the amorphous part a-Si(*p*) and "B" the crystalline part c-Si(*n*). The two individual materials A and B are joined together and the respective majority carriers of the materials A and B will diffuse into the other area via the concentration balance. Only the front side is shown

[4]Hetero derives from the Greek word Heteros meaning «different» or «other».

[5]Homo-, Greek prefix expressing the notion of "same, identical".

[6]Configurations in which the bandgap of one of the materials is lying within the bandgap of the other material are called «**heterostructures of Type 1**». However, if the bandgaps are «staggered» in the sense that the conduction band of the second material is lower than the conduction band of the first material, and at the same time, the valence band of the second material is also lower than the valence band of the first material, then we have a «**heterostructure of Type 2**».

3. those between Copper-Indium-Gallium-Diselenide and Cadmium Sulfide (CIGS/CdS),
4. those relating to Group[7] III–V tandem cells (GaAs, etc.).

These other contacts are not described in this chapter: (1) has been mentioned in Chap. 6, (2) and (3) will be mentioned in Chap. 8—whereas (4) will not at all be treated in this book. Indeed, although III–V tandem cells give the highest cell efficiencies ever obtained (over 40%), their fabrication cost is at present prohibitively high, so that they are only used in special situations (like concentrating photovoltaics, CPV, see Chap. 10, Sect. 10.3.2) and in space applications.

(b) ***Band Diagram and Tunnelling***

In Fig. 7.5 the construction of the band diagram of a heterojunction cell is shown [6]—for the sake of simplicity this is done only for **the front side, here for the *p*-side, where the *pn*-junction is.** The band diagram is essential to understand the benefits of passivated contacts, which ultimately characterize the HJT cell. Let us first look only at the individual layers—the individual components of the heterojunction before joining them together. Individual layers: The bandgap of the *p*-doped, amorphous silicon a-Si:H(*p*) layer is shown hatched on the left in Fig. 7.5, drawing A. Just next to it (drawing B), is the bandgap of the *n*-doped silicon c-Si (*n*) layer (wafer). This bandgap, dotted in drawing B, is within the first bandgap («type 1» heterojunction). When building the heterojunction, e.g. putting in contact the (*p*) amorphous silicon layer and the (*n*) crystalline silicon wafer, equilibration occurs, which will determine the band diagram as discussed below.

As in the dark—without sunlight—the **potential in a material has to be balanced**, the two Fermi levels will be adjusted to the same potential. Let us look at what happens in detail: When we bring the two individual materials A and B into contact,[8] the respective majority charge carriers begin to diffuse to the other side. In the *n*-doped silicon c-Si (*n*) layer, the electrons are the majority carriers. The electrons flow from the right side in Fig. 7.6 (position ②) to the left side (position ①).

This flow charges the right side (β) positively and the left side (α) negatively, see Fig. 7.6. This charge carrier exchange flows until the two Fermi levels have become equal and an equilibrium state for E_F prevails. In Fig. 7.6 the two Fermi levels are therefore on the same line.

Let us take a closer look to the region around the interface. The band bending and the ***gradient caused*** thereby in the Conduction Band Edge E_C and in the Valence Band Edge E_V constitute an electrical potential for the charge carriers that drives

[7]The term «Group» here refers to the numbering of the columns in the periodic system.

[8]The idea that we start with individual layers, which we afterwards join together is, of course, just a "Gedankenexperiment", an artifice we imagine in our minds, so as to understand better what happens. In reality the two individual layers are joined together right from the beginning—as the HJT cell is fabricated.

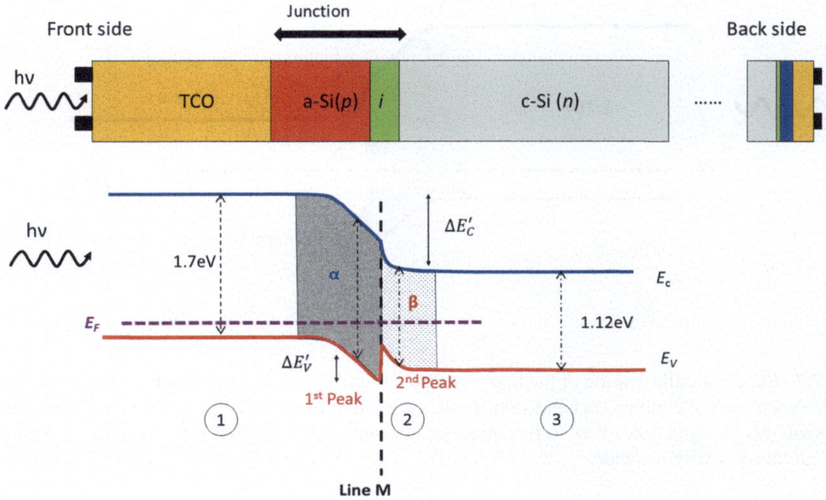

Fig. 7.6 Bandgap diagram after joining the two materials A and B. Shown is the *p*-side of the heterojunction cell (here the front side, with the *pn*-junction) without metallization. The band-bending of the valence band and the conduction band is illustrated. The conduction band stretches around $\Delta E_C' (= \Delta E_C + \text{band-bending})$, while the valence band stretches around $\Delta E_V' \; (= \Delta E_V - \text{band-bending})$ and its offset is therefore smaller. One can also clearly see the peak that occurs when the two different potentials of the two semiconductors are joined together. For simplicity's sake, we assume that the convergence takes place at line M. Above the band diagram the cross-section of the solar cell is illustrated. The light hits the solar cell on the left side; the back side of the cell is to the right

them in opposite directions to each other, electrons to the right hand side and holes to the left hand side in Fig. 7.6. [9]

In the area of the junction (see Fig. 7.6) the transition from the bandgap of amorphous silicon (~1.7 eV) to the bandgap of crystalline silicon (1.12 eV) takes place. The edges E_C of the conduction band and E_V of the valence band are bent. Initially, the valence band follows the conduction band at a distance of 1.7 eV, up to the transition; and then, after the transition the distance becomes 1.12 eV. The transition takes place within a few atomic layers, i.e. within a width of a nanometre or less, whereas the junction itself with the charged regions described above extends over tens of nanometres in the highly doped amorphous silicon and hundreds of nanometres in the lowly doped crystalline silicon. There is therefore a band offset of ~0.6 eV at the interface between the two materials (line M in Fig. 7.6), which is typically split into 0.45 eV as conduction-band offset and 0.15 eV as valance-band offset. The downward bending due to the junction formation followed by this band offset leads to the

[9] Those majority carriers that accumulate now on opposite sides constitute a concentration imbalance which is a chemical potential. The process is self-regulating with electrical and chemical potential neutralising each other at every point in space. In fact, this is the requirement which is postulated in the beginning, the combined electrochemical potential is identified with the Fermi level.

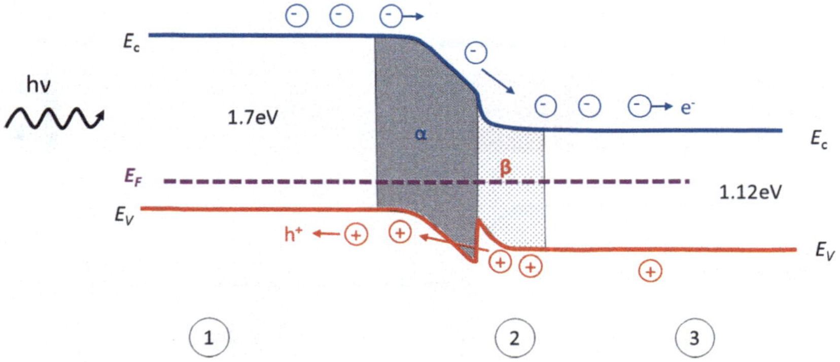

Fig. 7.7 Schematic illustration of the flow of electrons and holes. The negatively charged electrons "slide down" on the conduction band slope to the energetically lower, right side); on the other hand, the positively charged holes flow to the energetically higher side, the left side. They have to tunnel through the very narrow peak

spike in the valence band on the left of the Line M (1st peak). After that—shortly to the right of the vertical line M—the bandgap reduces completely to 1.12 eV. Again, the valence band follows the conduction band. The valence band makes a 2nd peak here, but this time upwards [6].

(c) ***Charge Transport***

Now what is the effect of these features on the carrier currents in the solar cell? **Photogeneration** of electrons and holes in the crystalline silicon c-Si(*n*) layer (wafer) leads to an excess of both electrons and holes in the wafer compared to the equilibrium state described above. Near the entry point of the light (on the left side, position ① in Fig. 7.7), the photogeneration will be at its highest value, and from their onward it will drop off exponentially to the right (see Chap. 4, Fig. 4.2). At the back side of the wafer (position ③), the photogeneration will be substantially lower.

Now, we will look at the electrons and holes separately: Electrons will be diffusing from left to right, as their concentration is very high on the left side of the c-Si (n) layer, (position ①, strong photogeneration) and relatively low on the right side (position ③, weak photogeneration); this is the point where the electrons leave the solar cell, see Fig. 7.7.

There will be no electrons "going back" from the c-Si (*n*) layer to the a-Si:H(*p*) layer, as there is a potential difference of $\Delta E'_C$ (Fig. 7.6) to be overcome. Holes, on the other hand, flow from the right side in Fig. 7.7 to the left side, because the valence band edge of c-Si is lower than that of a-Si. The holes can tunnel through the very narrow peak. The holes are prevented from flowing to the right by the peak,[10] and additionally, by the potential difference in $\Delta E'_V$(Fig. 7.6). Hence, the peaks in

[10]The probability for charge carriers to tunnel through a peak depends on the height and width of the peak, and also on the density of charge carriers present just before the peak; now, in ② there is a high density of holes, so there will be many holes tunnelling through the peak from left to right,

the valence band help in obtaining a more effective separation of charge carriers. Since holes can flow freely to the left-side contact, but electrons do not flow in that direction, an ideal contact—without any recombination at the electrode surface, is formed (selective contacts or passivated contacts).

However, the peak must not be too high, otherwise it constitutes a blockade and the holes cannot tunnel through. Conversely, if the peak is too small, cell efficiency and the fill factor are reduced [7]. Later in Sect. 7.2.3, we will also look at the back side of the HJT cell and see then that there is a peak at the back side, as well, through which the electrons can tunnel through.

Thanks to the fact that almost no electrons flow through the (*p*) amorphous silicon layer (and similarly on the back side, no holes flow through the a-Si(*n*) amorphous silicon layer), recombination losses are avoided. The contacts created in this manner are passivated contacts. They are selective and not Ohmic. They let either only the holes pass through or else only the electrons pass through. [8] In addition, these selective barriers are unidirectional: The charge carriers, once they tunnel through the barrier, cannot return. And finally, they are not Ohmic contacts and reduce recombination losses. This has an impact on the equivalent circuit diagram. In the equivalent circuit diagram of a homojunction cell with fired contacts (Al-BSF cell, PERC cell, Chap. 5) we have a two-diode model with saturation currents[11]: J_{01} and J_{02}. With the passivated contacts of the HJT cell, the saturation current J_{02} is minimized, because recombination is strongly reduced.

The saturation currents, which have already been discussed in Chap. 3 have a very high priority in the evaluation of a solar cell, because they determine the open circuit voltage V_{oc} of the solar cell and, thus, also its efficiency η. Figure 7.8 shows the equivalent circuit diagrams for homojunction and heterojunction cells. J_{01} denotes the leakage current caused by the surface recombination and the losses in the bulk.

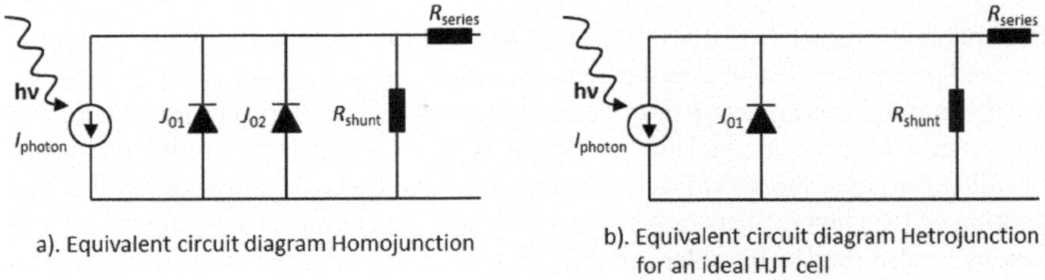

Fig. 7.8 **a** Equivalent circuit for homojunction cells with fired contacts (see Chap. 5 on PERC cell) and **b** equivalent circuit for HJT cells. J_{01} denotes the leakage current caused by the surface recombination and by the losses in the bulk. J_{02} denotes the losses caused by the space charge zones. In homojunction cells the losses in the space charge zones are higher than in HJT cells (The reason being that in the heterojunction cell, the *p*-region is separated from the *n*-type silicon by the intrinsic layer)

whereas in ① there is a very low density of holes (most holes there are siphoned off towards the contact)—thus, there will be very few holes tunneling through the peak from left to right.

[11] Also called dark or leakage currents.

J_{02} denotes the leakage current caused by the space charge zones. In homojunction cells a distinct space charge zone is present. In heterojunction cell, the p-region is separated from the n-region by a layer of intrinsic amorphous silicon, and the space charge region is not so pronounced, because of lower doping efficiencies and thinner layers, as compared to homojunction cells.

(d) ***Structure of Amorphous Layer***

Amorphous silicon consists of the same atoms as pure, crystalline silicon. However, it does not show a periodic crystalline structure. Although at the interface with the crystalline silicon it takes over the "crystalline" structure but already three to four bond lengths further, e.g. already after 0.5 nm, the bond angles deviate strongly, so that no periodic correlation as in crystalline silicon is recognizable. This leads to two effects: first, due to the absence of a crystalline structure, the electrical conductivity decreases and, secondly, many open bonds occur, e.g. many unsaturated defects, so-called "dangling bonds" (see also Chap. 6). These lead to increased recombination of the charge carriers. The dangling bonds can be neutralized with hydrogen; this step is called "saturation" in technical jargon. This is why hydrogen is added to amorphous silicon. The electrical conductivity of the amorphous material can be increased by doping, but remains always relatively low, due to its amorphous character.

In HJT cells the low conductivity of the amorphous layer is not an obstacle provided the amorphous layer is thin. The charge carriers can then cross this layer vertically. Since the layer is very thin, there is no appreciable loss due to electrical resistance. However, the low conductivity is not sufficient to enable the charge carriers to move laterally to the metal contacts. It is therefore necessary to cover the amorphous layer with an additional, highly conductive layer. This layer must also be transparent to light.

(e) ***TCO Layer***

Transparent Conductive Oxides (TCO) are suitable candidates, which incorporate all required properties. The charge carriers only have to flow now vertically through the two thin amorphous layers *i-p* or *i-n* to the cell surface. The lateral current conduction to the metal fingers is then taken over by the TCO layer. In addition to its function as a conductive layer, the TCO layer also takes over the function of an **A**nti-**R**eflection **C**oating (ARC) layer. Such an ARC layer reduces reflection losses—more light can thereby be absorbed by the solar cell. The TCO layer must hence be highly transparent and should ideally not absorb any light itself (parasitic absorption). As explained in Chap. 4, this layer has an optimal thickness of 80 nm, if it is used (as is the case here) as a ***single*** anti-reflective layer.

The two objectives that the TCO layer has to meet contradict each other: Either the TCO layer is very transparent but does not conduct well electrically, or it conducts very well electrically and is less transparent because it is thicker or it is doped to a higher extent. In order to determine the optimum for the cell, a compromise has to be found in which the mutual distance of the metal fingers and their thickness must also be taken into account. An optimum is usually found with a relatively low conductivity of the TCO layer combined with fine and narrow metal fingers, which are, typically, less than 40 μm thick. These approximately 60–80 metal fingers consist of highly conductive silver. Usually the metal fingers are printed on the TCO using a silver paste. In most cases, a screen-printing process is used.

The most common TCO material currently used in heterojunction cells is **I**ndium **T**in **O**xide (ITO). Depending on the application, the tin content varies between 5 and 10%. A high tin content increases conductivity, but—at the same time—it limits transparency. Further in ITO sputtering, one can have a whole range of transparency versus conductivity by just varying the oxygen content in the film. Production is about choosing the optimal parameters for the most suitable TCO layer. Since indium is a scarce raw material, efforts are currently being made to reduce its use—or in the medium term, even to replace it completely.

7.2.3 The Heterojunction Cell

(a) ***HJT Cells with the pn-Junction on the Front Side***

So far we have described the physical properties of the heterojunction cell; in the following we want to investigate the different possibilities for the cell structure. Basically we need three layers on both sides of the cell:

- an Intrinsic-amorphous layer
- a doped amorphous layer and
- a TCO layer.

All we have to do in order to create a Heterojunction cell is to coat both sides of the wafer with a double layer of intrinsic and doped amorphous silicon and then finish it on either side with a TCO layer. However, we have to make sure that the doping on the front side is opposite to the one used on the back side. We obtain a symmetrical cell in terms of the number and type of layers. Figure 7.9 illustrates this in detail. This is the structure, where the *p-n* junction (emitter)[12], e.g. the separation of electrons and holes is located on the front side. This can be recognized by the fact that the doping of the crystalline bulk material (c-Si) is *n*-type, whilst the doped amorphous layer on the front side (a-Si:H(p^+)) is *p*-type. The amorphous layer (a-Si:H(n^+)) on the back side is here *n*-type like the bulk. In short: the cell is *p*-type on the front side and *n*-type on the back side. On the back side there are the regions *n* (bulk c-Si:H(*n*) and a-Si:H(n^+)[13]: As can be seen in Fig. 7.9, through the peak $\Delta E_C''$ situated at the back side, the electrons can tunnel through, but not the holes.

If the barriers (peaks) are narrow enough, there is a certain probability of residence beyond the barrier and the charge carriers 'see' an energetically more favourable

[12]In photovoltaics, the term emitter is defined as the region where the minority carriers leave the solar cell. In *n*-type material, the minority carriers are the holes and the holes leave the solar cell on the *p* side (positive electrode). The reverse applies to *p*-type material. In *p*-type material the minority carriers are the electrons and these leave the solar cell on the *n*-side (negative electrode).

[13]n^+ means more doped than *n* and correspondingly p^+ means more doped than *p*.

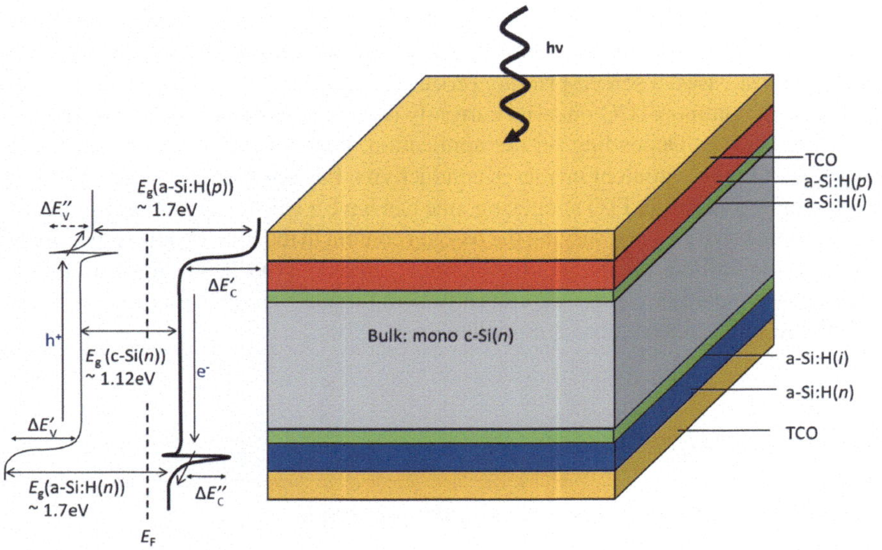

Fig. 7.9 Cross-section of a typical HJT solar cell with bilateral passivation and bifacial solar cell design, with the *p-n* junction on the front. Next to it the bandgap diagram [9] with valence band E_V and conduction band E_C. The Fermi level is shown as dotted line. There are two narrow peaks one in the conduction band and one in the valence band. Electrons can tunnel through the peak in the conduction band and holes through the one in the valence band. Thus, selective contacts are formed on both sides of the solar cell

area on the other side of the barrier. In a simplified way the carrier transport can be represented as follows: As soon as the charge carriers are on the other side of the peak, the way back is blocked, because the starting point is energetically less favourable. Furthermore: the holes can tunnel through the peak $\Delta E_V''$ on the front side, but not the electrons. The reverse is true on the back side. Through the peak $\Delta E_C''$ on the back side the electrons can tunnel through, but not the holes.[14]

This construction of the contact leads to a selective permeability for the charge carrier types. In addition, there is no direct Ohmic contact to the wafer and the electron-hole pairs cannot recombine. Such contacts are called passivated contacts. It should be noted that too much hydrogen can increase the height of the peak in the valence band to such an extent, that even the holes can no longer tunnel through and are blocked.

(b) ***HJT Cells with the pn-Junction on the Back Side***

The TCO layer on the front fulfills several functions: (1) It is an anti-reflective layer, (2) it is a protective layer for the very thin and highly sensitive amorphous layers. Furthermore, this layer has to be (3) highly conductive.

[14]In a more detailed view, the charge carriers flow through the peaks forward and backwards. However, the difference between forwards rates and backwards rates is very large, so that the backward flow can be neglected. See also footnote 10.

Achieving these three characteristics at the same time is not at all easy—in practice it means making compromises. If, however, the c-Si(*n*)/a-Si:H(*p*) interface is placed on the back side, a completely new situation arises, with interesting advantages. We are now collecting the electrons at the front contact. The electrons are the majority charge carriers in the *n*-type wafers we generally employ for HJT cells—therefore the electrons can also utilise the lateral conductivity of the bulk to reach the metal contacts. This allows us to reduce the thickness of the TCO on the front side by approximately 20% and to optimize the front TCO for its optical function, e.g. no compromises have to be made with respect to the anti-reflection properties. On the back side, however, reduced transmission of the TCO layer plays a minor role. This is because less light reaches the cell at the back. On the back, the TCO layer can be optimized for conducting and can be thicker than on the front side.[15] A thicker TCO layer on the back increases light trapping for long-wave light. In the case of bifacial cells, however, the TCO layer on the back cannot be as thick as otherwise—the albedo effect (see Chap. 10) must be taken into account in case of bifacial cells when determining the thickness.

Figure 7.10 illustrates the difference between an HJT cell with the *pn*-junction on the front (left) and an HJT cell with the *pn*-junction on the back (right).

If we compare both designs then we see that the TCO layer can be thinner in the case where the junction is on the back side.

There is another reason to move the *pn*-junction to the back. Indium-Tin-Oxide (ITO) is an *n*-type semiconductor. If the ITO layer lies on top of the amorphous *p*-type layer, which is doped with boron, a diffusion of boron atoms into the ITO layer can occur. As a result, the boron-doped *p*-type amorphous layer becomes "depleted".[16] This leads to a band bending and to performance losses [11]. To prevent this, the thickness of the a-Si:H(*p*)-layer must be relatively high, with all the disadvantages for the front side we have described above. This is an additional reason for moving the *pn*-junction to the back side.

Because the TCO layer is conductive, there is no PID (**P**otential **I**nduced **D**egradation) effect here (see Chap. 10). The sodium ions Na^{++} from the glass or the aluminum ions Al^{++} from the aluminum frame are neutralized by the TCO layer and cannot influence the space charge zone.

Figure 7.11 shows a cross-section of an HJT cell with the pn-junction on the back side.

(c) ***Illustration of the High Passivation in HJT Cells***

Unilluminated and Undoped Silicon Crystal

In an unilluminated and undoped silicon crystal, the equilibrium relationship follows the law of mass action [12]. The product of electron and hole densities is at

[15] The sheet resistivity of the TCO layer close to the amorphous layer is ~100 Ω/sq. Close to the glass of the module the sheet resistivity is much lower ~40 Ω/sq. Thus, with the thickness of the TCO layer, the resistivity can be optimized.

[16] "depleted" means **here**: it loses a part of the holes which are otherwise present in a *p*-type amorphous layer.

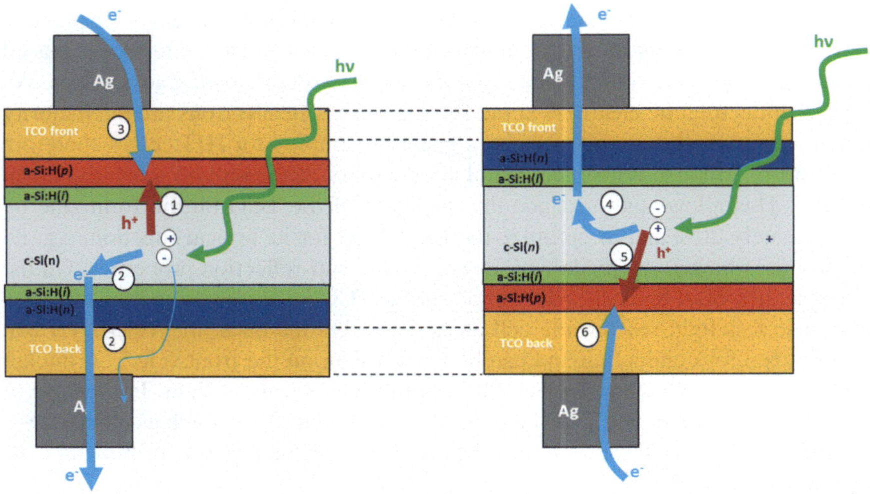

Fig. 7.10 **Left:** Cross-section of an HJT cell with the *pn*-junction on the front side. **Right**: Cross-section of an HJT cell with the *pn*-junction on the back side. The numbers denote the different paths for holes (h^+) and electrons (e^-)
pn-junction on the front side, left: ① Photons $h\nu$ generate electron-hole pairs; holes (h^+) travelling vertically through bulk c-Si(n) and at the a-Si interface they are hindered by the band offset, but are tunnelling through the junction (valence band). ② Electrons (e^-) have a high mobility and, thus, they can flow laterally through the bulk c-Si(*n*) and then after tunnelling through the amorphous stack they flow vertically through the back TCO where they leave the cell; ③ the electrons (e^-) re-enter on the front contact and they slide through the a-Si:H(*p*) and a-Si:H(*i*) layer into the bulk and finally recombine with the holes
***pn* junction on the back side, right**: ④ Electrons (e^-) travelling through bulk c-Si(*n*) and then after tunnelling through the amorphous stack they flow vertically through the a-Si:H(*n*) and through the front TCO. ⑤ Holes (h^+) travelling through bulk c-Si(*n*) up to the a-Si interfaces on the back ⑥ electrons (e^-) re-enter and reaching the bulk where they recombine with the holes

equilibrium a constant:

$$n \cdot p = n_i^2 \tag{7.1}$$

p Density of (free) holes in valence band
n Density of (free) electrons in conductive band
n_i Intrinsic charge carrier density in silicon at room temperature, approx. $10^{10}\ \text{cm}^{-3}$.

Unilluminated and Doped Silicon Crystal
If we dope silicon[17] with phosphorous atoms (*n*-type) having e.g. a density $N_D = 10^{16}\ \text{cm}^{-3}$, we obtain at room temperature, for holes, based on (7.1), assuming that all doping atoms are activated (ionized).[18]

[17]Silicon has 10^{23} atoms per cm^3.
[18]This means that we have $n = N_D + n_i \approx N_D$.

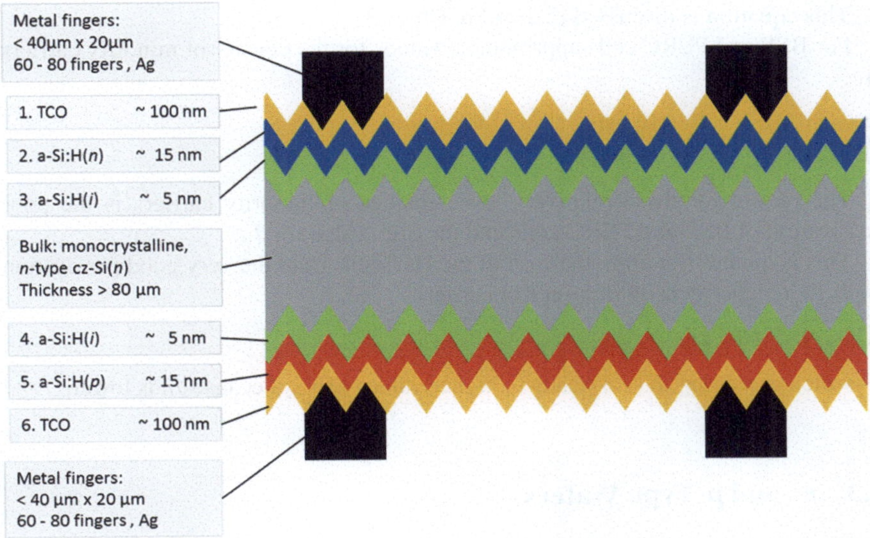

Fig. 7.11 Cross-section of a texturized typical bifacial HJT solar cell with *pn*-junction at the back and double-sided passivation as presented by CSEM/IMT/Meyer Burger [10]

$$p = \frac{n_\mathrm{i}^2}{N_\mathrm{D}} = \frac{10^{20}}{10^{16}} = 10^4\,\mathrm{cm}^{-3} \tag{7.2}$$

One immediately recognizes the fact that the holes are the minority carriers with a density of only 10^4 cm^{-3} whereas the majority carriers, here the electrons, have a density of 10^{16} cm^{-3}.

Illuminated and Doped Silicon Crystal
In the illuminated case, the photons generate electron-hole pairs. In our example, the density of minority charge carriers, which result from the generation of electron-hole pairs is 10^{16} cm^{-3}

We see that the density of minority carriers, which is only 10^4 cm^{-3} in the non-illuminated case, rises to 10^{16} cm^{-3} due to illumination: this increase corresponds to a factor of 10^{12}. The density of majority carriers is practically unchanged. In solar cells, losses due to saturation currents J_0 have to be taken into account; these are mainly influenced by the passivation quality of the surfaces. The open circuit voltage V_OC is given by (7.3):

$$V_\mathrm{oc} = \frac{nkT}{q}\mathrm{In}\left(\frac{J_\mathrm{ph}}{J_0} + 1\right) \approx \frac{nkT}{q}\mathrm{In}\left(\frac{J_\mathrm{ph}}{J_0}\right) \tag{7.3}$$

This equation is discussed in detail in Chap. 3.

For BSF and PERC cells approximate values for the density of minority carriers are:

Standard BSF cell 10^{14}cm^{-3}
PERC-cell 10^{15}cm^{-3}.

The reason for these relatively low densities of minority carriers is the poor passivation of BSF and PERC cells and the high value for J_0.

This is quite different in the case of the HJT cell: Here the very good passivation leads to a higher density of minority carriers

HJT cells 10^{16} cm^{-3},

which according to (7.3) leads to higher values of V_{oc} because J_0 is lower.

7.3 *n*- and *p*-Type Wafers

The entire development of standard solar cells is based on the use of *p*-type, e.g. boron-doped crystalline silicon. This development is historically due to the fact that in the early days of photovoltaics, wafers were manufactured from *p*-doped waste products of the semiconductor Industry. Over the years, the amount of waste was no longer sufficient and much effort was made to produce high-quality *p*-type silicon specifically for photovoltaics. In the past *p*-type material was cheaper to produce than *n*-type material. Today, both kinds of materials have more or less the same production costs. In principle, the question naturally arises why one generally uses *n*-type crystalline silicon for heterojunction (HJT) solar cells and not *p*-type silicon. There are several reasons which are listed in the next sections:

(a) Capture Cross-sections In the comparison of *n*-type and *p*-type silicon, the key quantity "*capture cross-section*" plays a decisive role. The capture cross-section indicates the size of the capture radius for electrons or holes, for capture by foreign atoms (impurities).[19] In the following example, we assume that iron is the dominant impurity in silicon. We will investigate the cross section capture of iron for electrons and holes more closely.

As an example we consider capture cross section of iron [13] for electrons (e) and holes (h):

$$\sigma_{\text{e}} = 3.5 \times 10^{-11}\,\text{cm}^{-2}$$
$$\sigma_{\text{h}} = 4.5 \times 10^{-16}\,\text{cm}^{-2}$$

[19] Actually the term «capture cross-section» is used not only in the case of impurities, but in the case of ***all*** recombination centres or defects—as an example it is also used in the case of "dangling bonds", which appear in amorphous silicon layers (see Chap. 6).

The capture cross section is 10^5 times larger for electrons than for holes regardless of whether we have *n*-type material or *p*-type material. However in *n*-type material the holes are the minority carriers and their density is less influenced by the capture process than the density of majority carriers (electrons).

Thanks to the smaller capture cross-section of the holes, the latter are captured less frequently by the impurities (here as an example, by the iron impurities) and have longer lifetimes. At the same time, they are the minority carriers, which contribute significantly to the open circuit voltage V_{oc}. This is one reason why *n*-type silicon leads to better cell properties than *p*-type silicon.

So, even in the hypothetical case where *n*-type material and *p*-type silicon would have the same number (e.g. the same density) of impurities, and the same impurities like iron (Fe), these impurities are less harmful in *n*-type silicon than in *p*-type silicon. Thus, *n*-type silicon does not have to be gettered (see section e below), at all, in order to obtain the material quality needed for solar cells.

(b) Carrier Lifetimes The lifetime of minority carriers is generally taken as a marker for the quality of a solar cell material. The lifetime of minority carriers indicates how long the minority charge carriers will, on an average, exist before they recombine with carriers of opposite charge (e.g. with the majority carriers). If the lifetime is too short, charge carriers cannot reach the junction from their point of origin in bulk silicon and recombine before. For *n*-type material the lifetime of holes (minority carriers) is up to approximately 10 ms, for *p*-type material the lifetime of electrons (minority carriers) is only up to approximately 2 ms. The probability that the minority charge carriers reach the junction and contribute to the solar cell current is, thus, two times higher for *n*-type material.[20]

(c) Carrier Mobilities In *n*-type material, the holes are the minority carriers and their mobility is only 450 cm^2/Vs. The mobility of the electrons is three times higher at 1400 cm^2/Vs. This then is the case in a cell with a back *pn*-junction: in the best case, the charge carrier generation takes place in a cell near the back surface and the hole diffuses to the back side by the shortest route. In the less favourable case, the hole has to take a longer path through the bulk to reach the junction. But since the mobility of the holes is relatively small, a high-quality bulk material is crucial, so that even in unfavourable cases the charge carriers have a high probability to reach the junction. *n*-type material satisfies these requirements, because the capture cross-section[21] σ_p for holes in *n*-type silicon (minority carriers) is smaller than the capture cross-section σ_n for electrons in *p*-type silicon (minority carriers).

(d) Boron Oxygen Complex and Degradation Effects Furthermore, boron in silicon tends to form together with oxygen so-called boron-oxygen complexes,[22] which

[20]Diffusion length $L = (\tau * D)^{1/2}$.

[21]Explanation of capture cross-sections: Low capture cross-section means that activity radius around impurities is small and recombination activity is low. High capture cross-section means that activity radius around impurities is large, and recombination activity is high.

[22]In addition to boron, iron and copper in combination with oxygen can also produce interference effects.

lead to the **LID (light induced degradation**, see Chap. 10) effect, which in its turn causes degradation in the performance of a solar cell of 5–15%. LID effects occur above all in monocrystalline *p*-type silicon, which is doped with boron and only to a lesser extent in multicrystalline solar cells; they hardly occur, at all in *n*-type silicon. When we bring more hydrogen into the bulk, we can reduce the LID effect and we have less oxygen in silicon.[23] LID has according to recent investigations [14] been observed to occur especially in solar cells that have undergone high temperature cell processes (see Chap. 10). Another way to reduce the LID effect is to dope with gallium (Ga). Because of patent restrictions, *p*-type gallium-doped wafers could not be offered commercially until 2018; it is at the moment not clear whether Ga-doped material will ever come onto the market.[24]

With hydrogen we can reduce the LID effect. But the higher the hydrogen content, the higher the **LETID (light and elevated Temperature Induced Degradation)** effect. LETID occurs mainly in multicrystalline material when solar cells are exposed to ambient temperatures higher than 50 °C and is less pronounced in monocrystalline solar cells. It is enhanced because of the high firing temperature (800–1000 °C) used for the metallisation paste during cell manufacturing (see Chap. 5). Due to these high temperatures, more hydrogen penetrates into the cell and, thus, the higher the LETID effect will be. The thinner the wafer, the less pronounced are LID and LETID.

If both effects, LID and LETID, occur simultaneously, this can lead to degradation, which is as high as 15%. Since HJT cells are manufactured with *n*-type material and additionally very low process temperatures are applied, LID and LETID effects are negligible.

(e) Gettering According to the above explanations a–d, *n*-type material has significant advantages over *p*-type material when both have the same density of impurities. This is also the reason why *n*-type material does in general not require gettering processes. On the other hand, it is not at all possible to apply gettering processes to HJT cells: Due to the temperature sensitivity of the amorphous layers, process temperatures for HJT cells must not reach more than approximately 200 °C, whereas for gettering one requires temperatures of more than 500 °C. Because of this limitation, gettering[25] cannot be used during the production of HJT cells, as it is used for the production of "standard" (homojunction) *p*-type crystalline silicon solar cells. Indeed, gettering is not at all necessary for HJT cells with *n*-type material. This is because today's *n*-type silicon is so pure that a high temperature gettering step (even if it could be applied) would no longer significantly contribute to cell improvement.

---

[23] Oxygen penetrates the silicon during crystallization and cannot be completely avoided.

[24] This is because it is very difficult to obtain homogeneous doping of silicon with Gallium (see segredation coefficient Chap. 5).

[25] When silicon is cooled after the crystal pulling process, thermal donors (TD) may be formed by oxygen clusters. These are negatively charged. Thermal donors influence the resistivity. In the *n*-type material the resistivity decreases, in the *p*-type material it increases. TD dissolve at over 500 °C during a gettering process. However, the influence on the improvement of the cell efficiency is not economically meaningful (0.2% abs.); this is the reason why gettering processes are not carried out with *n*-type material.

(f) Thin Wafer With a share of more than 50% in the total production costs, the production of wafer material is the biggest cost driver in the manufacturing of a solar cell. Therefore, wafer production also offers the greatest potential for savings. Due to the low process temperatures and the symmetrical cell structure of heterojunction cells, the use of very thin wafers becomes possible here. A thick metal layer (Aluminium) on the back side, as is usual with monofacial homojunction cells, leads to a warping of the cells—this is due to the different thermal expansion coefficients α of the metal and silicon; these differences in the values of α have a negative effect—in conventional cell production—because of the high firing temperatures (>800 °C) and the subsequent rapid cooling. This can lead to major problems during cell production such as warping and cell breakage during screen printing. In the case of the heterojunction cell, the metal layer is completely omitted, so that thinner wafers can be used for cell production. This leads to two opposite effects: A thinner wafer means that more light passes through the solar cell without being absorbed, so less light contributes to carrier generation. This reduces the short-circuit current density J_{sc} of the solar cell. At the same time, passivation of the wafer surface is becoming increasingly important, as more charge carriers reach the surfaces and do not recombine in the bulk. A high degree of passivation, as is the case with heterojunction cells, leads to an increase in the open circuit voltage V_{oc}. To a certain extent, these two effects (lower J_{sc} and higher V_{oc}) cancel each other out and cell efficiency remains the same. Nevertheless, in addition to the cost advantage, there is also a performance advantage for thin wafers. When the solar cells are integrated into a module, series resistance losses occur in accordance with (7.4).

$$P_{\mathrm{loss}} = I^2 R \tag{7.4}$$

Equation (7.4) means that the power loss P_{loss} increases quadratically with the total current in the module. A lower current means, thus, a reduction in module losses. It can be seen that with the same power class of individual cells, the power of the modules from thin cells is higher than that of the modules from thicker cells [15].

According to Fig. 7.12 the optimum wafer thickness for a heterojunction cell is 80–100 μm whereas for homojunction cells the minimum cell thickness is about >140 μm.

For the reasons a–f given above, we may conclude that the design of the silicon heterojunction cell optimally complements the properties of n-type material, and that n-type wafers are ideally suited for the production of highly efficient HJT cells.

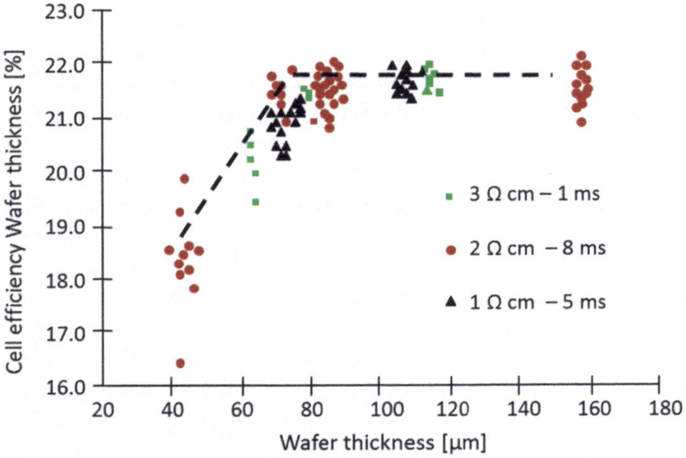

Fig. 7.12 Cell efficiency of HJT cells as a function of wafer thickness for three different silicon materials with 1, 2 and 3 Ω cm resistivity. Down to 100 μm there is no loss in efficiency. Below 100 μm there is a reduction of efficiency which is caused e.g. by surface damage through handling and loss due to reduced absorption in thin wafers. Above chart is provided by CEA-INES and investigations were done on their pilot line

7.4 Cell Process Steps

7.4.1 Wafer Cleaning and Texturization

Crystal Plane {100} and {111} From a large cylindrical silicon single crystal (ingot) of a length of approximately 2–4 m, square bricks of approximately 700 mm length are cut out. Finally, thin wafers with a thickness of 100–180 μm are produced from these bricks using diamond wire saws. The monocrystalline ingot is advantageously processed in the {100}-crystal plane, because this is the crystal direction which is the easiest for subsequent cutting. In this crystal orientation the atoms sit regularly in parallel planes between which one can easily cut. Furthermore the subsequent texturization of the surface is easier to realize. In contrast, the {111}-crystal plane is not easy to machine because it is harder to do cutting in this case.

Cleaning Sawing produces tiny silicon splinters and saw wire residues that adhere to the wafer surface. After cutting, the wafer must therefore first be cleaned of the so-called «kerf». This is done in various baths with a solution of a few % of potassium hydroxide KOH dissolved in distilled water.

Texturization During sawing, damage occurs to the wafer surface: so-called «saw marks and microcracks» are formed. They can extend up to approximately 5–10 μm into the wafer. Saw damages and microcracks lead to substantial efficiency losses, if untreated. Removing the saw damage and texturing the wafer surface, is usually

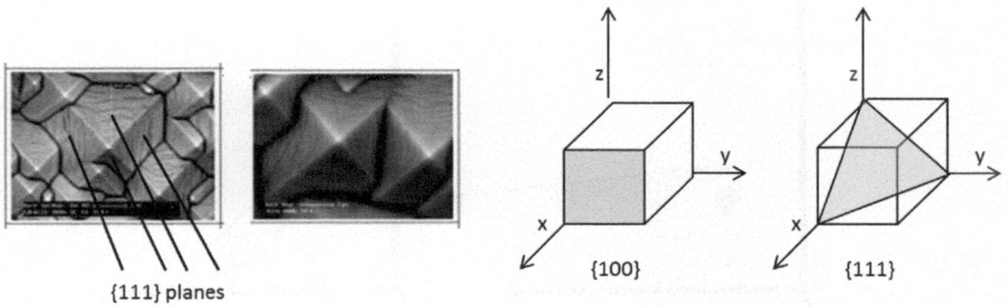

Fig. 7.13 Visual representation of pyramidal structures as suitable for highly efficient cells. The pyramids are formed when the wafers are **sawn in the {100} orientation**. A subsequent anisotropic texturing along the crystal planes creates pyramids with the **diagonal surfaces reflecting the {111} structure.** In the right half of the Figure, the two crystal orientations {100} and {111} mentioned above (in the text) are shown [9]

done together, in a sequence of different wet chemical etching processes. First, the organic residues are cleaned with KOH and hydrogen peroxide baths, followed by the texturing of the wafer surface, which takes place by anisotropic etching. In this alkaline process, KOH plus some special additives is generally used in present-day photovoltaic manufacturing units. Finally, the metal residues produced by the baths themselves, such as potassium (K), must be removed by means of an HF/HCl bath. Anisotropic etching leads to preferential etching along certain crystal planes. By exploiting this property, it is possible to texture the surface homogeneously and to obtain pyramidal structures, as illustrated in Fig. 7.13. The crystal lattice[26] determines the pyramidal structure. Wafers with a {100} orientation[27] form 3–6 μm high pyramids with a square base and diagonal planes according to the {111} orientation. This significantly contributes to the reduction of light reflection. In this way, reflection of visible light by the wafer surface as low as 11–12% is achieved. Of course, not all pyramids have the same size, irregularities occur. It is important that the diagonal planes are not broken, so that the intrinsic layer, which is only approximately 5 nm thick, can grow well on the pyramidal surfaces.

The binding energy is different in the individual crystal orientations. Figure 7.14 illustrates this fact. At the {100} crystal plane a silicon atom has two free bonds. At the {111} crystal plane, however, there is only one free bond. The bond energy in the {111} plane is therefore much higher and the etch rate is low. The preferred etching direction is therefore the {100} crystal plane.

---

[26]The term crystal structure describes the arrangement of atoms, molecules or ions in a crystalline material. In the silicon crystal, the silicon atoms are located in an ordered three-dimensional structure. In *x*, *y*, *z* direction the structures look different, but are always periodical. Therefore, the material properties are also different in *x*, *y*, *z* directions. This fact is exploited, for example, in sawing by selecting the direction {100}.

[27]The crystal structure of silicon is cubic-surface-centred. The silicon crystal can be imagined as a cube. At the 8 corners sit the atoms (cubic) and in the middle of the 6 surfaces sit one atom each (surface-centred).

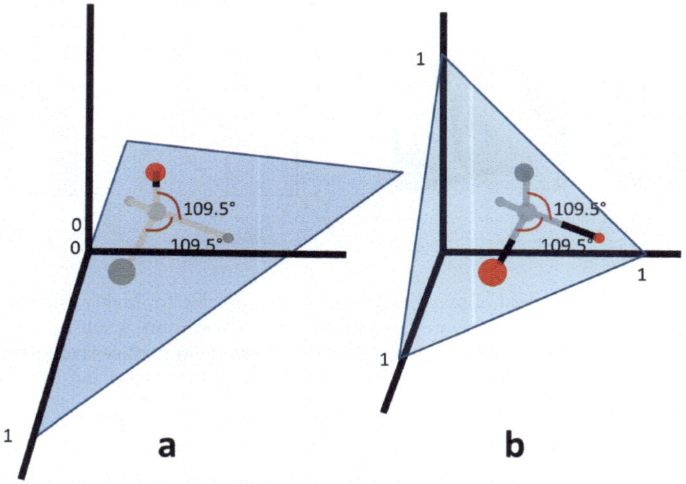

Fig. 7.14 Illustration of four bonds of silicon in different crystal planes: Left only one free bond in the {111} plane; Right two free bonds in the {100} plane

In the actual etching process the OH^- groups of potassium hydroxide KOH play the decisive role. In the {100} plane two electrons are needed to saturate the two free bonds and two electrons are injected into the conductive band. This allows the silicon atom to detach itself from the solid (anistropic etching). The released electrons form further OH^- groups with water (from the chemical bath) and the process continues.

$$\mathrm{Si} + 2\mathrm{OH}^- \rightarrow \mathrm{Si(OH)}_2^{2+} + 4\mathrm{e} \tag{7.5}$$

$$4\mathrm{e}^- + 4\mathrm{H_2O} \rightarrow 4\mathrm{OH}^- + 2\mathrm{H_2} \tag{7.6}$$

A more cost-effective and qualitatively equivalent process is based on ozone.[28]

7.4.2 *Determination of Depth of Saw Damages and Lifetime Measurements*

Measurement of Saw Damages During sawing, damage occurs to the wafer surface: so-called «microcracks» are formed; which can extend up to approximately

[28]The etching effect here is isotropic. Isotropic etching is independent of the crystal direction. The same amount is etched away everywhere on the wafer surface. So-called saw marks can therefore not be completely removed because regardless of the crystal structure, the same amount is removed everywhere. The saw marks are also etched and remain visually intact.

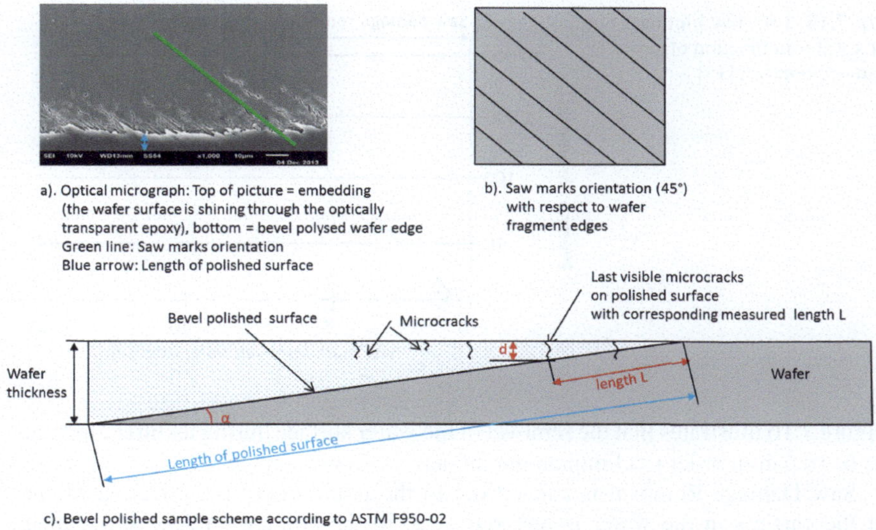

Fig. 7.15 **a** Shows a "Bevel" polished wafer. **b** Illustration of saw marks orientation of 45° for Bevel polishing. **c** Wafer edge being polished with a polishing angle of $\alpha = 8°$. From this, the depth of the saw damage can be derived according to the formulas in the text. *By courtesy of* Meyer Burger Technology AG

5–10 μm into the wafer. The depth of the saw damage depends on the cutting parameters and the size and shape of the diamonds on the diamond wire. Usually the diamonds are between 8 and 20 μm in size, the diameter of the saw wire being 40–80 μm thick. This results in saw damage with a depth of 4–6 μm. Before the wafers are textured, the depth of saw damage, especially the microcracks, are measured. For this the Bevel method is used. The Bevel procedure is described in Fig. 7.15a–c.

The scheme for Bevel polishing is given by the International Standard ASTM F950-02.[29] How much must actually be etched away from the wafer surface can be calculated using the formulas

$$\sin(\alpha) = \text{wafer thickness}/\text{length of polished surface} \tag{7.7}$$

$$\text{Depth of microcracks: } d = \sin(\alpha) * \text{length } L \tag{7.8}$$

Lifetime Measurement in Cell Production In order to optimize the process time, the minority carrier lifetime of the wafers is measured after the texturization. On the one hand, we want to remove all saw damages, on the other hand we do not want to etch too much away in order to keep production times as short as possible.

[29]"ASTM" means "American Society for Testing and Materials" but is now, an international standard organization that develops and publishes based on voluntary consensus technical standards for a wide range of materials, products, systems and services.

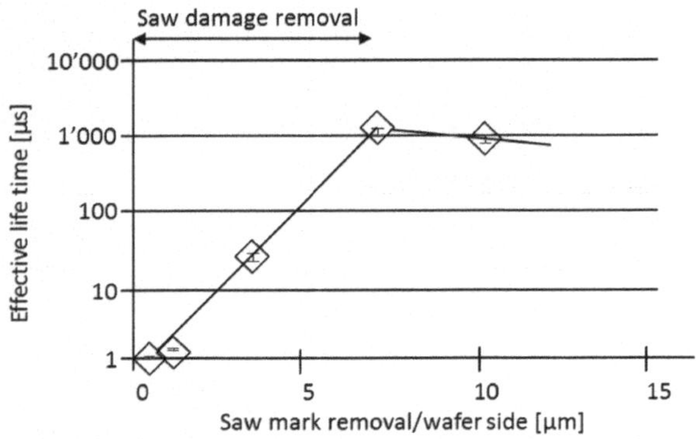

Fig. 7.16 Effective lifetime of a wafer in function of saw damage removal [15]

Figure 7.16 illustrates that the removal of the wafer surface during texturization must be 6–10 μm in order to eliminate the influence of saw damage.

Saw Damage Removal is carried out to the point where the effective lifetime of the carriers in the wafer is restored—e.g. to the point where the saw damage has no influence anymore on lifetime—and further etching would only reduce the wafer thickness without any positive effect—this means that from that point onwards the effective lifetime would now be reduced. In addition to the volume (bulk) lifetime τ_{bulk}, the effective lifetime τ_{eff} takes also into account the effective surface recombination speed S_{eff} and the wafer thickness w, according to (7.9)

$$\frac{1}{\tau_{\mathrm{eff}}} = \frac{1}{\tau_{\mathrm{bulk}}} + \frac{2S_{\mathrm{eff}}}{w} \tag{7.9}$$

For the measurement of the lifetime τ_{eff}, the wafer is passivated after saw damage etching, by applying a thin intrinsic amorphous silicon layer, which neutralizes the surface recombination effects and the bulk lifetime τ_{bulk} can now be checked.

7.4.3 Deposition of Intrinsic and Doped Amorphous Layers

After the surface has been cleaned, textured and cleaned again, the wafer surface has to be passivated. Usually one starts with the *n*-stack layer (a-Si:H(*i*) and a-Si:H(*n*)), either on the front side or on the back side. Hydrogen is then added to saturate the dangling bonds at the wafer surface. Hydrogen can also saturate the defects in the crystal. These occur, among other things, when interstitial oxygen is present, which sits between the lattice sites and leads to distortions of the crystal. Oxygen penetrates into the silicon crystal to the order of 10^{18} cm^{-3} during the production of the single crystals («monocrystals»), see Chap. 5. Since oxygen is larger than silicon, the lattice is strongly bent by these oxygen atoms and free bonds (dangling bonds) that constitute recombination centres are formed. Such free bonds can be passivated with hydrogen (hydrogenation).

The deposition of the a-Si layers is usually carried out by PE-CVD (Plasma Enhanced Chemical Vapour Deposition, see also Chap. 6). In a mixture of silane (SiH_4) and hydrogen a plasma is ignited. Within the plasma, the molecules decompose and the atoms are directed via an electric field onto the substrate, e.g. onto the wafer. There they are deposited as an amorphous layer. In this way, the intrinsic, undoped passivation layer a-Si(*i*) is formed without the addition of further gases. By adding phosphine (PH_3) or diborane (B_2H_6), *n*- or *p*-doping of the amorphous layer can be achieved. The a-Si:H(*p*) deposition causes more complications regarding impurities and thermal stability. This is the reason why the *p*-layer is deposited after the *n*-layer.

During the process, it must be ensured that no unwanted contamination by the doping materials boron (diborane) and phosphorus (phosphine) occurs, since such "cross-contamination" can lead to severe efficiency losses. The cell can be protected from this, by using a separate transport box (transport carrier) for each layer and the PECVD deposition equipment is equipped with separate plasma chambers. With parallel plate plasma reactors and a box-in-box vacuum chamber system, large areas can be deposited homogeneously with a thickness tolerance of less than 5%, at layer thicknesses of a few nm. The deposition takes place at 13.56 MHz, at 10^{-3} mbar and at temperatures of 150–250 °C. It is absolutely necessary to avoid the amorphous layers becoming too hot. Otherwise a restructuring of the originally amorphous layers up to crystallization occurs, whereby the layer loses its excellent passivation properties. The deposition of the four amorphous layers takes only approximately 60 s; despite this fast deposition, very high lifetime values of more than 4 ms for the minority carriers are achieved, with industrially produced silicon (measured at the reference excess charge carrier concentration of 10^{15} cm^{-3}). With high-quality floatzone (FZ) silicon, lifetime values of up to 10 ms can be achieved. Admittedly, FZ silicon is very expensive and is not suitable for mass production in the photovoltaic industry—however, this result indicates the basic potential for further improvement.

7.4.4 Coating of the TCO Layer

After the four amorphous silicon layers have been deposited, the front and back of the cell are each coated with TCO. With the ITO (indium tin oxide) generally used for HJT cells, the coating is usually done with sputtering. On the front side, the TCO layer is responsible for light trapping (e.g. it forms the ARC—the Anti Reflection Coating); furthermore, it has to provide an Ohmic contact to the amorphous silicon and to the metal contact grid. On the back, where the *p*-*n*-junction is preferentially located, the TCO layer can be slightly thicker (see Sect. 7.2.3) to ensure high electrical conductivity. In the case of bifacial HJT cells, the back TCO layer additionally acts as an ARC layer—just like the front one. Indeed, bifacial cells receive on their back side the scattered light of the environment and the light reflection of the ground, in order to generate additional charge carriers. The reflection of the ground naturally depends on its composition; it is quantified by the «albedo».[30] (Fresh snow has for example an albedo of 80%, whereas green grass has an albedo of only 25%.[31]) HJT cells can be manufactured up to 92% bifaciality.[32] With an albedo of 20%, for example, 18% of surplus energy (18% = 20% × 92%) is generated from the back side, and adds to the energy generated from the front side. A bifacial HJT module with, for example, 20% module efficiency[33] from the front side, 92% bifaciality and with 20% albedo, has an overall efficiency of 20% × (1 + 0.18) = 23.6%.

7.4.5 Metallisation and Contacting

The electrical conductance of the TCO layer of 15 Ω at 100 nm thickness is not sufficient to collect the current over the entire cell without large losses. It is therefore not sufficient to pick up the current only at the edge of the solar cell; rather, it is necessary to provide thin metal fingers to support the transport of the charge carriers to the edge of the cell. Of course, this is again a compromise between low electrical (resistance) losses obtained with many and thick metal fingers and low optical (shading) losses obtained with few and thin metal fingers. There are four different technologies for contacting HJT cells:

- Busbar Technology
- Multi wire technology
- SmartWire Connection Technology (SWCT)
- Adhesive Technology.

All four technologies are illustrated in Fig. 7.17. As a rule, about 60–90 fingers with a width of 35–50 μm and a height of about 5–20 μm are used; they are applied by screen printing onto the solar cell.[34] In HJT cells, the high temperature technology like for homojunction solar cells does not work because the amorphous layer cannot endure more than 200 °C. For the HJT technology so-called low-temperature pastes given by polymer-based screen printing pastes are used with a curing temperature

[30] The «albedo» is a dimensionless quantity indicating the ratio between the reflected and the incident global irradiance (see also Chaps. 2 and 10).

[31] These values are applicable in the case where no object obstructs the incoming sunlight. In a PV system values of albedo are considerably lower and depend on the layout of the system.

[32] This means that the conversion efficiency for the light coming in from the back side is 92% of the conversion efficiency of light coming in, from the front side.

[33] which is equivalent to a world-record monofacial module of 20% module efficiency.

[34] The Busbar technology is often realized with high temperature pastes, at which temperatures of approx. 800 °C are used to melt the silver flakes in the paste. This is how the fingers and the busbars are applied. High temperature pastes are also used in multiwire applications. Instead of the busbars, pads are printed to fix the wires. The conductivity of such pastes is high. The five ribbons (busbar technology) or the 2 × 7 wires (multi wire technology) are soldered crosswise to the fingers at approx. 240 °C.

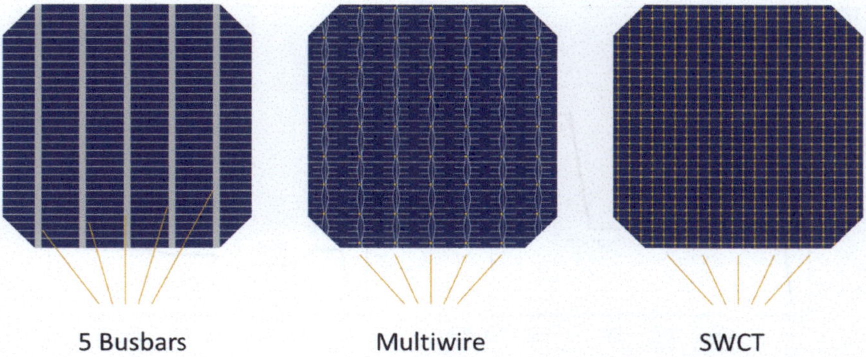

Fig. 7.17 Illustration of the three main metallization technologies, whereas Adhesive Technology is equal to sketch 5 Busbars. *Courtesy of* Meyer Burger Technology AG

of 200 °C. During the lamination process, the fingers are connected to the wires at approximately 120 °C. The conductivity of low temperature pastes is not quite as high as that of standard pastes, and they are more expensive. The SWCT^{TM} technology makes up for this disadvantage by using many wires (18–36) and, thus, reducing the current per contact point. This allows one to save up to 50% silver [16] and total metallization costs for the entire cell can be reduced compared to standard pastes. In addition, the round wires reflect obliquely incident light into the cell, resulting in 2–5% more current for the cell. During the curing time of 10–20 min, the silver pastes polymerize and achieve their adhesion to the TCO, as well as their conductivity. At the same time, the TCO layer oxidizes, making it more transparent. Another technology to contact HJT cells can be implemented with Electrically Conductive Adhesives (ECA). With this technology the ribbons are glued on the busbars at low temperature.

7.4.6 Process Temperature and Process Cycle Time

The simple structure of the HJT cell structure is characterized by only six coating processes, two of which are identical. On the front and back side, the intrinsic amorphous layer is deposited, followed by a doped amorphous silicon layer and, finally, by a TCO layer. With regard to equipment, only two different coating machines are required for these six coating processes.

The layers are deposited in a heterojunction cell at low process temperatures of less than 250 °C (see Fig. 7.18). These low process temperatures are interesting in terms of energy and open the way to the use of thin wafer material.[35] The process times for HJT cells are approximately 33 min and about 50% shorter than those used

[35]Standard cells use process temperatures of 800–1000 °C.

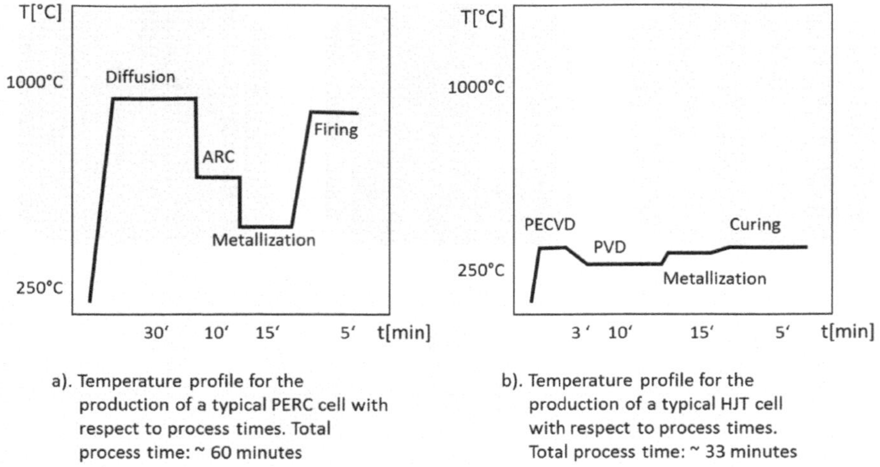

Fig. 7.18 **a** Temperature profile for manufacturing a typical PERC cell. **b** Temperature profile for manufacturing a typical HJT cell. The production of a HJT cell has a production time of approximately 33 min compared to 60 min for a typical PERC cell. The temperature budget for a HJT cell is about 4 times lower than the temperature budget for a PERC cell

for standard PERC cells: the latter take 60 min to manufacture. The HJT cell requires 8 process steps, while a standard PERC cell requires 11 process steps.

7.5 Temperature Coefficient of HJT Cells

The temperature coefficient TC of a solar cell is mainly given by the temperature dependence of the open circuit voltage V_{oc}—which in turn depends on the bandgap of the cell (see Chap. 3). The temperature coefficient *TC* of a solar cell depends strongly on the losses in the cell and consists of three parts [17]:

$$TC_{\text{MPP}} = TC_{V_{oc}} + TC_{J_{sc}} + TC_{\text{FF}} \tag{7.10}$$

In (7.10) $TC_{V_{oc}}$ is the predominant term and accounts for 80–90% of TC_{MPP}.

Table 7.1 Indicative overview of the temperature coefficients (TC) and bifacial factors for a few different types of solar cells

Technology	*TC* power @MPP (%/K)	Bifacial factor (%)
AL-BSF	−0.4	0
PERC	−0.38	0
PERT	−0.35	76
HJT	−0.3	92

$TC_{V_{oc}}$ is examined in detail in Chap. 3. Table 7.1 shows values of the temperature coefficients (TC) and bifacial factors obtained for various c-Si cell types:

An interesting aspect for the operation of solar modules under natural weather conditions is their performance at different temperatures. For Al-BSF c-Si solar cells, the temperature coefficient of the MPP point is ~−0.40/°C, which means that the module's output decreases sharply with increasing module temperature. The reference temperature is 25 °C according to STC (Standard Test Conditions). In contrast, the temperature coefficients of heterojunction solar cells are ~−0.3%/°C and the power decreases less with higher module temperatures. Haschke et al. [18] shows that solar modules with high V_{oc} and lower J_{sc} perform better in hot environments than modules with the same power but lower V_{oc} and higher J_{sc}. The explanation lies in the fact that in Al-BSF cells the temperature coefficient of the fill factor is high, so that the high current cannot be dissipated and high losses result. In this respect, modules equipped with cells having passivated contacts and high V_{oc} are the most advantageous choice in hot areas.

7.6 Levelized Cost of Electricity (LCOE) of HJT Cells

For all the reasons mentioned above, such as simplicity of the manufacturing process, low process temperatures, favourable temperature coefficients, bifacial cell architecture, use of thin wafers, advantageous properties with respect to LID, LETID and PID degradation, excellent passivation, the LCOE[36] values of HJT cells are very interesting compared to those of other cell architectures [19]. Detailed indications about the Levelized Cost of Electricity (LCOE) can be found in Chap. 13, Sect. 13.1.2.

References

1. R,V.K. Chavali, S. De Wolf, M.A. Alam MA, Device physics underlying silicon heterojunction and passivating-contact solar cells: A topical review. Prog. Photovolt. Res. Appl. 1–20. https://doi.org/10.1002/pip.2959
2. W.G.J.H.M. van Sark, L. Korte, F. Roca (eds.), *Physics and Technology of Amophous-Crystalline Heterostructure Silicon Solar Cells* (Springer, Berlin, Heidelberg, 2012)
3. S.W. Glunz, E. Schneiderlöchner, D. Kray, A. Grohe, M. Hermle, H. Kampwerth, R. Preu, G. Wille, Laser-fired contact silicon solar cells on p- and n-substrates, in *Conference 19th European Photovoltaic Solar Energy Conference* (2004)
4. M. Taguchi, M. Tanak, T. Matsuyama, T. Matsuoka, S. Tsuda, S. Nakano, Y. Kishi, Y. Kuwano, Improvement of the conversion efficiency of polycrystalline silicon thin film solar cells, in *5th International Photovoltaic Science and Engineering Conference* (Kyoto, Japan, 1990), p. 689
5. M. Tanaka, M. Taguchi, T. Matsuyama, T. Swada, S. Tsuda, S. Nakano, H. Hanafusa, Y. Kuwano, Development of New a-Si/c–Si heterojunction solar cells: ACJ-HIT (Artificially

[36] A common method is to calculate the energy price using the Levelized Cost of Electricity (LCOE). This allows different technologies to be compared on the basis of a standardized calculation basis, both within the PV and with other energy production sources.

constructed junction-heterojunction with intrinsic thin-layer). Jpn. J. App. Phys. **31**, 3518 (1992)
6. G. Faraone, Electronic structure of hetero-crystalline superlattices. Dipartimento di Fisica, Università degli Studi di Milano (2012)
7. R.V.K. Chavali, S. de Wolf, M.A. Alam, Device physics underlying silicon heterojunction and passivating-contact solar cells: a topical review (Wiley Photovoltaics, 2017). https://doi.org/10.1002/pip.2959
8. U. Würfel, A. Cuevas, Fellow, IEEE, P. Würfel, Charge carrier separation in solar cells. IEEE J. Photovolt. **5**(1) (2015)
9. S. De Wolf, A. Descoeudres, Z.C. Holamn, C. Ballif, High-efficiency silicon heterojunction solar cells, A review. Green **2**(1), 7–24 (2012). https://doi.org/10.1515/green-2011-0018
10. C. Ballif, S. de Wolf, A. Descoeudres, Z.C. Holman, Amorphous silicon/crystalline silicon heterojunction solar cells, in *Semiconducteurs and Semimentals*, vol. 90 (Elsevier 2014)
11. M. Bivour, S. Schröer, M. Hermle, Numerical analysis of electrical TCO/a:Si:H(p) contact properties for silicone heterojunction solar cells. Energy Procedia **38**, 658–669 (2013)
12. D.A. Neamen, Semiconductor physics and devices, in *Basic Principles*, 3rd edn. (McGraw-Hill Higher Education, 2003)
13. F. Rougieux, C. Sun, D. Macdonald, Determining the charge states and capture mechnisme of defects in silicon through accurate recombination analysis: a review. Solar Energy Mater. Solar Cells (2018) (Elsevier B.V.)
14. J. Lindroos, H. Savin, Review of light-induced degradation in crystalline silicone solar cells. Solar Energy Mater. Solar Cells **147**, 115–126 (2016)
15. D.L. Baetzner, J. Heiber, M. Lanz, C. Heiniger, F. Debrot, R. Kramer, L. Andreaetta, D. Lachenal, W. Frammelsberger, B. Legradic, J. Meixenberger, P. Papet, B. Strahm, G. Wahli, Reducing wafers surface saw damage with diamond wire sawing technology, in *29th European Photovoltaic Solar Energy Conference and Exhibition* (2014)
16. T. Söderström, P. Papet, J. Ufheil, Smart wire connection technology, in *Proceedings of the 28th EU-PVSEC, European Photovoltaic Solar Energy Conference Paris* (2013), pp. 495–499
17. O. Dupré, R. Vaillon, M.A. Green, Physics of temperature coefficients in solar cells. Sol. Energy Mater. Sol. Cells (2015). https://doi.org/10.1016/j.solmat.2015.03.025
18. J. Haschke et al., The impact of silicon solar cell architecture and cell interconnection on energy yield in hot & sunny climates. Energy Environ. Sci. (2017)
19. A.H.A. Richter, S. Leu, Cell and module design from the LCOE perspective in *Manufacturing the Solar Future, The 2014 Production Annual, Photovoltaics International* (2014)

Sylvère Leu was from 2008 to 2017 with the Meyer Burger Group, where he became a member of the executive board, as Chief Innovation Officer and Technology Officer (CIO/CTO). Since his retirement in 2017, he is active as technology consultant, for the Meyer Burger Group. Sylvère graduated from the Swiss Federal Institute of Technology in Zürich (ETHZ). He additionally obtained a Master's Degree in Business Administration at the University of St. Gallen (HSG). As a Swiss pioneer, Sylvère Leu started to work in photovoltaics 30 years ago. He constructed an industrially relevant laminator and sun simulator in his own company. Beside his job he worked as an associate lecturer at University of St. Gallen (HSG), for several years, in the field of industrial production. At the end of 2005 he was charged with building up an integrated 250 MW_p photovoltaic facility, including wafer-, cell and module manufacturing, in Germany

Detlef Sontag is Senior Technologist for Solar Cells at Meyer Burger (Germany) GmbH. since 2015—one of the world's leading PV equipment manufacturing companies. He graduated as a physicist at the University of Konstanz in 2000, and obtained his Ph.D. degree in 2004 working on novel wafer materials for Photovoltaics. In 2004 he joined Deutsche Cell GmbH, a subsidiary of Solar World AG, as quality assurance engineer with insights in the industrial crystalline silicon PV manufacturing processes along the whole value chain from block casting via cell processing to module assembling. Detlef Sontag joined the equipment manufacturer Roth & Rau AG as Senior Principal for cell technology in 2009. There he supported the planning and commissioning of the Technology centre and worked actively on the development of cell technologies like PERC and HJT including project coordination of corresponding R&D-projects.

Chapter 8
CdTe and $CuInGaSe_2$ Thin-Film Solar Cells

Alessandro Romeo

Abstract Chapter 8 is divided into seven sections:

- Section 1. Describes the peculiarities of polycrystalline thin film materials with emphasis on CdTe and CIGS. It highlights what are the pros and cons in fabricating photovoltaic devices with these materials.
- Section 2. Describes in detail CIGS-based thin-film solar cells: the structure of the cell with the different layers; the different fabrication processes, record efficiencies and industrial applications.
- Section 3. Describes in detail CdTe-based thin-film solar cells: the structure of the cell with the different layers; the different fabrication processes, record efficiencies and industrial applications.
- Section 4. Describes the flexible configuration that thin film solar devices can assume if special substrates are used.
- Section 5. Describes the all in-line fabrication process: from glass to finished module in one production line.
- Section 6. Shows the behaviour of thin film photovoltaics at low light irradiation and at high temperatures.
- Section 7. Shows the environmental impact of the devices in both fabrication, operation and end-of-life.

8.1 Thin-Film Polycrystalline Materials

Thin film solar cells are based on materials, which show an extraordinary high absorption coefficient so that there is no need to build thick solar cells to absorb all the light.

Using this kind of materials allows one to fabricate devices with an overall thickness of less than 10 μm and a clear advantage in terms of material supply and fabrication energy.

A. Romeo (✉)
University of Verona, Verona, Italy
e-mail: alessandro.romeo@univr.it

A. Shah (ed.), *Solar Cells and Modules*, Springer Series in Materials Science 301,
https://doi.org/10.1007/978-3-030-46487-5_8

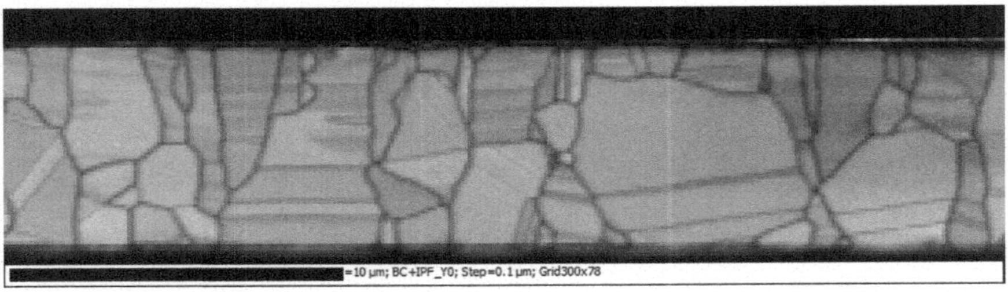

Fig. 8.1 Cross-section of CdTe thin-film layer in electron backscatter diffraction mode. (Courtesy of Daniel Abou-Ras, Helmoltz Zentrum Berlin)

Thin-film solar cells offer a wide variety of choices in terms of device design, fabrication methods and substrates (flexible or rigid, metal or insulator). The deposition of different layers (contact, buffer, absorber, reflector, etc.) can be done using several techniques, which will be described later. Indeed, such versatility allows for tailoring and engineering of the layers, in order to match the solar spectrum and to improve device performance.

Typically, the thin films used in these devices are polycrystalline materials, where the layer is a pattern of small crystals, whose width can range between 0.1 and 5 μm. As shown in Fig. 8.1, the layer is a patch of differently sized grains with different orientations. This configuration looks very disordered and irregular considering that it has to allow carriers to move through the material; however the films, when properly prepared, have the required conductivity, and devices can reach very high efficiencies up to 25%.

On the other hand, the advantage of such a disordered structure is that it does not need a very precise control of crystal growth; neither does it need high energy for crystallization: this is an advantage compared to other technologies, such as crystalline silicon.

So, considering the very small amount of material used and the relatively low fabrication temperatures involved, thin-film PV devices have held the promise since the 1970s to provide a photovoltaic technology with lower investment of fabrication energy, as compared to the more traditional photovoltaic technologies.

However, the very large scaling of crystalline silicon (c-Si) has changed the perspectives in the last 10 years. Chinese manufacturers have heavily invested in crystalline silicon fabrication plants, thereby strongly reducing the price of crystalline silicon wafer-based (c-Si) solar cells, while thin-film module manufacturers are still in the stage of early industrialization; resulting thus, in a higher market price for thin-film modules. However, thin-film modules would have the long-term advantage to be able to provide lower production costs than those for c-Si modules, once the same high level of manufacturing volume is reached.

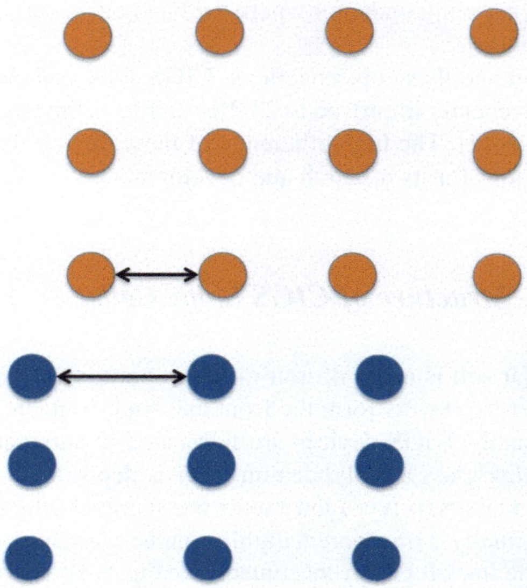

Fig. 8.2 Sketch showing two structures (orange atoms and blue atoms) with different lattice constants (indicated with arrows)

Thin-film solar cells are based on hetero-junctions[1] where the *n* and the *p* semiconductors consist of different materials. This adds more complication to the device: as a matter of fact the semiconductor layers are polycrystalline materials with different lattice constants, and the consequent lattice mismatch (see Fig. 8.2) gives place to defects at the interface which reduce the conversion efficiency by increasing carrier recombination.

8.2 CIGS Solar Cells

8.2.1 Introduction

$Cu(In,Ga)Se_2$ or CIGS, is a semiconductor with a high absorption coefficient and a direct bandgap that can be varied from 1.0 eV for $CuInSe_2$ to 1.67 eV for $CuGaSe_2$ by subsequent substitution of In atoms with Ga atoms and addition of sulphur. These properties, in combination with a large tolerance to compositional variations and low recombination of carriers in grain boundaries, are important in order to reach high efficiencies with thin, polycrystalline films and fast processing.

CIGS gives the researcher the possibility to play with the ratio of copper, indium and gallium in the material. In this way being able to change the bandgap and allowing

[1] A heterojunction is the interface that occurs between two dissimilar crystalline semiconductors. These semiconducting materials have unequal band gaps as opposed to a homojunction.

to design a semiconductor where the bandgap is varied from the junction area to the back contact.

Thanks to these optimizations, CIGS solar cells have reached efficiencies up to 22.9% (recently improved to 23.3%) and have the target of reaching 25% in the next few years [1]. The high efficiency of these devices has always been one of the key motivations for its research and development.

8.2.2 Structure of CIGS Solar Cells

The solar cell is made of four different parts: two layers form the *p-n* junction and the other two layers form the front and back contacts.

Typically, CIGS devices are fabricated in substrate configuration where on the supporting glass a molybdenum layer is deposited as back contact, then the CIGS absorber layers (*p*-type) and a semi-transparent buffer layer (*n*-type) are subsequently grown; finally, a transparent highly conductive semiconductor (TCO) is deposited on top of the stack for the front contact (see Fig. 8.3). This configuration gives the highest efficiency for CIGS cells, due to favourable process conditions and compatibility of the materials, but it requires an additional encapsulation layer and/or glass to protect the cell surface, the latter is not needed in the superstrate configuration (explained later). However, the superstrate configuration delivers CIGS solar cells with lower efficiencies.

Now we will consider the single layers that compose the device.

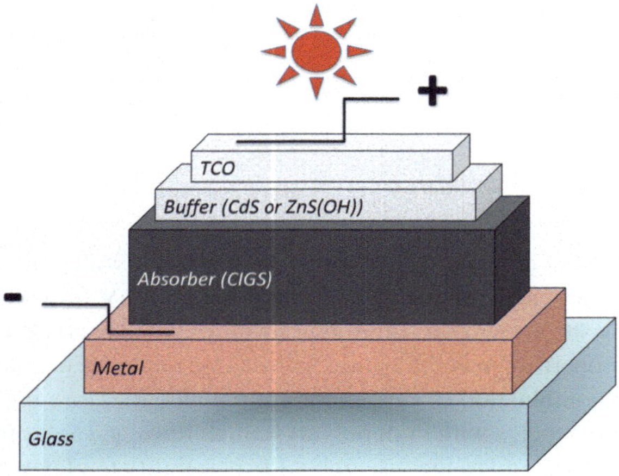

Fig. 8.3 Schematic structure of a typical CIGS solar cell (substrate configuration)

8.2.2.1 Back Contact and Absorber Layer

CIGS solar cells can be fabricated on soda lime glass; the first layer, grown on the supporting substrate, is the back contact and typically molybdenum (Mo) is the most common material for this purpose.

Mo is generally grown by sputtering or by e-beam evaporation and its influence on the Ohmic contact behaviour at the CIGS/Mo interface makes $MoSe_2$ formation an important issue. $MoSe_2$ is generated after molybdenum deposition during CIGS growth, especially when co-evaporation of the elements in a selenium atmosphere is done. It was for a long time considered as a side effect, but it has been proven that it improves the performance of the back contact.

The CIGS absorber material can be prepared with a large variety of deposition techniques that go well beyond the scope of this chapter. In this sense we will concentrate on the two most successful processes [2]:

(a) Co-evaporation.
(b) RF-sputtering of precursors and subsequent heating of the stack in a selenium atmosphere (selenization).

Co-evaporation

It consists of evaporating each single element (Cu, In, Ga and Se) at the same time, by using a crucible for each element and controlling the evaporation flux. Using co-evaporation it is possible to reach very high efficiencies at the laboratory scale, particularly by the application of an evaporation profile with constant rates. In fact by varying evaporation rates of the single elements it is possible to profile the bandgap and/or to adjust the composition.

For example by varying the Cu flux, the bulk copper content can exceed the standard composition in a part of the process, enhancing grain growth. Moreover it is possible to control the process by varying the composition from Cu-rich to Cu-poor or vice versa.

On one hand, excess copper is beneficial during the deposition process but on the other hand, a Cu-rich CIGS film also leads to segregation of Cu_xSe that shunts the cell.

Already during the early eighties, the design of a deposition process with a balance between Cu-rich and Cu-poor was known and introduced by Boeing (the so-called Boeing process). This has subsequently resulted in a co-evaporation process divided into three-stages: a sequential process which starts with a $(In,Ga)_2Se_3$ growth stage, continues with a deposition of Cu and selenium and ends with another $(In,Ga)_2Se_3$ stage.

A very important feature of co-evaporation is the possibility to control in situ[2] the ratio between gallium and indium. Changing the Ga/(Ga+In) ratio during the process is one of the key factors for high efficiency, since a high Ga content near the back contact improves the minority carrier lifetimes, reducing recombination.

---

[2]In situ: in the same evaporation chamber and during the evaporation process.

In the three-stage process bandgap grading is obtained both at the back contact and at the front contact, resulting in a double bandgap grading. In this way, the gallium concentration increases both at the back and at the front of the CIGS layer with a minimum bandgap in the CIGS bulk. This structure was formed spontaneously delivering high efficiencies and reaching the conclusion that a front Ga-rich layer is beneficial for device performance.

Sputtering + selenization (+sulfurization) of precursor materials

This kind of process consists of the subsequent deposition of a precursor material (that can be an element or a compound), followed by thermal annealing in a selenium and/or a sulphur environment, for the final formation of the absorber.

The most used precursor materials are metallic and metal selenide layers.

Precursors can be deposited by thermal evaporation, sputtering or electrodeposition. However, the most successful process is the deposition of elemental layers by DC magnetron sputtering: well established for the production of large-area solar modules up to $60 \times 120\ cm^2$ size [3]. In this particular case the chalcogenide[3] formation is made by the so-called SAS process: sulfurization after selenization by sequential selenization in H_2Se (99.99%) and subsequent sulfurization in H_2S (99.99%) with Argon (Ar) in a furnace.

The addition of sulphur into the CIGS matrix is made here, in order to increase the open-circuit voltage of the solar cell and to tune the bandgap-grading.

In sputtering deposition, a plasma[4] is generated with an inert gas (typically Argon), and then the ions are accelerated on to a target material by a potential difference that can be provided by a direct (DC) or radiofrequency (RF) current. The collisions generated on the target transfer enough energy to the target atoms or molecules, which are then able to move in form of neutral particles and reach the substrate.

In RF sputtering, the accelerating voltage on the target is typically being switched with a frequency of a few kHz; while in DC sputtering, the voltage is constant. For semiconductors or insulators, RF sputtering is the natural choice to avoid electrical charging of the target. However, it is also possible to use a DC pulsed technique where the voltage is fixed but pulsed, to avoid charging and consequently arcing.[5]

It has been found that the beneficial effect of using soda-lime glass substrates could be assigned to the supply of sodium. Similar or even better results can be obtained by using an Na-precursor layer, either deposited before/after or during the CIGS process. This leads in its turn to higher values of open-circuit voltage and higher fill factors. All these effects could be explained by a higher doping density, as Na, K and other alkali elements enhance substantially the performance by increasing the absorber doping.

[3] Chalcogenides are materials containing one or more chalcogen elements (e.g. S, Se or Te) as a substantial constituent.

[4] Plasma consists of a gas of ions and atoms which have some of their orbital electrons removed, and free electrons.

[5] For arcing we mean the formation of an electric arc, that would shunt the DC circuit.

Moreover, alkali treatment (among them Rb and Cs) increases the bandgap at the surface of CIGS improving the junction quality. This has allowed one to reach efficiencies above 22%.

8.2.2.2 Buffer Layer

CIGS solar cells were originally made with CdS as buffer layer deposited by evaporation; but this limits the current density since with its bandgap of 2.4 eV, CdS absorbs visible light from 400 to 500 nm.

A significant improvement has been obtained with the strong reduction of CdS thickness, obtained by replacing thermal evaporation with chemical bath deposition (CBD) that assures a higher conformal coverage; and by adding a highly resistive and transparent thin ZnO layer. On the other hand, attempts to replace CdS completely by ZnO did not give satisfying results. Therefore CdS by CBD is mainly used to protect CIGS against sputter damage; it etches the CIGS surface and passivates the grain boundaries.

One of the most important improvements for the buffer layer is the application of ZnS as buffer layer, which was obtained by substituting Cd with Zn in the chemical bath. Increase of the bandgap from 2.4 eV for CdS to 3.6 eV for ZnS has been shown. This also reduces the lattice mismatch between CIGS and (Zn,Cd)S.

ZnS and ZnO are II–VI semiconductors with wide bandgaps: 3.8 and 3.3 eV, respectively. The most successful buffer layers from this group of materials contain oxygen and sulphur in combination with hydrogen in the form of hydroxides.

The bandgap of ZnO_{1-x},S_x exhibits a large bowing (modification of bandgap by intermixing of two materials, in this case ZnO and ZnS) with a minimum bandgap of 2.6 eV for $x = 0.5$. At the moment this solution is applied in large-scale module production.

8.2.2.3 Front Contact

Today, CIGS solar cells are mainly fabricated with a double layer of ZnO:Al/ZnO, generally deposited by RF-sputtering. Doping of the conducting ZnO layer is achieved by group III elements (B, Al, Ga, In, Tl), particularly with aluminium. The presence of an intrinsic ZnO layer increases the open-circuit voltage and prevents the solar cell from shunts.

8.2.3 Performance and Degradation

Nowadays, CIGS modules are commercially produced by different companies, such as Solar Frontier (Japan), Avancis (Germany-China), Solopower (USA), Solibro-Hanergy (Sweden-China), Global Solar (USA-China), MiaSolé (USA-China), Flisom (CH).

Efficiencies of completed modules can exceed 19% as produced by Solar Frontier, with an area of 841 cm^2 and 15% for larger modules as produced by MiaSolé, with an area of 9703 cm^2 [1], moreover it has been demonstrated that the stability of the modules is comparable, if not higher than the stability of c-Si modules.

Main remaining issues can come from shading-induced hot spots, metastabilities (induced by light soaking after dark storage) and potential induced degradation that can be enhanced by Na migration from glass and impurity diffusion through grain boundaries and for this reason strongly depends on the quality of the module.

8.3 CdTe/CdS Solar Cells

8.3.1 Introduction

CdTe, is a semiconductor with a high absorption coefficient and a direct bandgap of 1.45 eV this value is very near to the optimum bandgap value. It is a robust and reliable material that can be deposited at low or high temperatures; it evaporates congruently (Cd and Te evaporate with the same rate) and it is basically very easy to keep a 1:1 Cd-Te composition with most of the deposition techniques.

When deposited at high temperatures on a transparent buffer layer, CdTe cells have reached efficiencies up to 22.1% [1].

8.3.2 Structure of CdTe Based Solar Cells

From the early eighties until 2015, the typical CdTe solar cells in superstrate configuration[6] were composed of (see Fig. 8.4): a transparent conductive oxide (TCO), CdS as *n*-part of the junction and window layer, CdTe as *p*-part of the junction and absorber layer and back contact. The front and the back contacts can be composed of different layers; this will be described later.

In this configuration, the supporting substrate obviously needs to be transparent. Considering that in a thin-film device more than 98% of the material is glass, the choice of soda lime glass as substrate allows keeping the final price low. However, soda lime glass has two disadvantages: (1) the presence of sodium and potassium that

[6]Superstrate configuration: when light shines through the glass to reach the underlying solar cell.

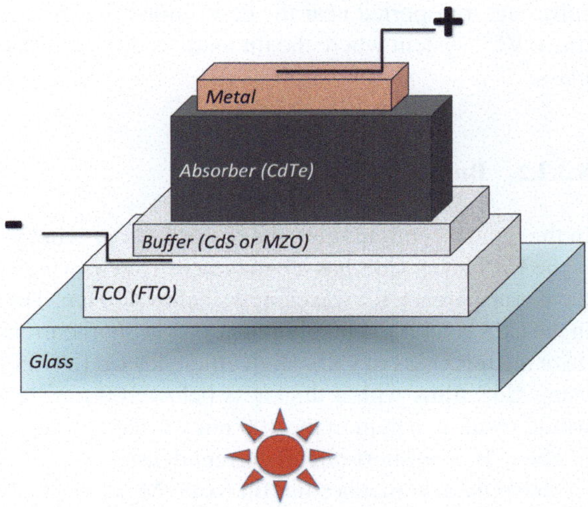

Fig. 8.4 Schematic structure of a typical CdTe solar cell (superstrate configuration)

diffuse into the layers during depositions and annealing, (2) the process temperature is limited to the 500 °C range.

8.3.2.1 Front Contact

The front-contact has to be a highly conductive semiconductor. A layer that allows at the same time to contact the cells and to let the light pass. For this reason the choice is to employ a transparent conductive oxide (TCO).

However, oxides are generally materials with very low conductivity, so it is necessary to dope them with specific elements. In this way, TCOs become highly conductive (with a free carrier concentration between 10^{18} and 10^{20} cm^{-3}).

Most used TCOs are: fluorine tin oxide, SnO_2:F (FTO); zinc oxide, ZnO:Al (AZO); and indium tin oxide In_2O_3:Sn (ITO). Insulating oxides (high resistant transparent layers: HRT) can also be applied in addition between TCO and CdS in order to prevent shunts due to pinholes in the thin CdS layer and, additionally, improve band alignment.

Since TCOs are the first layers to be deposited on the substrate, they can affect the structure and properties of the layers above them. Impurities from the TCO might diffuse into the device junction and into the absorber. Moreover, the morphology of TCOs affects the CdS and the CdTe grain structure and, consequently, the electrical properties of these layers. Industrially the most used TCO for CdTe solar cells is SnO_2:F (FTO). Commercially available tin oxide thin films are fabricated by Chemical Vapour Deposition technique (CVD), where the thin film grows as consequence of a gas reaction of the inorganic and metal organic precursors, occurring on the substrate, transported near the deposition zone. Nowadays, FTO is made in an "in-line CVD" system where the tin oxide is deposited on the "just manufactured" float glass.[7]

8.3.2.2 Buffer Layer

In the typical configuration, the buffer layer or window layer (*n*-part of the junction) is the CdS layer. CdS has a bandgap of 2.4 eV, which does not allow having a complete transparency for wavelengths below 400 nm. The "parasitic" optical absorption in the CdS layer is the predominant loss in photocurrent, making it necessary to minimize the thickness of CdS. High efficiency CdTe/CdS solar cells have been achieved using CdS films with a thickness below 100 nm. A reduction from 120 to 60 nm would result in a gain in the 400 nm wavelength region, with a subsequent increase of about 10% in short-circuit current density.

Recently also magnesium tin oxide $Mg_xZn_{1-x}O$ (MZO) has been introduced as a buffer layer. MZO results in higher transparency compared to CdS. Moreover, the band-gap of MZO can be tuned by changing the substrate temperature during the deposition, and by doing so it is possible to control the Mg content. This results in an increase in efficiency, in particular in terms of short-circuit current density and of fill factor.

This new buffer layer works very well when a thin layer of CdSe is introduced in the CdTe structure near the junction. Comparison of spectral response measurements of solar cells made with a CdSe buffer, between CdS and CdTe layers suggests that the improved current density can be attributed to a specific design of the junction.

The additional insertion of CdSe improves spectral response both in the blue region (between 300 and 500 nm) and in the long wavelength region (825–930 nm): this results in an increase in short-circuit current density. In fact CdSe has a lower band gap than CdTe, allowing for the absorption of photons with lower energy.

In summary the insertion of an intermediate buffer layer of CdSe and substitution of CdS with MgZnO, allows avoiding the "parasitic" absorption due to CdS, improves the junction quality and drastically increases efficiency.

8.3.2.3 CdTe

In the standard configuration, the absorber consists of a CdTe layer, with a thickness ranging from 1 to 8 μm. Its crystal structure and quality affect both the carrier concentration and the carrier lifetime, having an effect on both the open-circuit voltage and on the short-circuit current density.

[7]Float glass is a sheet of glass made by floating molten glass. Generally float glass has a melting temperature around 500 °C and contains alkali elements.

Close space sublimation (CSS) is probably the most popular technique; here the material is sublimated[8] at high pressure, with high substrate temperatures: resulting, thus, in fast deposition. This, on an industrial scale, is transformed into the process called vapour transport deposition (VTD).

Substrate temperature influences grain growth: the higher the substrate temperatures are, the larger the grain sizes will be with a consequent reduction in the number of grain boundaries.

Larger grain size with low substrate temperature can be obtained by the standard recrystallization treatment: the so-called $CdCl_2$ annealing treatment. This consists of a heat treatment at 380–420 °C under a chlorine atmosphere (that can be provided by a chlorine containing gas or by a deposition of $CdCl_2$ or of $MgCl_2$ on the CdTe layer). This can be done in argon or in air atmosphere. This process step re-crystallizes the grains and removes structural defects in both CdTe and in the intermixed layer of CdTe-CdS, which is generally formed at the junction. Devices without this post-deposition annealing obtain conversion efficiencies not larger than 10%.

In the past CdTe has been deposited by a large amount of techniques, however the most successful ones are CSS [4], RF Sputtering [5], thermal evaporation [6] and VTD [7].

We will describe here only the VTD process in more detail, since it is the method that is mostly applied for large-scale production.

Vapour Transport Deposition (VTD)

Vapour transport deposition is a technique that combines the features of being an atmospheric pressure deposition and having a high substrate temperature like in the close space sublimation technology (but in VTD the material is delivered on the substrate by carrier gases).

In this way, the features of a high pressure and a non-vacuum low-cost process are obtained, but, at the same time, it is much easier to refill the material sources, since the crucibles are far away from the deposition region. This makes VTD particularly suitable for industrial processes, so there is no wonder that Solar Cell Inc. (now First Solar), in collaboration with the National Renewable Energies Laboratory, has introduced it for their fabrication line.

The process of CdTe deposition consists of three different stages:

(1) Heating of the material from a powder feeder into the vaporizer where CdTe is introduced in gas phase, (2) transporting CdTe into the deposition chamber through a distributor by means of inert carrier gases, (3) depositing the layer on the substrate at high temperature (coalescence[9] and growth of the crystals).

Up to now, this is the process that leads to the highest efficiency for CdTe solar cells, 22.1% [1] fabricated at the batch CdTe facility at First Solar.

---

[8]Sublimation is the transition of a substance directly from the solid to the gas phase, without passing through an intermediate liquid phase.

[9]Coalescence is the process by which two or more small grains merge during growth to form a single larger grain.

8.3.2.4 Back Contact

The back contact is the part of the solar cell, which is opposite to the sunny side. With many other technologies, as in the case of $CuInGaSe_2$, a metal such as molybdenum would be sufficient for contacting the cell. On the contrary, for CdTe the back contact is a complicated issue, due to the high electron affinity[10] of this material: this requires a metal with a work function[11] greater than 5.7 eV. No metals available have this value and different solutions have to be applied.

Generally, CdTe forms a Schottky contact with metals, presenting an energy barrier. The height of this can be reduced by generating a heavily *p*-doped CdTe surface and by applying a buffer layer with a high carrier concentration.

This has been obtained in the past by applying a Cu/metal double layer; but it leads to a high instability of the device due to copper diffusion.

On the other hand, applying copper above a diffusion barrier, with a molybdenum layer on top of this stack, makes a more stable back contact. In this way, copper has the predominant task of contacting the cell and reducing the series resistance. A large variety of combinations have been proposed with this structure: ZnTe:Cu, As_2Te_3:Cu, Bi_2Te_3:Au. ZnTe is, at the moment, the variant, which is the most used.

8.3.3 Performance and Degradation

Nowadays, CdTe modules are commercially produced by a few companies, mainly First Solar (USA), which is the largest producer of thin-film modules in the market, and by Calyxo-TS Solar (Germany).

Efficiencies of finished modules can exceed 18% (by First Solar with an area of 7038 cm^2) [1]. In the past there has been a problem with stability of the solar cell efficiency; this was connected with the diffusion of copper, which is needed to provide a suitable back contact, as previously explained. However, in the last years the embedding of Cu into a ZnTe layer has practically solved this problem, as demonstrated by the high reliability of First Solar modules [8].

8.4 Flexible Thin-Film Solar Cells

As mentioned above, conventional polycrystalline thin-film solar cells are usually manufactured on (TCO-coated) glass substrates.

---

[10]The electron affinity of an atom or molecule is defined as the amount of energy released or spent when an electron is added to a neutral atom or molecule in the gaseous state to form a negative ion.

[11]In solid-state physics, the work function is the minimum energy needed to remove an electron from a metal or semiconductor and place in the space (vacuum) immediately outside the solid surface.

This, despite the extremely reduced thickness of the entire device, does not allow for any weight advantage or for shape adaptability to different surfaces.

On the other hand, crystalline silicon solar cells generally have thicknesses above 100 μm, resulting in heavier and rigid structures that cannot be bended and certainly not rolled.

Now, if thin film solar cells are fabricated on flexible substrates several advantages result for space as well as for terrestrial applications. Since the absorber can be 2–10 μm thick, if the solar cell is deposited on a thin and flexible substrate (with a thickness between 10 and 100 μm) one would obtain significant weight savings compared to glass substrates.

Moreover, specific power is a very important factor not only for space applications, but also for terrestrial applications: cells mounted on flexible foil have a special advantage since they can be wrapped onto any suitably oriented structure reducing the installation costs.

In this sense it is useful to think in terms of the power to weight ratio (*P*/*W*), the higher the *P/W,* the lower is the cost per Watt. At the same time a high *P/W* ratio simplifies the embedding of the modules in building structures.

So with thin films it is possible, by choosing a suitable substrate, to obtain lightweight and adaptable modules. But what kind of substrates can be used?

Typically metal foil or polymer substrates offer the possibility of fabricating flexible CdTe and CIGS solar cells.

The procedure to fabricate the modules is very similar to the one used for rigid modules: starting from the flexible substrate the different layers are deposited with the deposition techniques mentioned above. An important difference is that with flexible polymers and/or metal foils it is possible to design a special fabrication process (roll-to-roll), where the substrate is unrolled into the deposition line and rolled back again after going through all the deposition steps (see Fig. 8.5). This improves once more the scalability of the process, since it becomes possible to use a single roll to fabricate a very large number of modules, thereby reducing the fabrication costs. On the other hand, if special polymers are used as substrates, this contributes to increasing the module production costs.

The main difference between CIGS and CdTe, when deposited on flexible substrates, is that CIGS gives better results in the substrate configuration whilst CdTe performs better in the superstrate configuration. This gives a completely different panorama of possibilities for the fabrication of flexible devices. CIGS, when

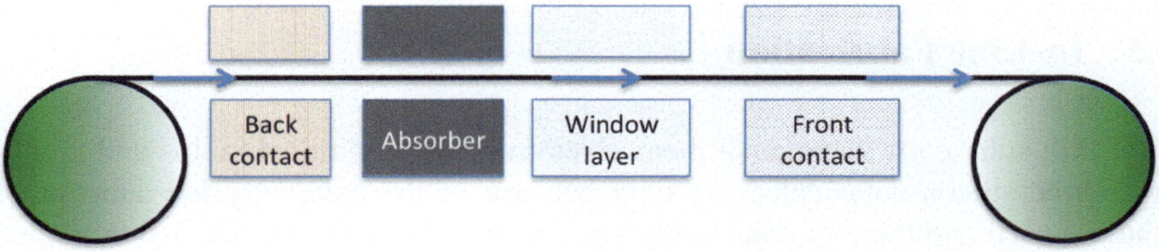

Fig. 8.5 Schematic representation of a roll-to-roll process

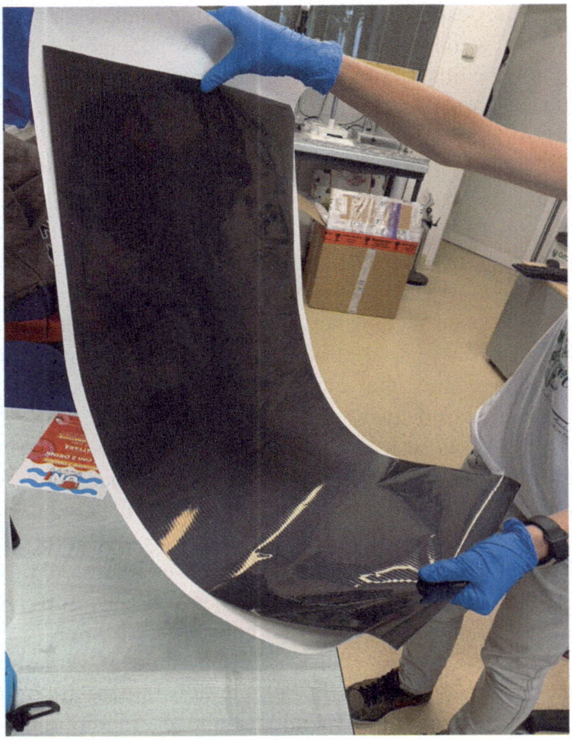

Fig. 8.6 An example of a CIGS solar module on a flexible substrate (from Flisom AG)

deposited in the substrate configuration, requires a flexible substrate, which withstands high temperatures and that does not release impurities into the active materials. On the other hand, CdTe when deposited in the superstrate configuration requires for the substrate not only high stability at high temperatures but also excellent transparency.

In case of CIGS, high-efficiency solar cells with efficiencies exceeding 19% have already been presented and commercial modules as well (see Fig. 8.6). On the other hand, flexible CdTe solar cells are still in the prototype stage and their efficiencies are lower (13.6%) [9].

8.5 In-Line Fabrication

Thin-film solar cells distinguish themselves from the traditional multi- and monocrystalline silicon solar cells, not only because of the extremely low amount of material used and the low processing temperatures they require, but also through a difference in the nature of module fabrication.

In particular there is no need, in the case of thin-film solar cells to prepare first small-size wafers, which are later assembled into modules, as in the case of crystalline silicon module fabrication.

On the contrary for thin film modules, the deposition of the whole stack, is done in the same fabrication line. Front contact, buffer layer, absorber and back contact are directly deposited on large-size glass panels: this means that here it is possible to start from pure glass and end up with a complete, encapsulated module in one single fabrication line. This is called "in-line fabrication".

If instead of glass a flexible substrate is used, then, such an in-line fabrication method can be implemented with a roll-to-roll system, taking the automation of the production line even further, with continuous supporting foil that can be cut at the end into different modules.

In Fig. 8.7, a CdTe fabrication line from former CdTe module producer Arendi S.p.A. is depicted. In this case the glass was first washed automatically, prior to insertion in the deposition line. Similarly, in Fig. 8.8, a sketch of an industrial production for CIGS-based modules is shown: the main difference with the CdTe fabrication process being the process sequence, due to the inverted structure of the layers. The high scalability of these technologies is proven by the "Takt time" (average time between the start of production of one unit and the start of production of the next one), which is typically less than a minute per module.

To avoid low voltages and high currents (which would lead to high Ohmic losses), CIGS and CdTe solar modules generally consist of a series of stripes connected in series, these are fabricated by a sequence of laser scribing and layer deposition steps, as shown in Fig. 8.9. The laser scribing steps are:

- Scribing of the front contact
- Scribing of the absorber

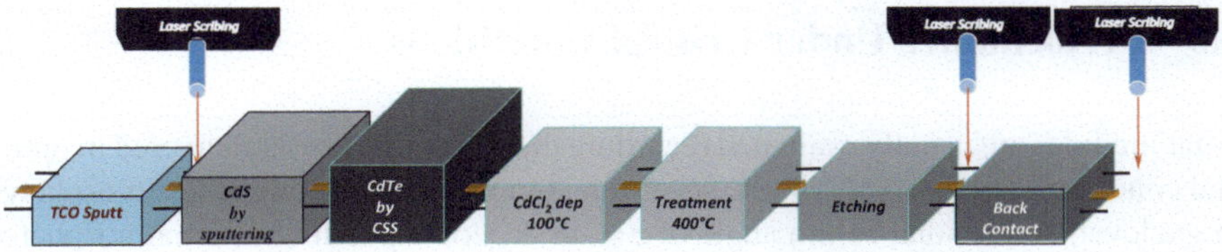

Fig. 8.7 Schematic illustration of a fabrication sequence for CdTe-based modules (Arendi S.p.A.)

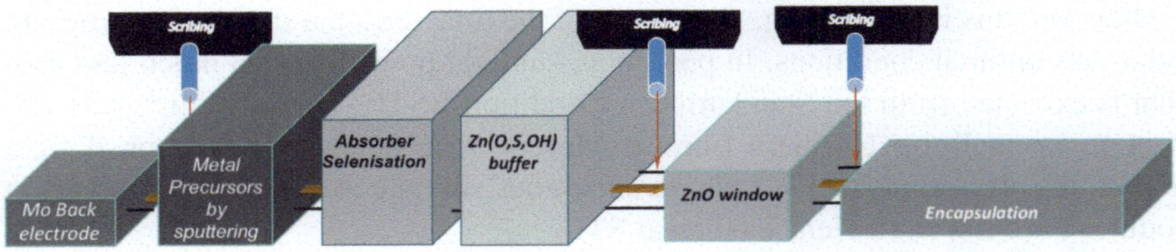

Fig. 8.8 Schematic illustration of a fabrication sequence for CIGS-based modules

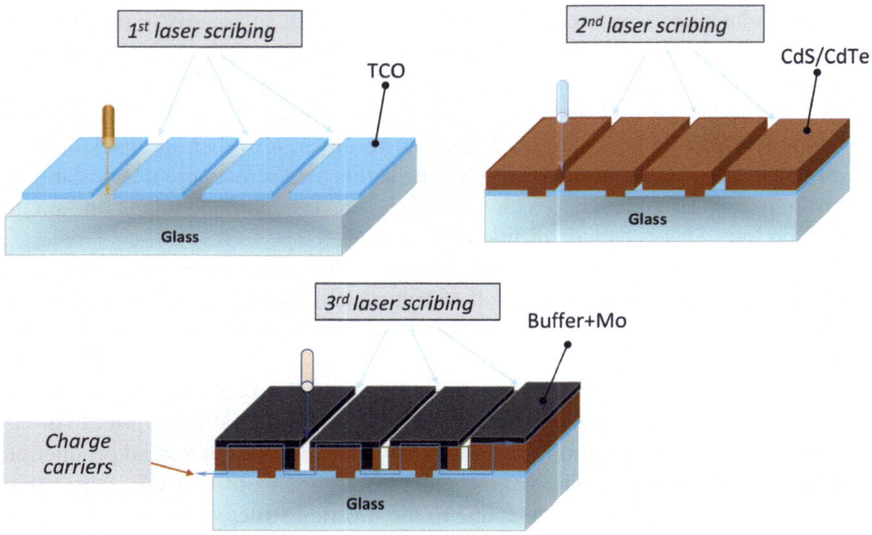

Fig. 8.9 Sequence of laser scribing steps for CdTe solar cells

- Scribing of the back contact.

Thanks to these in-line fabrication processes both CIGS and CdTe modules are now fabricated with very large areas, scaling from 60 × 120 cm to 120 × 200 cm (Series 6 from First Solar is achieving with 60 × 120 cm area, 18% conversion efficiency).

8.6 Performance Under Critical Conditions

Solar modules are usually evaluated regarding the maximum power delivered in optimal conditions such as AM 1.5 (the spectrum after travelling through the atmosphere to sea level, with a solar zenith angle of $z = 48.2°$, corresponding to 1.5 atmosphere thickness) with 1000 W/m^2, and light perfectly oriented on the devices. Thus, efficiency and power generation of the different technologies are usually compared only in these conditions.

However, this is misleading, since most of the time the solar devices are working under non-optimal conditions. In particular, sunlight power is often much less than what is expected from the Standard Test Conditions (STC = 1000 W/m^2, AM 1.5, 25 °C), especially in Northern Europe. Moreover, in hot countries, even if light irradiation corresponds to STC, temperature goes very much above 25 °C and again modules have lower efficiencies than at STC.

This has created a large misunderstanding when choosing the right technology and discussing about efficiency and power generation, because thin-film modules have the capacity of higher performance at conditions which are not optimal, compared to crystalline silicon devices.

8.6.1 Low-Light Conditions

CdTe as well as CIGS are direct-band-gap materials with high absorption coefficients.

In diffused light conditions, the spectrum changes its range of wavelengths towards the blue region, so the distribution of light power is more concentrated in the 300–500 nm range. This is very important if we consider that different absorber materials work at different wavelength ranges. For diffused light solar cell materials, like CdTe, that have their working range centred in the UV and the visible part of the spectrum are preferable, in these conditions.

It has been demonstrated [10] that mono-crystalline and multi-crystalline silicon technologies when installed in Southern Europe (Cyprus) have very large seasonal variations in module efficiency, with higher efficiencies reached in winter periods and lower ones in summer time. Also CIGS and CdTe thin film technologies have demonstrated highest efficiencies in winter time but with a reduced seasonal variation in comparison to crystalline silicon.

8.6.2 High Temperature Conditions

The operating temperature of a PV module can strongly affect its performance. As mentioned before, PV modules are measured at STC, but their actual operating temperatures are generally higher.

For this reason a Nominal Operating Cell Temperature (NOCT) was introduced to give an indication of the PV device operating temperature; it is therefore a useful parameter for PV system designers. According to IEC 61215 and 61646 regulations, NOCT is measured in open circuit at 45° inclination, 800 W m^{-2} with a wind speed of 1 m s^{-1} and at an air temperature of 20 °C. This results in typical NOCT values around 50 °C. However, still real PV operating temperatures exceed NOCT values especially in summer days.

Special attention to the operating temperature must be given in the case of BIPV, as the operating temperature of the PV module is given by the heat exchange mechanisms at the back of the modules. If the PV installation is designed and assembled without proper ventilation, high module operating temperatures will result. During summer days thin film glass/glass modules can reach temperatures up to 70 °C, 79 °C, and 92 °C respectively for open-rack mounted, for building added photovoltaics and for building integrated photovoltaics.

Now due to the different temperature coefficient of each absorber material, the loss of power due to higher operating temperatures—compared to STC conditions—will be more pronounced for some technologies and less for others.

Table 8.1 Temperature coefficients for different materials (see Chap. 5 for explanation of PERC)

Technology/product series	Manufacturer	Power temperature coefficient (%/°C)
IBC-crystalline silicon	Sunpower	−0.29
Multi-PERC crystalline silicon	Canadian solar	−0.37
Mono-PERC crystalline silicon	Longi solar	−0.38
Bifacial n-type PERT	Jinko solar	−0.38
CdTe-series 6	First solar	−0.32
CdTe-series 3 black plus	First solar	−0.29
CdTe-CX3Pro	Calyxo	−0.25
CIGS	Solar frontier	−0.33
CIGS	Avancis	−0.39
CIGS	Nice solar	−0.36
CIGS	Solibro	−0.32
CIGS	Flisom	−0.35

In particular, as shown in Table 8.1, CdTe shows a low magnitude of the power temperature coefficient, suggesting that their power reduction with increasing temperature would be rather limited compared to other technologies.

Tests of power delivered on modules of different technologies were done by putting the device into a thermostatic chamber and by measuring the power with a pulsed solar simulator (10-ms pulse duration) in the range of 25–60 °C [11]. These show clearly the superior behaviour of CdTe devices whose power is less dependent on temperature. Recently, as shown in Table 8.1, also certain crystalline silicon based solar cells have demonstrated a low magnitude of *TC* (see Chap. 3 for more details on this topic).

8.7 Environmental Aspects

The presence of cadmium gives a negative perception of CdTe solar cells, since cadmium is a carcinogenic element in its elemental form. Cadmium is in fact banned for devices under the regulations of the European RoHS (Restriction of Hazardous Substances) directive. This has been often used as a strong weapon against the widespread use of this low-cost and efficient PV technology.

However, a large variety of studies on the impact of CdTe in the environment have been presented over the years [12–15]. Contrary to cadmium, CdTe is not labeled as carcinogenic, as it is a stable compound. Studies on acute inhalation and oral toxicities of CdTe in rats found that the both median lethal concentration and lethal dose are more than 9 times higher than that of elemental Cd [15]. Moreover, CdTe showed no acute adverse reproductive effects in rats and low aquatic toxicity [16]. Now, if CdTe as a compound, does not represent a real harm for the environment, then the question arises if and how this compound might release pure toxic Cd.

Different studies have been made to clarify this issue. Early works from broken modules left under rainwater showed that cadmium concentration stayed within the limits of the law for drinking water, and similar results were obtained for concentration in the soil when the damaged modules were left on the ground [14].

A research laboratory at the University of Arizona very recently did further studies on the subject. Their studies concerned non-encapsulated CdTe, by examining 1.5 g of "snipped" CdTe film together with 300 g of crushed glass, in order to maintain the ratio between active material and encapsulant, at the same ratio as conditions within a solar module. In this way an acidic column and a methanogenic column were applied to test the leachate[12] produced in a municipal landfill. Even though the material was not encapsulated, the results showed that the Cd concentration was 6.7-fold lower than the threshold limit established in the TCLP (toxic characteristic leaching procedure) conditions and 4.6-fold lower than the one for the WET (waste extraction test) conditions [17].

Another high risk could be in case of fires, but it has to be considered that the melting point of CdTe is 1050 °C, a temperature normally not reached during fires of residential houses, so that one can consider that during a fire in a residential location, no cadmium is released into the atmosphere.

Higher temperatures are reached in fires of industrial plants. However, when modules were subjected to temperatures of 1100 °C, the melted cadmium telluride was captured and embedded in the melted encapsulating glass [13].

Also for CIGS the environmental impact is low, as most of the elements present in a CIGS solar cell are not toxic, with the exception of selenium and a small amount of Cd in the buffer layer.

Indeed, different studies have been conducted with CIGS solar cells and no significant release of toxic materials has been measured during fires, the main compound released being CO_2 (as usual in fires).

Another issue is the presence, within CIGS and CdTe solar cells, of rare elements, such as indium and tellurium, respectively: These elements may constitute a bottleneck for the development of these technologies on a very large scale.

However, calculations have shown that a production of several TW_p of CIGS and CdTe solar cells would be possible with the present reserves of Indium and Tellurium—Now such a production volume may only be reached by 2050, even in the most aggressive scenarios for the development of renewable energy.

Material-related sustainability issues can also be faced by developing recovery during primary production, by reducing the thickness of the semiconductor layers below 4 μm, and, most importantly by recycling modules at their end-of-life.

[12] Water that has percolated through a solid and leached out some of the constituents.

For this reason, a substantial part of the research in CIGS and CdTe solar cell technologies is directed towards the reduction of layer thickness.

References

1. M.A. Green, Y. Hishikawa, E.D. Dunlop, D.H. Levi, J. Hohl-Ebinger, M. Yoshita, A.W.Y. Ho-Baillie, Solar cell efficiency tables (version 53). Prog. Photovoltaics Res. Appl. **27**(1), 3–12 (2019). https://doi.org/10.1002/pip.3102
2. A. Romeo, M. Terheggen, D. Abou-Ras, D.L. Bätzner, F.-J. Haug, M. Kälin, D. Rudmann, A.N. Tiwari, Development of thin-film Cu(In, Ga)Se_2 and CdTe solar cells. Prog. Photovoltaics Res. Appl. **12**(23), 93–111 (2004). https://doi.org/10.1002/pip.527
3. K. Kushiya, CIS-based thin-film PV technology in solar frontier K.K., Sol. Energy Mater. Sol. Cells **122**, 309–313 (2014). https://doi.org/10.1016/j.solmat.2013.09.014
4. D. Bonnet, The CdTe thin film solar cell—an overview. Int. J. Sol. Energy **12**(1–4), 1–14 (1992). https://doi.org/10.1080/01425919208909746
5. J.M. Kephart, R.M. Geisthardt, W.S. Sampath, Optimization of CdTe thin-film solar cell efficiency using a sputtered, oxygenated CdS window layer. Prog. Photovoltaics Res. Appl. **23**(11), 1484–1492 (2015). https://doi.org/10.1002/pip.2578
6. I. Rimmaudo, A. Salavei, E. Artegiani, D. Menossi, M. Giarola, G. Mariotto, A. Gasparotto, A. Romeo, Improved stability of CdTe solar cells by absorber surface etching. Sol. Energy Mater. Sol. Cells **162** (2017). https://doi.org/10.1016/j.solmat.2016.12.044
7. G.M. Hanket, B.E. McCandless, W.A. Buchanan, S. Fields, R.W. Birkmire, Design of a vapor transport deposition process for thin film materials. J. Vac. Sci. Technol., A **24**(5), 1695–1701 (2006). https://doi.org/10.1116/1.2214689
8. J.Y.Ye, T. Reindl, A.G. Aberle, T.M Walsh, Performance degradation of various PV module technologies in tropical Singapore. *IEEE J. Photovoltaics* **4** (5), 1288–1294 (2014). https://doi.org/10.1109/JPHOTOV.2014.2338051
9. L. Kranz, C. Gretener, J. Perrenoud, R. Schmitt, F. Pianezzi, F. La Mattina, F. Bloesch, E. Cheah, A. Chirila, C.M. Fella, H.T. Hagendorfer, S. Jaeger, S. Nishiwaki, A.R. Uhl, S. Buecheler, A.N. Tiwari, Doping of polycrystalline CdTe for high-efficiency solar cells on flexible metal foil. Nat. Commun. **4**, 1–7 (2013). https://doi.org/10.1038/ncomms3306
10. G. Makrides, B. Zinsser, G.E. Georghiou, M. Schubert, J.H. Werner, Outdoor efficiency of different photovoltaic systems installed in Cyprus and Germany, in *Conf. Rec. IEEE Photovolt. Spec. Conf.* 7–12 (2008). https://doi.org/10.1109/PVSC.2008.4922830
11. A. Virtuani, D. Pavanello, G. Friesen, Overview of temperature coefficients of different thin film photovoltaic technologies, in *25th Eur. Photovolt. Sol. Energy Conf. Exhib./5th World Conf. Photovolt. Energy Conversion, 6–10 Sept. 2010, Val. Spain*, (January), 4248–4252 (2010). https://doi.org/10.4229/25thEUPVSEC2010-4AV.3.83
12. V.M. Fthenakis, E.A. Alsema, M.J. de Wild-Scholten, Life cycle assessment of photovoltaics: perceptions, needs, and challenges, in *Conf. Rec. Thirty-first IEEE Photovolt. Spec. Conf. 2005* 1655–1658 (2005). https://doi.org/10.1109/PVSC.2005.1488464
13. V.M. Fthenakis, H.C. Kim, E. Alsema, Emissions from photovoltaic life cycles emissions from photovoltaic life cycles. Environ. Sci. Technol. **42**(6), 2168–2174 (2008). https://doi.org/10.1021/es071763q
14. H. Steinberger, Health, safety and environmental risks from the operation of CdTe and CIS thin-film modules. Prog. Photovoltaics Res. Appl. **6**(2), 99–103 (1998). https://doi.org/10.1002/(SICI)1099-159X(199803/04)6:2%3c99:AID-PIP211%3e3.0.CO;2-Q
15. J. Zayed, S. Philippe, Acute oral and inhalation toxicities in rats With cadmium telluride. Int. J. Toxicol. **28**(4), 259–265 (2009). https://doi.org/10.1177/1091581809337630
16. V.M. Fthenakis, S.C. Morris, P.D. Moskowitz, D.L. Morgan, Toxicity of cadmium telluride, copper indium diselenide, and copper gallium diselenide. Prog. Photovoltaics Res. Appl.

7(6), 489–497 (1999). https://doi.org/10.1002/(SICI)1099-159X(199911/12)7:6%3c489:AID-PIP287%3e3.0.CO;2-N

17. A. Ramos-Ruiz, J.V. Wilkening, J.A. Field, R. Sierra-Alvarez, Leaching of cadmium and tellurium from cadmium telluride (CdTe) thin-film solar panels under simulated landfill conditions. J. Hazard. Mater. **336**, 57–64 (2017). https://doi.org/10.1016/j.jhazmat.2017.04.052

Alessandro Romeo is professor of Physics at the University of Verona and research group leader, working on a whole spectrum of chalcogenide solar cells (CdTe, Sb_2Se_3, CZTS and CIGS solar cells). Alessandro obtained his PhD at the Swiss Federal Institute of Technology Zurich (ETHZ). He has been working in the field of thin film solar cells for more than 20 years having done postdoctoral research in different laboratories (University of Parma, Swiss Federal Institute of Technology, University of Verona). He is a member of the scientific committee for the European Photovoltaic Solar Energy Conference since 2010. Alessandro was one of the organizers of the E-MRS Spring Meeting "Thin Film Chalcogenide Solar Cells" in 2010, 2012 and 2014.

Chapter 9
Solar Module Technology

Alessandro Virtuani

Abstract In a solar photovoltaic module, a number of individual solar cells are electrically connected to increase their power output. Cells and interconnects are then packaged in order to: (1) protect the electrical circuit from weathering, (2) provide structural stability and protect the mechanical integrity of the cells, (3) isolate the electrical circuit from the environment, protecting operators against electrical shocks. In this chapter, we focus initially on the electrical layout of cells/modules and on the typical sandwich and encapsulation structures used in solar module manufacturing, by having a close look at the materials and processes used in the industry, and at innovative concepts that are increasingly entering the market. We then review the topics of performance, reliability and durability of solar modules and the related subjects of industry qualification standards, service lifetime and warranties. A closer look at failure modes experienced by modules in operation will be given in Chap. 10, next up. In this chapter, our major focus will be on the dominant technology of wafer-based crystalline silicon (c-Si) solar cells/modules. Peculiarities of module technology for thin film modules will be reviewed briefly.

9.1 Electrical Layout of Solar Modules

9.1.1 Cell Interconnections

In a PV module, a number of individual solar cells are electrically connected to increase their power output. In wafer-based crystalline solar (c-Si) solar cells, the **busbars** present on the top of the cell (see Fig. 9.1) are connected directly to the rear contact of the adjacent cell, by means of **cell interconnect ribbons**, generally tin-coated copper ribbons (or silver-coated ones). This is shown in Fig. 9.2 for the case of cells with 2-busbars (2BB); this figure shows as well how sub-strings are connected by so-called **string interconnects**. Most often, the ribbons are soldered directly onto

A. Virtuani (✉)
EPFL (PV-Lab), Neuchâtel, Switzerland
e-mail: alessandro.virtuani@epfl.ch

A. Shah (ed.), *Solar Cells and Modules*, Springer Series in Materials Science 301,
https://doi.org/10.1007/978-3-030-46487-5_9

Fig. 9.1 c-Si solar cells with (from the top left) two (2BB), three (3BB), five (5BB) busbar and cell without busbars (below right). Current-collecting fingers are placed perpendicular to the busbars

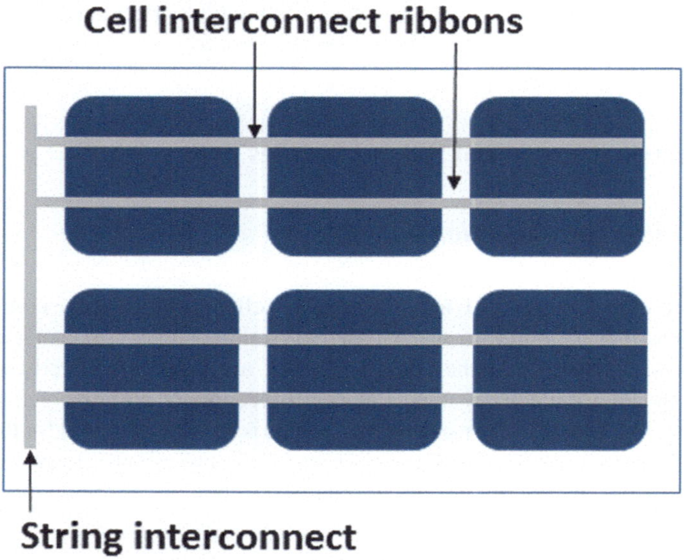

Fig. 9.2 Simplified sketch of a solar module with two-busbar (2BB) cells. *Cell interconnect ribbons* and *string interconnects* are shown. *Busbars* are hidden below the ribbons

the cell busbar. In some cases, as for HJT cells that require low-temperature processes to avoid damaging the passivating layers, *electrically conductive adhesives* (ECA) are used in the place of soldering. As of today, most solar cells have four or five busbars (4BB-5BB). An increased number of busbars or even multi-wire concepts with no busbars can be expected in future PV modules, as this will lead to reduced silver consumption; it should also lead to a reduction in the vulnerability of module performance to cell cracks and breakage (see Sect. 9.3). **IBC (Interdigitated Back Contact)** cells as well, where both positive and negative contacts are at the rear side of the cell, do not need busbars. To connect neighbouring cells, the ribbons are placed on the rear side of the cells.

Similarly, thin film modules can be manufactured by contacting adjacent solar cells with ribbons. However, this is usually done in a similar way by using the so-called **monolithic interconnection** of cells. This method constitutes one of the advantages of thin film technologies as compared to wafer-based ones. In this process, semiconductor and metal layers are deposited over the full surface of a substrate. The solar cells are then separated by means of laser or mechanical scribes, which are performed sequentially. The front side of a solar cell is generally connected to the rear side of the adjacent cell by means of the transparent conductive oxide (TCO) layer, which covers the solar cells and provides the front contact. This process, schematically shown in Fig. 9.3, creates some dead areas (approximately 100–200 μm) between neighbouring solar cells. A poor quality of scribing lines can lead, as well, to shunts between cells. On the other hand, this patterning process offers a great flexibility in terms of cells design (size and shape); it can be fully automatized and can easily be integrated into thin film solar cell manufacturing lines.

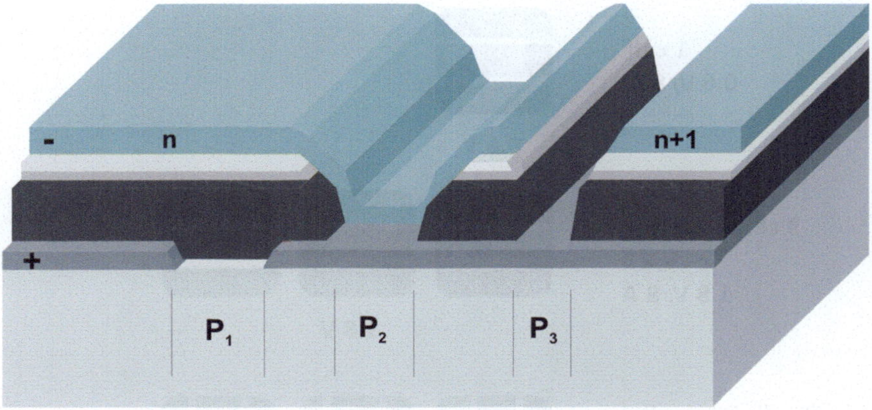

Fig. 9.3 Monolithic interconnection of CIGS solar cells. Sequential laser and/or mechanical patterning cuts (*P*1, *P*2, and *P*3) are used to define the size/form of the solar cells. The front side (the negative pole of the cell) of cell *n* is connected to the rear side (the positive pole) of the adjacent solar cell *n* + 1 by means of a transparent conductive oxide layer [1]

9.1.2 Series and Parallel Connections of Cells

If we connect solar cells in series (**series connection**), voltages add-up, while the overall current corresponds to the current of a single solar cell. If a single cell in the string is delivering a lower current (lower performance, cell breakage, shaded cell, etc.), this cell sets the current of the whole string. This is why cells must be *current-matched* when sorted in module manufacturing lines. Otherwise, mismatches will lead to a reduction in module performance.

On the contrary, in a **parallel connection**, currents add-up and the overall voltage is equal to the voltage of a single solar cell. However, in a parallel connection, if voltages of cells/strings are not balanced, the overall voltage of the connection will be close to the voltage of the cell/string with the lowest voltage.

In general, if n and m are, respectively, the number of the cells in series and in parallel, the following relations hold for the open-circuit voltage V^m_{oc} and short-circuit currents I^m_{sc} of the modules:

(1) In a series connection: $V^m_{oc} = V_{oc} \times n$, $I^m_{sc} = I_{sc}$;
(2) In a parallel connection: $V^m_{oc} = V_{oc}$, $I^m_{sc} = I_{sc} \times m$;

where V_{oc} and I_{sc} are the parameters for a single solar cell.

Figure 9.4 illustrates an example for the relationship between the current and voltage characteristics of a single solar cell and that of three solar cells connected once in series and, the other time, in parallel.

In general, commercially available c-Si solar modules are composed of 60 or 72 series-connected solar cells. In some cases, particularly for off-grid applications

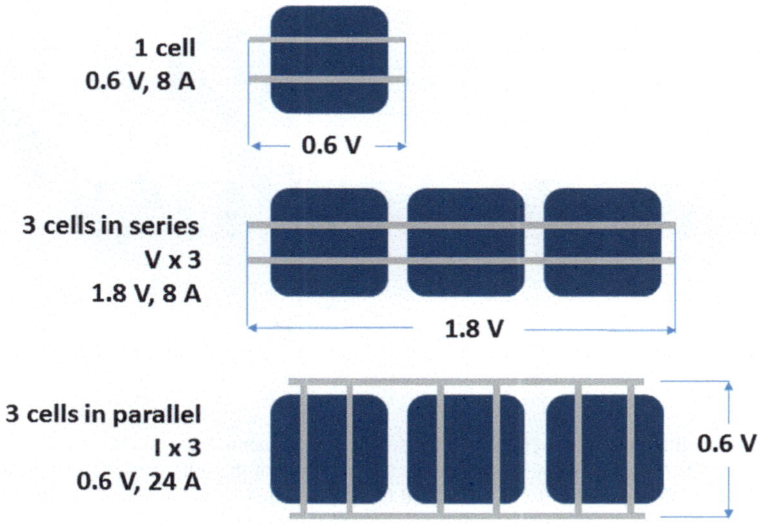

Fig. 9.4 Relationship between the current and voltage relationship of a single solar cell and that of three solar cells connected in series and in parallel

that may require larger currents for faster battery charging, modules can contain paralleled strings. This has the advantage of reducing shading losses. However, the resulting higher currents will lead to higher resistive losses.

Thin-film solar modules, irrespective of what technology, are in general characterized by a higher number of cells connected in series, and by smaller cell active areas. They are, thus, characterized by higher voltages and lower currents.

9.1.3 Cell-to-Module Losses

Just as in the case for solar cells, the performance of modules is characterized by the power P_{max} and efficiency η measured at Standard Test Conditions (STC = 1000 W/m^2, AM1.5, 25 °C). On a module level, we can distinguish between *total-area efficiency* η_{tot} and *aperture-area efficiency*[1] η_{apa}. The former is determined using the total area (including the edges and also the frame, if present); the latter using only the surface of the active-area of the cells (which corresponds to a smaller surface). Understandably, $\eta_{tot} < \eta_{apa}$. When comparing the efficiency of different technologies and/or products we need to make sure that we are referring to the same parameter. In general, η_{tot} tends to be a more reliable parameter for a direct comparison.

When assembling solar cells in a module, the overall efficiency of a module (i.e. both η_{tot} and η_{apa}) is generally lower than the average efficiency η_{cell} of the cells used.

The difference in the efficiency (or power output) between cells and modules is described by a parameter called the "**Cell-to-Module (CTM) ratio**", which accounts for the so-called CTM losses.

For example, a conventional c-Si module (with a power of 293 W_p and an area: ~1.63 m^2 = 1.65 m × 0.992 m), made of 60 cells (each 156 mm × 156 mm) with an average efficiency η_{cell} of 21%, would have a typical total-area efficiency η_{tot} of 19% with a CTM ratio (calculated for the efficiency) of ~90%. Typical CTM values lay in the range of 85–95%, even if novel concepts (see Sect. 9.2.3) can lead to higher CTM ratios.[2]

The difference in efficiency between cells and modules depends on several factors, mainly: *geometrical factors* (e.g. module margins, cell spacing, etc.), *optical factors* (e.g. glass/air interface and front cover reflections, encapsulant absorption, etc.) and *electrical losses* (increased series resistance due to ribbons, interconnects,

[1]For c-Si *aperture-area* or *active-cell area* are sometimes used interchangeably. For thin films, the *active-cell area* is generally smaller than the *aperture-area*, which includes the portion of the cell covered by grid fingers and interconnects.

[2]It should be noted that CTM ratios may be determined using either power or efficiency values for the module and the cells. These ratios do not correspond, because when efficiency values are used, the cell and module areas are factored as well in the computation. CTM ratios are higher if calculated using power values, rather than efficiency values.

connectors, etc.). Losses due to *mismatched* cells (i.e. cells with different electrical properties), can be, as well, a critical loss mechanism, but can be minimized by proper sorting of the cells.

On the other hand, the integration of cells into modules leads as well to optical *gains* (partly offsetting CTM losses), mostly thanks to an improved optical coupling (glass/encapsulant/cell) that reduces the relatively high reflectivity of the cell at the air/cell interface. Additional optical gains can be related to internal reflectance and coupling promoted by the reflection of the light over the metallization and the rear cover, particularly if a white backsheet is used [2, 3]. If the angle of incidence of the light is not normal part of the light reflected by the cells, the interconnects or the backsheet in fact will undergo multiple reflections in the encapulant/glass structure and eventually impinge over the active surface of an adjacent cell.

9.1.4 By-Pass Diodes

The electrical layout of a solar module generally contains by-pass diodes, which are used to minimize the effects of partial shading on cells. Such partial shading can seriously affect the output of a module (*small shadows, big losses!*), and can even cause damages to the materials composing the module.

Partial shading can easily occur due to the presence of trees, leaves or bird droppings. It can be particularly severe in the built environment, where it can originate from the presence of chimneys, antennas, neighbouring buildings, etc.

If a single cell (or a group of cells) is shaded, it will become reverse-biased, not generating power any longer, but becoming a power sink that dissipates the power generated by the other cells of the string. This will result for the shaded cell (or the shaded portion of a cell) in local overheating (in so-called "**hot-spots**"). The same effect will take place, if one of the cells is partially or fully damaged. Temperatures $\geq$100 °C, depending on weather conditions and the number of cells in a string, can be reached and, in turn, lead to destructive effects (e.g. encapsulant browning, backsheet and glass cracking, loss of adhesion, etc.), which in the long run can compromise the performance and electrical insulation of the module, as will be described in the next chapter. Additionally, a large reverse bias applied to a solar cell (which will depend on the number of cells connected to the shaded cell in the string) can lead to *junction breakdown* completely destroying the cell.

The effects of a *hot-spot* can be prevented through the use of a by-pass diode connected in parallel with an opposite polarity to that of the solar cell, as Fig. 9.5 illustrates. Under normal operating conditions, each solar cell is forward biased. The by-pass diode will then be reverse biased and carries no current. When a solar cell becomes reverse biased, due to shading or cell damage, the by-pass diode will then allow the current from the illuminated solar cells to flow in the external circuit, *by-passing* the shaded/damaged cell and preventing hot-spot formation. The maximum reverse bias across the shaded/damaged cell is limited to a single diode drop (i.e. ~−0.6 V), irrespective of the number of cells connected in series.

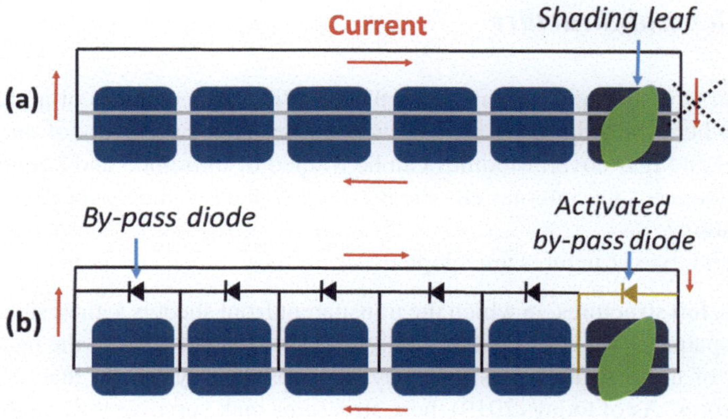

Fig. 9.5 **a** A string of six short-circuit connected solar cell of which one is shaded. **b** By-pass diodes placed in parallel to each cell in the string: when the cell is shaded, the by-pass diode is activated, preventing reverse-biasing of the shaded cell and dissipation of the power generated by the other cells in the string. In commercial solar modules a single by-pass diode protects whole groups of cells, rather than a single cell

Ideally, each cell should be protected by a by-pass diode. In practice, for cost reasons and to avoid over-complexity in manufacturing, commercial modules generally make use of three by-pass diodes. For a typical module with 60/72 cells in series, each by-pass diode protects three sub-strings made of 20/24 cells each. If a single cell is completely shadowed, the entire sub-string of the cell will be by-passed and the overall module power will be reduced by 1/3 and not entirely, as would be the case if no by-pass diodes were present and all cells were connected in series.

9.2 Module Architectures, Materials and Processes

As mentioned in the introduction, cells and interconnects are packaged in order to: (1) protect the electrical circuit from weathering, (2) provide structural stability and protect the mechanical integrity of the cells/interconnects, (3) isolate the electrical circuit from the environment protecting operators against electrical shocks. A careful selection of the proper materials, a careful design and optimized manufacturing processes are crucial to ensure the long-term performance of the modules and prevent safety threats.

9.2.1 General Structure

The typical structure of a module, as depicted in Fig. 9.6, consists of a multi-stack structure where the solar cells are sandwiched between two layers of encapsulant and a front and rear cover. Modules can be framed or unframed and a junction box is used to connect the internal electrical circuit to cables and connectors used for module cabling.

In general, two structures are adopted:

(1) **Glass/foil** structures, in which the transparent front sheet is a glass sheet and the rear opaque cover is formed by a polymeric **backsheet**. One of the main advantages of these structures is that they are considerably lighter than glass/glass structures. As of today (2019) these structures make up over 90% of the market share and overall cumulative installed capacity;
(2) **Glass/glass** structures make use of a glass sheet as a rear cover, too. Compared to the glass/backsheet variant, these structures are more stable from a mechanical standpoint (which makes glass/glass modules more robust against wind, snow loads and hail impacts); they also provide a more solid packaging with respect to moisture ingress. For crystalline-silicon modules, glass/glass structures are largely adopted in **building-integrated PV (BIPV)** products. Presently, however, glass/glass structures are experiencing a revival for ground-mounted installations, too, due to a considerable market interest for bifacial PV modules (see Sect. 9.3), for which a transparent backsheet is required. Projections forecast market shares for glass/glass modules in the range of 30% by 2028 [4]. Additionally, glass/glass structures are generally adopted for thin-film modules, which in some cases can be extremely sensitive to moisture and for

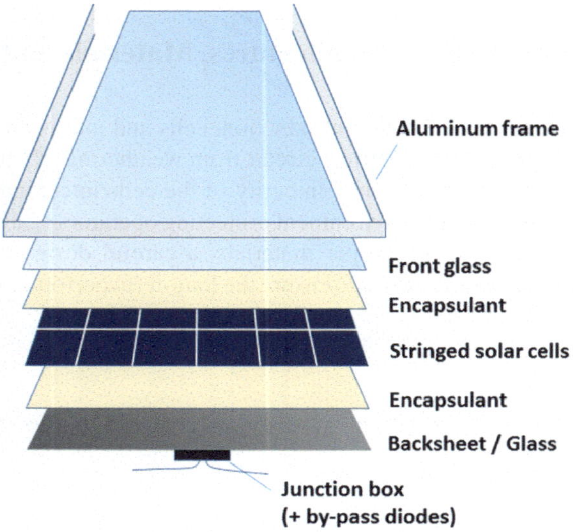

Fig. 9.6 Typical sandwich structure of a c-Si solar PV module

which one glass sheet is usually adopted as a substrate for the deposition of the semi-conducting/absorber layers. Thin film modules generally make use of a single layer of encapsulant rather than two layers as for c-Si.

9.2.2 *Materials*

9.2.2.1 Front Covers

Most modules make use of a transparent glass sheet as front cover. In general, a soda-lime glass with a low iron (Fe^{2+}) content (so-called "*solar-grade glass*"), which reduces the absorption of light, is used. To improve its mechanical stability and protect the module against impacts (hail, transport, handling, etc.) the glass sheet needs to be tempered (or *heat-strengthened*[3]). A standard glass thickness is 3.2 mm; although, for cost and weight reasons,[4] efforts are posed to reduce the thickness of the glass. BIPV products, which may be exposed to high wind loads make use of thicker glass (4–6 mm or thicker). Depending on its composition, glass blocks a considerable part of the UV-B radiation (280–320 nm wavelength), but only a part of the UV-A radiation (320–400 nm wavelength).

Some special products (e.g. flexible thin film modules) use transparent polymers, such as ETFE (ethylene tetrafluoroethylene) as front covers. These polymers are highly transparent and can provide an excellent barrier against water ingress, but are, at the moment, also very costly.

9.2.2.2 Rear Covers

Most solar modules have a composite polymer sheet as **backsheet** generally formed by a multi-stack structure of three layers with an overall thickness of 280–400 μm.

A typical backsheet structure is made of PVF/PET/PVF where PET (Polyethylene terephthalate) is a polyester, which isolates the module electrically and provides mechanical stability, whereas PVF (polyvinyl-fluoride, also known with the commercial brand Tedlar®) protects the internal circuit (and PET layer) from weathering agents (moisture, dust, UV, etc.). Several other materials are used to manufacture backsheets. PVF (in the outer and inner layers) is often replaced by PA (polyamide), by other fluoropolymers, such as PVFD (polyvinylidene difluoride), or

[3]There are two different types of *heat-treated glasses*: tempered and heat-strengthened. With tempered glass, the cooling process is accelerated to create higher surface compression and/or edge compression in the glass. This is the process that makes the glass four to five times stronger and safer than annealed or untreated glass. With heat-strengthened glass, the cooling process is slower, which means the compression strength is lower. Heat-strengthened glass is approximately twice as strong as annealed, or untreated, glass.

[4]Glass weighs approximately: ~2.5 kg/mm/m^2.

by PET. Double layers or even single layers of polymers (so-called all-PET, all-**PO** (**polyolefin**) backsheets)—have also been used as backsheet and are presently under further development.

With their low permeability, polymeric backsheets protect the solar cells to a certain extent by delaying the ingress of moisture, oxygen and other atmospheric gases or pollutants, but are partly "breathable" allowing an atmospheric exchange between the internal module structure and the environment. This can be an advantage, as for example the acetic acid generated by the photo-degradation of EVA (ethylene-co-vinyl acetate) will be partly released to the external atmosphere (see Sect. 9.2.2.3).

In the presence of semiconducting and other layers, which are particularly sensitive to humidity—such as the transparent conductive oxides (TCO), used in thin film cells and in silicon-heterojunction technology (HJT)—generally glass/glass structures, which provide a more solid packaging, are used. Alternatively, polymer backsheets containing an intermediate metal sheet (aluminium or steel) are sometimes employed [5]. These latter are used only by a limited number of manufacturers, but combine the lightweight properties of polymer backsheets with the reduced permeability of glass/glass structures.

Glass/glass structures need to be symmetrical and balanced, with the rear cover glass having similar mechanical properties (thickness, tempering, etc.) as the front one, to avoid unnecessary stress and bending (or breakage) of the module.

9.2.2.3 Encapsulants

In a sandwich structure like that of solar modules, encapsulants serve multiple purposes: they are needed to bond different materials and layers, to hold module components (cells, interconnections, etc.) in the right position, to accommodate mechanical stresses and to transfer mechanical solicitations trough the module structure. In addition, they isolate the solar cells electrically, and, to a certain extent, protect them from the external environment (moisture, oxygen, other gases, etc.), preventing corrosion and other degradation mechanisms. Besides, they improve the optical coupling of the different materials, reducing reflection losses. A good encapsulant must provide all these functions over a long time, even after exposure to UV radiation, humidity, mechanical stress, and high temperatures.

In the past, several encapsulants have been deployed by the industry. These include thermo-plastic polymers (or thermo-plastics) and thermo-setting polymers (or elastomers). The main differences between the two classes of polymers are:

1. **Thermoplastic polymers** are chemically linear structures (they do not cross-ink) while **thermosetting polymers** have three-dimensional cross-linked structures.
2. Thermoplastics polymer become soft and melt on heating, whereas thermosetting polymers do not soften with heating but rather become hard, due to cross-linking;
3. Thermoplastic polymers can be reshaped, whereas thermosetting polymers cannot be reshaped.

Thermoplastics include PVB (polyvinil butyral), TPU (thermoplastic polyurethane), TPO (thermoplastic polyolefins), and Ionomer. Materials as EVA (ethylene-co-vinyl acetate) and POE (polyolefin elastomers) used by the solar industry are specially formulated to become thermosetting polymers.

Over the years, EVA has become the best choice for the industry, because—without being an optimal material–it combines overall satisfactory properties at a reasonable cost and a long-track record of deployment in the field.

As of today, in fact, over 90% of the solar modules manufactured and deployed globally contain EVA as encapsulant. On the other hand, the attempt to shorten process times [6] and the current interest in glass/glass structures (for bifacial modules) is opening up doors to the adoption of alternative materials. Some market analysts see promising trends for polyolefins (as an alternative material to EVA) with an estimated market share of ~30% of the market in 2030.

One drawback of EVA in fact is that, upon exposure to UV (in combination with heat and moisture), it tends to photo-degrade, generating acetic acid. This acid will reduce the transparency of the encapsulant and may lead to the corrosion of the interconnects (ribbons and fingers) and of the solar cell. Due to the "breathability" of polymer backsheets, conventional glass/foil structures tend to release acetic acid, partly mitigating the potential impact of acetic acid generation.

In more solid sealing structures (i.e. less permeable structures) such as glass/glass ones, acetic acid will remain trapped in the inner structure potentially posing a serious threat to the long-term performance of the modules. For glass/glass modules, therefore, the replacement of EVA by polyolefins (or other encapsulants) can potentially lead to longer service life-times.

Additionally, in a small market share (<5%) of modules, PVB (polyvinylbutyral) is employed as encapsulant material. PVB is a material used to manufacture safety glass and is widely employed in the automotive and building industry [7]. Because of its mechanical stability and safety properties, PVB is mainly used to manufacture BIPV (Building Integrated PV) modules consisting of glass/glass structures.

Over the long run, the resistance to weathering and the stability of an encapsulant largely depend on the additives used in its formulation. If we focus on EVA, as an example, several additives are added to the base material or resin. These include:

a. Thermally-activated *peroxydes* used to promote cross-linking at elevated temperatures;
b. *Silanes* used to promote adhesion between EVA and inorganic surfaces such as glass;
c. *UV absorbers* that reduce the effects of UV radiation;
d. *Antioxidants*.

Over the last thirty years, the formulation of EVA (and of other encapsulants) has improved considerably. Consequently, the stability of these polymers has increased, especially with regards to problems of discoloration (yellowing or browning), photo-thermal degradation, hydrolysis and reduced adhesion.

When selecting a polymer foil as an encapsulant for the manufacturing of solar modules, the more relevant properties are:

1. **Water Vapor Transmission Rates (WVTR, [g/m²/day])**: it gives a measure of the ability of a polymer to prevent moisture ingress into the module sandwich. Typical encapsulants do not entirely prevent water vapour ingress, but delay it. On the other hand, a strong long-term adhesion to the solar cells, glass or other materials is more important than low moisture ingress, as strong adhesion will prevent the accumulation of water, avoiding corrosion and loss of electrical insulation.
2. **Electrical volume resistivity (ρ, [Ω cm])**: it is an intrinsic property of a material quantifying how the material opposes the flow of an electrical current when exposed to an electric field. Relatively low resistance values in encapsulant materials have been linked to high leakage currents and to the occurrence of potential-induced degradation (PID) and electro-chemical corrosion of transparent conductive layers in thin-film technologies. The resistivity of a polymer decreases at higher temperatures or if moisture is trapped in the polymer. Materials with a high ρ, such as Ionomer, some polyolefins (PO) or high-ρ EVA have been shown to prevent or strongly reduce the insurgence of PID [5, 8].
3. **Optical transmittance (T, [%])**: a high optical transmittance of the polymer is required to avoid a reduction in efficiency of the solar cell. To ensure a high optical transmittance of the polymer over time, (the latter can be seriously affected by yellowing or browning), the UV stability of the encapsulant is critical. As previously mentioned, in fact, glass will generally block most UV-B radiation, but only a part of UV-A radiation.

Other relevant mechanical properties are the adhesion to a foreign substrate (glass or foil) and the polymer stiffness. Table 9.1 provides a schematic overview of some of these properties for the different classes of encapsulants mostly used in the solar industry. Data are taken from the literature [9, 10] and from manufacturers' data-sheets and do not pretend to be exhaustive. Additionally, it should be noted that, within one specific class, these parameters can vary considerably (sometimes even over order of magnitudes, as in the case of resistivity) based on the exact formulation of the polymer.

Table 9.1 Overview of the most relevant properties for a given class of encapsulants

Encapsulant (thickness 0.45–0.5 mm)	WVTR (g/m²/day)	Water absorption (%)	Volume resistivity at 23 °C (Ω cm)	Light transmittance (%)
EVA	15–25	0.2–0.3	10^{13}–10^{14}	>92
PVB	20–25	0.4–0.5	~10^{12}	>92
POE/TPO	≤5	<0.1	10^{14}–10^{17}	>92
Ionomer	≤1	<0.1	~5×10^{16}	>94

Data are taken from the literature and manufacturers' data-sheets. Within a specific class of materials, these parameters can vary considerably based on the exact formulation of the polymer

9.2.2.4 Other Module Components

Edge seal

In conventional glass/backsheet structures, moisture will diffuse through the permeable backsheet and equilibrate with the external environment in a time frame of even less than a year [9]. In more solid structures (glass/glass or foils with metal barriers), water can still diffuse in from the edges, depending on the diffusivity of the encapsulant and on the specific climatic conditions. A hot and humid climate will promote the diffusion of water vapour.

For cell materials that are extremely sensitive to moisture (as CIGS, transparent conductive oxides, etc.), one can prevent the diffusion of moisture into the module sandwich by using edge seals with an extremely low permeability, such as polyisobutylene (PIB) filled with desiccants. Figure 9.7 shows the results of power variation of a solar module after exposure to a damp-heat (DH) test: 85 °C, 85% relative humidity (RH). The example is for CIGS (a material extremely sensitive to moisture), for glass/glass mini-modules laminated with EVA, with and without an edge sealant. After 1000 h (i.e. 41 days) in damp heat (DH), the unsealed samples start to exhibit a pronounced decrease in performance. The sealed sample exhibits an excellent stability up to 5500 h [11].

Frame

The mechanical stability of solar modules is improved by employing an anodized aluminium frame that is generally used to mount the modules to fixed racks or

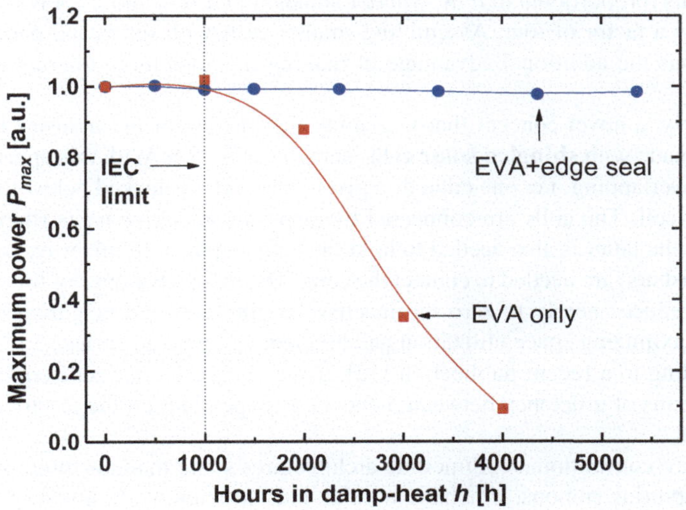

Fig. 9.7 Power variation after exposure to damp-heat (85 °C, 85% RH) test for CIGS glass/glass mini-modules laminated with EVA with and without edge sealant. See [11]

trackers. Glass/glass modules, having a more rigid structure, can be made unframed. Clamps are then used to fix the modules to the mounting structures. Alternatively, modules may use back-rails as fixation systems.

Junction box
A junction box is generally attached at the back (or edge) of the module. Inside the junction box, the internal module circuit is bonded to the wires that are used to connect—through special connectors—the module in series with other modules. Junction boxes generally accommodate one or more by-pass diodes as described in Sect. 9.1.4.

9.2.3 Advanced Module Concepts

Advanced modules concepts generally target (i) increased module performance by reducing Cell-to-Module (CTM) losses, (ii) increased energy-yield and/or (iii) increased reliability.

9.2.3.1 Increased Module Performance

A concept that is slowly becoming popular is using cells cut in half (**half-cell modules**) to reduce cell interconnection losses. In this way, the current of each module is reduced by a factor of two and the power losses due to resistive losses and interconnections (proportional to $I^2 \cdot R$, where I stands for current and R for resistance) is reduced by a factor of four. Assembling smaller cells with series and parallel connections has the additional advantage of making modules more tolerant to partial shading.

Similarly, a novel concept that is gaining the interest of the industry is that of modules made with **shingled solar cells**, shown in Fig. 9.8. With this approach, the cells are overlapping, i.e. one edge of a given solar cell is stacked below that of the next solar cell. The cells are connected by using a *conductive paste* (they are not soldered); the latter is also needed to hold the cells in place. In this way, no ribbons (and no busbars) are needed to contact the cells. The main advantage of this structure is that it reduces nearly to zero the inactive spacing between neighbouring cells, thereby maximizing space utilization and efficiency of a given surface.

According to a recent publication [12], using shingled solar cells could lead to gains in terms of efficiency between 5 and 12%, depending on the electrical layout of the cells.

For more conventional cell/module architectures some manufacturers make use of light capturing ribbons, which are not flat, but "structured". In this way, the light impinging over the ribbons is reflected and scattered at different angles, increasing the chances of total internal reflection in the sandwich structure, and the collection of light by the neighbouring cells. With this approach, shading losses due to the

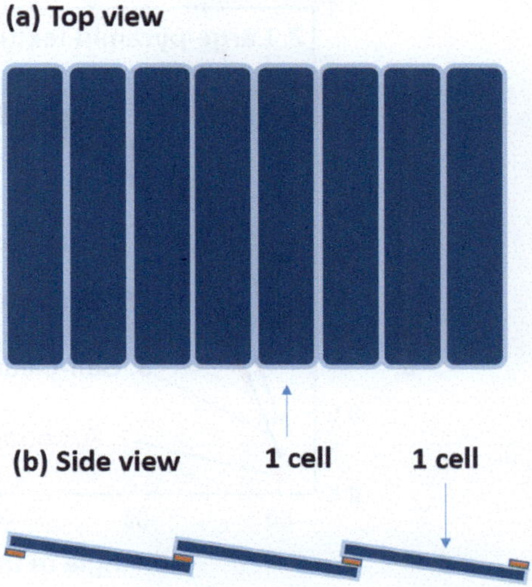

Fig. 9.8 Shingled solar cells: **a** top view of ten shingled solar cells in series, and **b** side view of three cells in series

presence of ribbons/busbars over the cell surface, which may account for up to 3%, are considerably reduced. The use of light capturing ribbons can lead to a gain in I_{sc} (and power) of +1.5% [13].

9.2.3.2 Increased Energy Yield

To increase the module's energy yield several approaches exist: one of the approaches is to apply **anti-reflection coatings (ARC)** on the front surface of the front glass, to reduce reflections at the front glass/air interface. Another approach is the use of a front *textured glass*, which increases the collection of light at low angles.

The effect of these two approaches is given in Fig. 9.9, which shows the angular dependence of the performance of various solar modules: one module with a standard flat front glass manufacture, two modules with textured glasses (with fine and large pyramids), a module with a self-cleaning glass, and, finally, a module with an AR coating on the glass.

The application of ARC or the use of textured glasses may increase the annual energy yield by 2–4%. On the other hand, textured glass can promote the adhesion of particles or dust and can become more prone to soiling losses. ARC or self-cleaning coatings may easily be scratched or removed during cleaning or field exposure.

A novel approach to increase the energy yield of modules, presently witnessing a considerable interest in the market, is that of using **bifacial cells (bifacial modules)**. In this way, the sunlight reflected by the ground can be collected by the rear side of a solar cell, increasing the module's efficiency (by +10–20%) and the annual energy yield of the module (by +10–30%). Compared to a mono-facial module, the

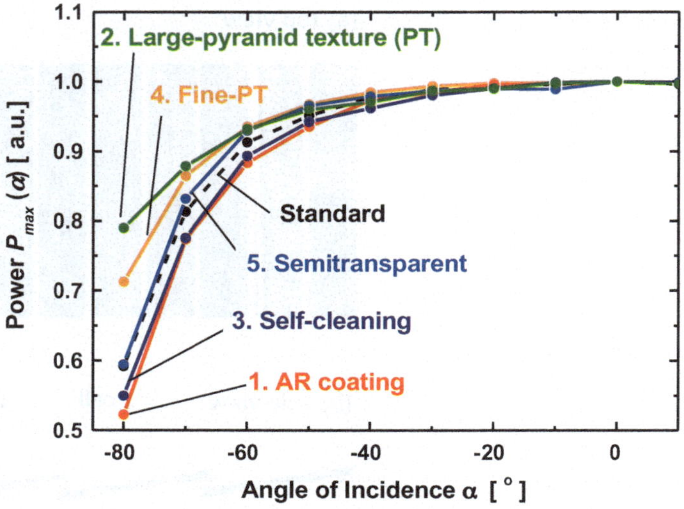

Fig. 9.9 Angular dependence of the performance of CIGS mini-modules manufactured with a standard flat front glass, two textured glasses (with fine and large pyramids), a glass coated with a self-cleaning coating and one with an anti-reflecting (AR) coating. See [14]

increased energy yield of a bifacial module will depend on the ground albedo and on possible rear shading from mounting structures (see Chap. 10).

9.2.3.3 Increased Reliability

Conventional high quality **glass/foil** modules have demonstrated service lifetime of 25 years or more. On the other hand, **glass/glass** structures may offer an additional protection for solar cells in harsh environments, particularly in locations with the presence of high mechanical solicitations that can originate from heavy snow, hail or wind loads.[5]

To minimize, in the long, run the impact of cracks on the performance of cells/modules, two other promising concepts are: **smart-wire technology (SWT)** and **multi-ribbon (or multi-wire) technology** [15]. In both cases the conventional flat ribbons (and busbars) of the cells are replaced by round wires with small diameters (300–400 μm, slightly lower than the thickness of the encaspulant) that are directly joined to the silver fingers of the cell. In these approaches, the ribbons (3 to 6) are replaced directly by 20 or more wires. An increased number of current-carrying ribbons reduces the current distribution per wire: this allows us to work with wires of lower conductance (i.e. both the diameter and the conductivity of the wire material can be reduced). Thus, silver consumption can be reduced, with a cost-saving

[5]Providing additionally an increased resistance to the generation of micro-cracks that may originate during the transport, handling, and installation phases.

Fig. 9.10 Two encapsulated solar cells: **a** c-Si solar cells with 2BBs and ribbons (left), and **b** busbar-less solar cell with no ribbons and with smart-wire technology (right)

potential at the cell level. Additionally, the use of multi-wire concepts, may lead to a considerable reduction of the sensitivity of cell/module performance to cell cracks or breakage, thereby increasing the durability of the modules (Fig. 9.10).

9.2.3.4 Modules with Integrated Inverters (Micro-inverters)

In general, strings of multiple solar PV modules are connected to an **inverter**, which converts direct current generated by the module to alternating current (i.e. a DC/AC converter—see Chap. 12). Particularly in the USA, and for some specific applications (such as small-scale installations in buildings) there is a tendency to integrate the Inverters, in the form of **micro-inverters**, directly into the module. With a micro-inverter a single panel can feed its output directly to the grid. Micro-inverters can be installed adjacent to a module (attached to the frame or replacing the junction box) or integrated directly into the module.

The main advantage of using micro-inverters is that they electrically isolate the modules from each other, so that in the case of module failure or shading, the output of the full string is affected only in a limited manner.

With a conventional string or central inverter, in fact, the complete or partial shading of a single panel may reduce the output of the full string to zero, with a considerable impact on the overall plant performance. For this reason, micro-inverters may be very beneficial in solar plants that are affected by persistent or frequent shading. As for example in the built environment, where the presence of several objects (antennas, chimneys, adjacent buildings, etc.) can cast shadows on the solar plant.

Compared to conventional inverters and electrical layouts, the use of micro-inverters requires a higher capital expenditure per watt-peak. Service and maintenance (including replacement costs) may also be higher.

9.2.4 Module Manufacturing Processes

The typical manufacturing line for c-Si modules (sometimes called *backend*, to differentiate from the *frontend* where the solar cells are manufactured) is generally composed of the following major processing steps (and of the corresponding equipment), which are, at present, in most cases fully automated:

1. **Pre-stringing**: where the glass is unpacked, cleaned in a glass washer and dried. All other materials (including cells and encapsulants) are prepared and inspected.
2. **Stringing**: a stringer is used to assemble and connect solar cells in a string. In this step the ribbons are soldered (or glued) to the busbars of the cell. Proper soldering and high adhesion of the ribbons is critical in ensuring the long-term performance of the cells/modules. Electro-luminescence (EL) imaging can be performed to verify the integrity of the solar cells after the stringing process. In thin-film manufacturing lines, the interconnection of cells is generally achieved by three sequential laser or mechanical patterning steps (*monolithical interconnection*, see Sect. 9.1.1).
3. **Lay-up station**: the module sandwich is formed by embedding the interconnected strings into two layers of encapsulants, and into the front and rear covers.
4. **Lamination**: this process is generally performed using a flatbed single-membrane vacuum laminator. The typical *pressure-temperature temporal profile* for EVA is represented in Fig. 9.11. A short *evacuation* (or *degassing*) step (1–2 min) is used to evacuate the air trapped within the different layers composing the module sandwich. The sandwich is then laid onto a heating-plate inside the laminator, and the module temperature increases in approximately 3 min to 50–60 °C (*pre-heating*). At this temperature, EVA generally softens and the **curing** step can start. A pressure is applied to the sandwich trough a membrane to avoid

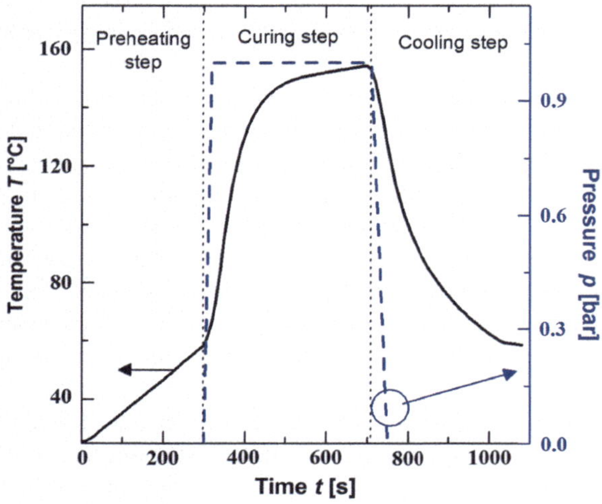

Fig. 9.11 Typical temperature-pressure (*T-p*) temporal profile used in a standard lamination process of solar modules containing EVA as encapsulant material. Three main processing steps are illustrated: (1) preheating, (2) curing, and (3) cooling (optional)

the formation of bubbles and voids and to prevent the misalignment of the cells. The real *curing* (i.e. the polymer cross-linking) of the encapsulant generally takes place at temperatures between 130 and 160 °C, during a few minutes. The membrane is then removed and the lamination chamber vented, by letting air enter the chamber. An optional *cooling step* can be applied to attenuate mechanical stress that may arise due to the mismatch in the **thermal expansions coefficients** of the different components of the module. Thereby the durability properties (and for some polymers the optical transmittance, too) of the module are improved. Overall, the lamination process can take between 10 and 20 min. Given the relatively high operating temperatures of the heating plate (145–160°), the energy consumption of the process can be significant. The quality of the lamination step is critical in ensuring long-term performance and durability of the modules. The lamination process parameters (temperature, pressure, time and heating ramps) need to be carefully optimized and tailored to the polymer used. On the other hand, the most critical quality parameter for a laminator is having a homogeneous heat distribution over the heating plate. In a high-quality laminator, temperature homogeneity fluctuations should be constrained to $\pm 2\%$ at most. An uneven temperature distribution may lead to different levels of cross-linking of the EVA, which, in the long-term, could lead to lack of adhesion and delamination in parts of the module.

5. **Finishing**: after the lamination, polymer material in excess is removed from the edges (*trimming*), an anodized aluminium frame is applied to the edges of framed modules (*framing*), and the electrical terminations of the modules are soldered to a junction box. Modules in a glass/foil structure are usually framed. Glass/glass modules, which generally have a superior mechanical stability, can be manufactured unframed.
6. **Quality inspection**: in-line routine quality checks generally include power measurements at STC (1000 W/m^2, 25 °C), low-irradiance performance (200 W/m^2, optional), and electro-luminescence (EL) imaging.

9.3 Module Testing, Reliability and Lifetime

9.3.1 Electrical Performance

In a similar way as for individual solar cells, the performance of modules is characterized by the short-circuit current I_{sc}, by the open-circuit voltage V_{oc}, by the current at maximum power point (MPP) I_{mpp}, by the voltage at MPP V_{mpp}, by the Fill Factor FF, by the power P_{max}, and by the efficiency η measured at Standard Test Conditions (STC = 1000 W/m^2, AM1.5, 25 °C). As modules are sold in terms of €/W$_p$ (or \$/W$_p$), a correct STC (i.e. W$_p$ Watt-peak) characterization of the module power (*power rating*) is critical, as it can be affected by several errors and measurement artifacts. These include spectral mismatch errors (deviations from the reference

AM1.5 spectrum), temperature and irradiance deviations, and slow signal-response of the device under test (e.g. *capacitive effects*) that can seriously affect specifically modules consisting of high-efficiency c-Si solar cells [16, 17].

The actual energy yield of a solar module in real operating conditions is influenced by several other parameters that can be determined in a laboratory - under controlled laboratory conditions—or outdoors. These include the module's angle-of-incidence (AOI) response (see Fig. 9.9), the low-irradiance performance, the spectral response and the temperature behaviour (temperature coefficients), shown in Figs. 9.12 and 9.13, respectively, for different technologies. Due to the fact that different PV technologies have different **temperature coefficients** (*TC*), the loss of power due to higher operating temperatures—compared to STC conditions—will be more pronounced for some technologies (e.g. conventional Al-BSF crystalline silicon) and less for others (amorphous silicon or CdTe). Temperature coefficients for solar cells are listed in Sect. 3.5.2.[6]

Figure 9.13 shows the **spectral response** of the same technologies. The spectral response is a measure of the spectral sensitivity of a given technology to the spectral distribution of the light and is useful in determining how the outdoor performance

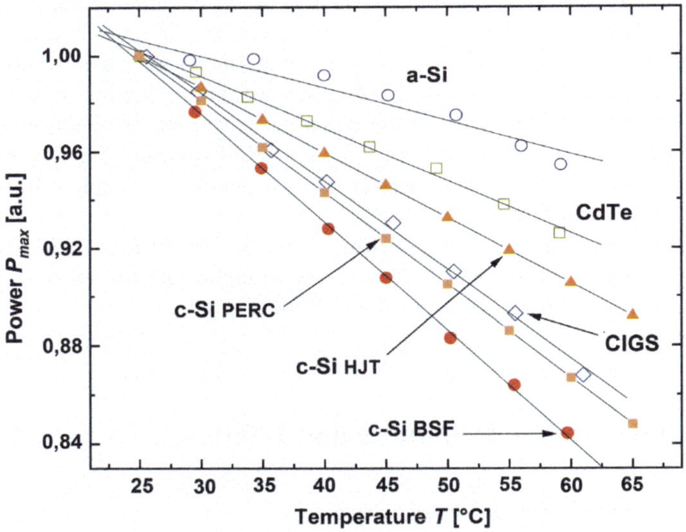

Fig. 9.12 Module maximum power $P_{\max}$ as a function of the temperature for different PV technologies. Empty symbols refer to thin film technologies: amorphous silicon (a-Si), CdTe and $Cu(In,Ga)Se_2$ (CIGS). Full symbol refer to wafer-based crystalline silicon (c-Si) technologies (BSF: back surface field, PERC: passivated emitter rear cell, HJT: hetero-junction technology). A linear fit to the data provides the power temperature coefficient *TC* of the modules. See [18]

[6]Temperature coefficients for solar cells and modules may differ to some extent in absolute terms. This because of the contributions of the interconnections, increased series resistance, etc. to the temperature coefficients measured for solar modules.

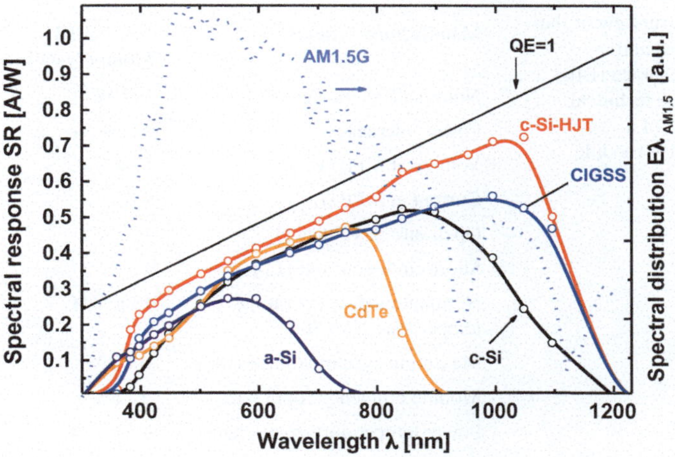

Fig. 9.13 Spectral response measurements (solid lines) of large-area commercial modules (c-Si, a-Si, CdTe, CIGSS or Cu(In,Ga)(Se,S)$_2$). Besides conventional crystalline silicon (c-Si), the SR of a c-Si Silicon hetero-junction module is shown too (c-Si HJT). In addition, the spectral distribution of the reference solar spectrum (AM1.5G) is shown (dashed line), together with a black solid line indicating the quantum efficiency QE. (A solar cell's **quantum efficiency** (QE) value indicates the amount of current that the cell will produce when irradiated by photons of a particular wavelength. If all (100%) incident photons are converted to electron/hole pairs, QE = 1. In reality, due to optical and recombination losses (or other loss mechanisms), QE is lower than this limit value (QE < 1)). See [17]

of a given technology will be affected by changes in the spectral content of the light (over the year or during the day) under real operating conditions [19].

The **energy yield** (and *energy rating*) of solar modules/systems will be the subject of the following Chap. 10.

The main electrical and mechanical parameters that represent the typical values found on the nameplate and technical datasheet of a commercial solar module are listed in Table 9.2.

The **Nominal Module Operating Temperature** (**NMOT**) is a parameter that provides a general indication of the operating temperature of a module in the field under specific reference conditions (i.e. an irradiance of 800 W/m^2, with a wind speed of 1 m/s, and an ambient temperature of 20 °C). It is a useful parameter for PV system designers when comparing the performance of different module designs. However, the actual operating temperature at any particular time is affected by the mounting structure, distance from ground, irradiance, wind speed, ambient temperature, and reflections and emissions from the ground and nearby objects. For accurate performance predictions, these factors need to be taken into account. Typical NMOT values lay around 47–50 °C and may be a few degrees lower for devices with a polymer/glass structure and slightly higher for glass/glass modules [20].

Table 9.2 Overview of the main electrical and mechanical characteristics as taken from the technical specifications of a commercial PV module

Manufacturer—model type	AAA-BBB Mono-crystalline, framed
Nominal STC power (P_{max})	345 W
Power tolerance	0 to +3%
Voltage at MPP (V_{mpp})	38.4 V
Current at MPP (I_{mpp})	9 A
Open-circuit voltage (V_{oc})	46.7 V
Short-circuit current (I_{sc})	9.5 A
Nominal module operating temperature NMOT	46 ± 2 °C
Maximum system voltage (DC)	1000 V
Module efficiency (optional)	17.8%
Power temperature coefficient	−0.4%/°C
Dimensions	1956 × 992 × 40 (mm)
Module area	1.94 m^2
Weight	19 kg
Maximum load	2400 Pa
Product warranty (workmanship)	10 years
Performance guarantee	80% after 25 years (linear)
Maximum system voltage (DC)	1000 V
Product certification/qualification	IEC 61215, IEC 61730

9.3.2 Module Reliability and Long-Term Performance

A solid encapsulation is needed to protect the active components of the module against aging and weathering. Recent evidence suggests that modules can reach service lifetimes of 25–30+ years [7, 21, 22]. A definition of **lifetime** is, however, arbitrary, depending on the application and on system use. A frequently used definition of lifetime for modules/systems refers to a threshold (power loss beyond a definite limit) corresponding to 80% of the initial nominal power of the device.

Module performance, in fact, declines over the years and may strongly be influenced by the general climatic conditions and by the type of installation (e.g. open-rack mounting versus full building integration). We can distinguish between:

1. **"Natural" degradation**: due to aging of the different materials/components (polymers, glass, seals, etc.) constituting the module. Degradation has different causes, which include effects of humidity, temperature, solar irradiation, UV and exposure to thermal stresses. Most often, a single degradation mechanism is promoted by a combination of causes (as for example the interplay of exposure to UV, high temperatures and humidity for encapsulant browning), rather than just one single cause.

2. **Accelerated degradation**: generally due to unexpected phenomena (e.g. potential-induced degradation (PID), snail-trails, etc.). These phenomena can be triggered by a poor selection of the materials constituting the module, by a poor module or system design, by damages occurred during the installation or transport phase, or by specific climatic conditions (heavy snow loads, high UV contents, marine environment, etc.).
3. **Sudden degradation following a catastrophic event/failure**: as for example a fire, a lighting strike, or an earthquake.

Typical **long-term annual degradation rates** have been reported (from statistical analyses of data given in the literature) to be in the order of -0.5 to -1%/y for conventional crystalline silicon modules and somewhat higher for thin film modules [7, 19, 21, 23–26]. Higher degradation rates occur in modules approaching their end of life [21, 26]. The most relevant aging and failure mechanisms for c-Si modules that may occur in the early period (so-called "infant" failures), during the middle of operating life, and at the wear out (end of life) phase will be described in Chap. 10, next up. These include:

- Effect of the environment on the surface of the module (e.g. pollution, soiling, etc.);
- Discoloration or haze of the encapsulant or glass;
- Delamination of encapsulant leading to water ingress;
- Cell interconnect corrosion, loss of adhesion, increase in series resistance;
- Wiring degradation;
- Delamination or water ingress in the junction box;
- By-pass diode failure;
- Glass breakage;
- Loose frames;
- Potential-induced degradation (PID);
- Cracked cells;
- Backsheet foil defects: cracking, brittling, burn-marks,[7] etc.

Figure 9.14 shows an example of both "*minor*" and "*major*" failures for two modules of the Tiso-10-kW solar plant connected to the grid in 1982 and operating for 35 years in a temperate climate [7, 21]. "Soft" or "minor" failures have, according to their definition, a moderate impact on performance. "Hard" or "major" failures are failures, which substantially reduce module performance; they may in parallel affect the operational safety of the module.

[7] A typical example of *burn-mark* is a localized dark colouring of the backsheet. This is generally due to localized overheating or arcing.

Fig. 9.14 Examples of a "*minor*" *failure* (left, i.e. partial yellowing of the encapsulant and possibly part of the back sheet), and of a "*major*" *failure* (right). For the latter module, a combination of multiple failure modes can be observed: delamination in proximity of the edge, leading to moisture ingress; burn marks between edge and cell; severe oxidation of contacts; loss of electrical insulation of the module. The modules are part of the Tiso-10-kW plant, which has been operating for over 35 years in a temperate climate [7, 21]

9.3.3 Accelerated-Aging Testing and Warranties

As it is not possible to wait 20+ years to replicate the failure modes observed in the field, a**ccelerated-aging testing** of PV modules has been introduced to replicate, under controlled laboratory conditions, the most common failure modes observed after many years of outdoor operation (such as those shown in Fig. 9.14); thanks to these tests, the module manufacturer can optimize his product design.

There are international standards such as IEC 61215 [27] for silicon and thin film modules, which address the long-term performance of PV modules; these standards are generally known as **qualification** (*design qualification and type approval*) **or performance standards**. They contain sequences of accelerated aging tests aimed at stressing the modules and assessing their durability in general open-air climates, most specifically in temperate climates. Electrical and mechanical safety requirements are addressed in the so-called **safety standard**: IEC 61730 [28].

Conformity to these qualification and safety standards is generally considered as a minimum quality requirement in most countries. Table 9.3 lists some of the major testing requirements addressing climatic, mechanical and electrical stresses contained in IEC 61215 (and IEC 61730) and the corresponding failure mechanisms observed in the field that these tests attempt at replicating.

Table 9.3 Selection of relevant accelerated-aging tests contained in IEC 61215 (*) and IEC 61730 (**) [27, 28] divided per different categories

Category	Test	Test conditions	Potential effects and field failure modes that the test attempts at replicating
Climatic	Addressing thermo-mechanical mismatches in the selection of materials and weak encapsulation structures		
	Thermal cycling (TC)	50/200 cycles, T: −40° to +85°	Loss of adhesion of interconnects, solder joint failures, cell cracks, glass breakage, delamination, etc.
	Humidity-freeze (HF)	10 cycles −40° to +85° 85%-RH when $T \geq 85°$	Loss of adhesion of interconnects, solder joint failures, cell cracks, glass breakage, delamination, etc.
	Damp-heat (DH)	1000 h 85 °C, 85%-*RH*	Delamination, corrosion of interconnects, EVA discoloration, degradation of TCO (in thin-film modules), etc.
	UV	15 kWh/m^2 (*) 2 × 60 kWh/m^2 (**)	Encapsulant/backsheet discoloration and browning
Mechanical	Assessing the resistance to mechanical impacts (hail/stones) and mechanical loads (snow/wind)		
	Mechanical Load (ML)	3 cycles, ±2.4 MPa	Glass and cell brakeage, loss of electrical continuity, solder joint failures, frame deformation, etc.
	Hail impact (H)	Hail-stone: Ø: ≥ 25 mm, velocity: ≥23 m/s	Glass breakage
Electrical	Electrical insulation		
	Dry insulation	(1000 V + 2x max. system voltage), 1 min	Loss of insulation in dry conditions
	Wet leakage	Max system voltage, 2 min	Loss of insulation in wet conditions

T temperature, *RH* relative humidity, Ø diameter

The lifetime, degradation or occurrence of failure modes for PV modules depends upon the severity of the climate, where the module is installed; it also depends on the type of installation. Increased **climatic stress** (e.g. higher temperatures, high levels of humidity, increased exposure to UV, etc.) may, in fact, be responsible for certain degradation modes, which are not specifically addressed in the existing qualification standards that explicitly refer to the general conditions encountered in temperate climates. In a similar way, as of today, the existing IEC standards do not differentiate between **different types of installations** (e.g. open-rack mounting versus building integration)—this means (as an example) that they do not take into account the higher operating temperatures prevailing in modules fully integrated into the envelope of a building.

These, and other considerations, have led over time to **criticism** related to existing qualification standards. However, one generally admits that these test sequences have been quite successful in identifying and eliminating module types suffering from large degradation rates early in their lifetime. Therefore, the existing standards have certainly contributed to an overall more sustainable diffusion of solar electricity.

Presently, module manufacturers generally compliment the basic IEC 61215/61730 certification testing with additional testing sequences. A typical example is testing according to IEC 62804 [29] for Potential Induced degradation (PID, see Chap. 10), to IEC 61701 [30] for salt-mist corrosion testing, and to IEC 62716 [31] for ammonia-corrosion testing.[8]

To overcome these limitations, **extended durability testing** (based on IEC) have been designed and proposed by several research groups and certification bodies. They include increasing stress levels and duration (time periods and number of cycles), multiple combined stress factors (e.g. humidity and UV or humidity and high applied voltage), and in some cases newly designed tests (e.g. dynamic mechanical load). It should be noted, however, that, as of today, no consensus exists among industrial players or certification bodies on the effectiveness of such extended testing.

Manufacturers use accelerated-aging testing on modules to release **warranties** for their products. Terms and conditions of warranties are always manufacturer-specific and usually cover two different aspects:

(a) **Defects in materials and/or workmanship**, usually with a duration of 5–10 years. Ten years is increasingly becoming an industry standard and should presently be considered as a minimum requirement.

[8]The salt-mist corrosion test is recommended if the modules are to be operated in coastal areas. Similarly, the ammonia corrosion test is recommended if the modules are to be operated in rural areas in proximity to livestock, as in farming areas.

(b) **Performance (i.e. power) degradation**. Typical threshold values are, respectively, minimum peak powers ≥90% of the initial nameplate value at 10 years, and ≥80% at 20–25 years. Some manufacturers, as a means for product differentiation, go beyond this and offer clauses on the trajectories of performance losses (e.g. stepwise versus linear degradation). As of today, some manufacturers are considering to offer performance warranties of 30 (or 30+) years.

References

1. M. Mittag, M. Ebert, Systematic PV-module optimization with the cell-to-module (CTM) analysis software "SmartCalc.CTM". Photovolt. Int. **36**, 97–104
2. H. Hanifi, C. Pfau, D. Dassler, S. Schindler, J. Schneider, M. Turek, J. Bagdahn, Investigation of cell-to-module (CTM) ratios of PV modules by analysis of loss and gain mechanisms. PV-Tech Mag. 90–98
3. International Technology Roadmap for Photovoltaic (ITRPV)—Results 2017, 9th edn, Mar 2018. www.itrpv.net/Reports/Downloads
4. A. Virtuani, E. Annigoni, C. Ballif, One-type-fits-all-systems: Strategies for preventing potential-induced degradation in crystalline silicon solar photovoltaic modules. Prog. Photovolt. **27**(1), 13–21 (2019)
5. G. Cattaneo, F. Galliano, V. Chapuis, H.Y. Li, C. Schlumpf, A. Faes, T. Söderström, Y. Yao, R. Grischke, M. Gragert, C. Ballif, L.E. Perret-Aebi, Encapsulants characterization for novel photovoltaic module design, in Proceedings of the 29th EUPVSEC conference (Amsterdam, 2014), pp. 147–151
6. E. Annigoni, A. Virtuani, M. Caccivio, G. Friesen, D. Chianese, C. Ballif, 35 years of photovoltaics: analysis of the TISO-10-kW solar plant, lessons learnt in safety and performance—Part 2. Prog. Photovolt. Res. Appl. 1–19 (2019)
7. W. Luo, Y. Sheng Khoo, P. Hacke, V. Naumann, D. Lausch, S.P. Harvey, J. Prakash Singh, J. Chai, Y. Wang, A.G. Aberle, S. Ramakrishna, Potential-induced degradation in photovoltaic modules: a critical review. Energy Environ. Sci. **10**, 43–68
8. Kempe M., "Modelling of rates of moisture ingress into photovoltaic modules", (2006), Sol. En. Mat. Sol. Cells, 90, 2720–2738
9. M. Kempe, Overview of scientific issues involved in selection of polymers for PV applications. Conference records of 37th IEEE PVSC conference, pp. 85–90 (2011)
10. A. Virtuani, L. Manni, D. Pic, J.L. Allary, E. Saucedo, A comprehensive protocol for assessing the quality of the lamination process of thin film glass/glass modules. In: Proceedings of the 28th EUPVSEC conference (Hamburg, 2015), pp. 2444–2448
11. A. Mondon, N. Klasen, E. Fokuhl, M. Mittag, M. Heinrich, H. Wirth, Comparison of layouts for shingled bifacial Pv modules in terms of power output, cell-to-module ratio and bifaciality. In: Proceedings of the 35th EUPVSEC conference (2018, Brussels), pp. 1006–1010
12. W. Muehleisen, L. Neumaier, C. Hirschl, T. Maier, M. Schwark, S. Seufzer, R. Battistutti, M. Pedevilla, J. Scheurer, R. Lorenz, Comparison of output power for solar cells with standard and structured ribbons. EPJ Photovolt. **7**, 70701 (2016)
13. A. Faes, M. Despeisse, J. Levrat, J. Champliaud, N. Badel, M. Kiaee, T. Söderström, Y. Yao, R. Grischke, M. Gragert, J. Ufheil, P. Papet, B. Strahm, G. Cattaneo, J. Cattin, Y. Baumgartner, A. Hessler-Wyser, C. Ballif, SmartWire solar cell interconnection technology. In: Proceedings of the 29th EUPVSEC conference (Amsterdam, 2014), pp. 2555–2561
14. E. Dunlop, F. Fabero, G. Friesen, W. Herrmann, J. Hohl-Ebinger, H.D. Mohring, H. Müllejans, A. Virtuani, W. Warta, W. Zaaiman, Z. Zamini, Guidelines for Pv power measurements in

industry (performance fp6 integrated project). EUR Sci. Tech. Res. Ser. (2010). ISBN 978–92-79-15780-6. https://doi.org/10.2788/90247, *publications.jrc.ec.europa.eu/repository/.../eur-24359-en.pdf*
15. A. Virtuani, H. Müllejans, E. Dunlop, Comparison of indoor and outdoor performance measurements of recent commercially available photovoltaic solar modules. Prog. Photovolt. Res. Appl. **19**, 11–20 (2011)
16. A. Virtuani, D. Stepparava, G. Friesen, A simple approach to model the performance of photovoltaic solar modules in operation. Sol. Energy **120**, 439–449 (2015)
17. T. Sample, A. Virtuani, Modification to the standard reference environment for nominal operating cell temperature (NOCT) to account for building integration. In: Proceedings of the 24th EUPVSEC conference (Hamburg, 2009), pp. 3332–3337
18. A. Virtuani, M. Caccivio, E. Annigoni, G. Friesen, D. Chianese, C. Ballif, T. Sample, 35 years of photovoltaics: analysis of the TISO-10-kW solar plant, lessons learnt in safety and performance—Part 1. Prog. Photovolt. Res. Appl. **27**, 328–339 (2019)
19. A. Skoczek, T. Sample, E.D. Dunlop, H.A. Ossenbrink, The results of performance measurement of field-aged crystalline silicon photovoltaic modules. Prog. Photovolt. Res. Appl. **17**, 227–240 (2009)
20. D.C. Jordan, S.R. Kurtz, Photovoltaic degradation rates—analytical review. Prog. Photovolt. Res. Appl. **21**(1), 12–29 (2011)
21. D.C. Jordan, S.R. Kurtz, K. VanSant, J. Newmiller, Compendium of photo-voltaic degradation rates. Prog. Photovolt. Res. Appl. **24**(7), 978–989 (2016)
22. D.C. Jordan, T.J. Silverman, B. Sekulic, S.R. Kurtz, PV degradation curves: non-linearities and failure modes. Prog. Photovolt. Res. Appl. **25**(7), 583–591 (2017)
23. D.C. Jordan, T.J. Silverman, J.H. Wohlgemuth, S.R. Kurtz, K.T. VanSant, Photovoltaic failure and degradation modes. Prog. Photovolt. Res. Appl. **25**(4), 318–326 (2017)
24. International Electro-technical Commission (IEC), IEC 61215:2016 Series, Terrestrial photovoltaic (PV) modules—Design qualification and type approva. IEC Geneva
25. International Electro-technical Commission (IEC), IEC 61730:2016 Parts 1 and 2 "Photovoltaic (PV) module safety qualification". IEC Geneva
26. International Electro-technical Commission (IEC), IEC 62804-1:2015 Ed. 1 "Photovoltaic (PV) modules—Test methods for the detection of potential-induced degradation—Part 1: Crystalline silicon". IEC Geneva
27. International Electro-technical Commission (IEC), IEC 61701:2011 Ed. 1 "Salt mist corrosion testing of photovoltaic (PV) modules". IEC Geneva
28. International Electro-technical Commission (IEC), IEC 62716:2013 Ed. 1, "Photovoltaic (PV) modules—Ammonia corrosion testing". IEC Geneva
29. A. Virtuani, Optimisation of $Cu(In,Ga)Se_2$ thin film solar cells and modules for low-irradiance conditions. Ph.d. thesis (2006, Shaker Verlag)
30. A. Virtuani, H. Vogt, P.C. Mogensen, S. Auvray, D. Dupuy, E. Mahe, Modified visual appearance of $Cu(In,Ga)(Se,S)_2$ thin film solar modules for building integrated photovoltaics. In: Proceedings of the 25th EUPVSEC conference (Valencia, 2010), pp. 4248–4252
31. A. Virtuani, D. Pavanello, G. Friesen, Overview of temperature coefficients of different thin film photovoltaic technologies. In: Proceedings of the 22nd EUPVSEC conference (Milan, 2007)

Alessandro Virtuani is a senior researcher at the PV-Lab of EPFL (Neuchâtel) where he is responsible for the *PV module and reliability* group. He is co-founder of a start-up *Officina del Sole* (www.o-sole.eu), which offers testing/consulting services in the field of PV. His research interest covers performance/reliability testing, development of innovative module concepts, energy yield modelling of PV devices/systems, Building-Integrated PV (BIPV), and technology transfer to industry. In 2016, he has served as technical advisor to IFC (*World Bank group*) with a project aimed at promoting a sustainable diffusion of rooftop solar systems in Sub-Saharan Africa. Alessandro has worked at ESTI (Joint Research Centre of the European Commission), the European centre of reference for PV testing. He is a physicist with a PhD in Materials Sciences (University of Milano-Bicocca, 2004), obtained carrying out research activity on the optimization of CIGS thin film solar cells/modules at ZSW (Stuttgart). In 2014, he obtained a Master's Degree in Business Administration from SDA-Bocconi. Alessandro is the inventor of the *DragonBack* measurement method for the testing of high-efficient silicon devices, and is a member of the ETIP-PV (European Innovation and Technology PV Platform).

Chapter 10
Module Deployment and Energy Rating

Mauro Pravettoni

Abstract This chapter will explain in detail how to calculate, from the power rating of a photovoltaic module, the total energy that the module is expected to generate in one year of installation (energy rating). Afterwards, special attention will be given to cases where this expected energy yield is not achieved due to power losses and failure modes (including, but not limited to, known degradation mechanisms such as LID and PID), looking first at the time scale of these failures, then investigating how and in which cases these failures are recoverable. Finally, the tools to accurately perform the energy yield measurement via monitoring of PV plants will be introduced, to assist the plant owner in understanding when a power plant is underperforming.

10.1 Preliminary Remarks

Assume you have a 3 kW power meter and you want to get rid of your electricity bill using the energy of the Sun hitting your house roof. A friend of yours, a solar PV installer, quotes for you a good deal for a 3 kW installation and you decide to go through with it. Will it be sufficient to supply enough energy to your gadgets and appliances? «No», or maybe at this stage you are simply not able to give an answer.

Well, then look at the problem in another way: you have checked the history of your electricity bills and done some calculations, which tell you that your consumption is roughly 5 kWh per day. You live in Central Europe and you expect that, on the average, you have 9–10 sunny hours a day all over the year (well, you know: a little bit more in summer, a little bit less in winter…): that would render a 1.5 kW installation well enough to supply electricity to your gadgets and appliances all over the year, correct?

M. Pravettoni (✉)
SERIS, National University of Singapore, Singapore, Singapore
e-mail: mauro.pravettoni@nus.edu.sg

A. Shah (ed.), *Solar Cells and Modules*, Springer Series in Materials Science 301,
https://doi.org/10.1007/978-3-030-46487-5_10

Obviously, the answer is «No» again. The reason is the following: with solar PV installations the *power rating* of power plants is related to their *energy rating* by a remarkably more complex relation than the simple and "naïve" Equation

$$\text{energy} = \text{power} \times \text{time}.$$

Therefore, let us look here at all the factors that influence, in a PV plant, the relationship between power rating and energy rating.

In the previous chapter the physical parameters that influence the module power have been described, together with the ***Standard Test Conditions (STC)*** at which the module power can be rated. Power rating at STC is an important tool in comparing various products from different manufacturers. Indeed, performance at STC can be reliably assessed, even artificially indoors on solar simulators, with a high degree of reproducibility within laboratories and between different laboratories. (The differences are typically of the order of 2–3% in most laboratories worldwide.) However, STC refer to idealized conditions, which hardly reflect the variety of physical and environmental conditions that modules experience outside the laboratory in the field and which reduce the power generation during operation. Therefore, energy rating is needed to give more confidence to the estimated energy yield during the lifetime of modules within power plants.

This chapter is on module deployment in real installations and explores what are the factors that transfer the module power at STC to the energy that the module is expected to produce in one year: these factors are analysed in Sect. 10.2. However, there are at least three important cases where the calculation procedure of Sect. 10.2 cannot be applied easily, and which are worth noting: these are the cases of installations on trackers, either for flat-plane modules (1) or for concentrating photovoltaics (2), and of bifacial modules (3); these three cases are presented in Sect. 10.3. Furthermore, certain loss mechanisms occur when transferring electricity from the module to the grid. Also, certain failure modes can occur when modules are put into operation: these failure modes can be "recoverable" (either partially or fully) or "unrecoverable". Together with the loss mechanisms they will be explained in Sect. 10.4, with some examples of failure modes on modules from real installations.

Finally, Sect. 10.5 is dedicated to the tools (hardware and software tools) that may be needed to correctly perform energy rating of PV power plants.

10.2 From Power Rating to Energy Rating

In this section we will guide the reader through the various physical and environmental effects that affect energy generation of PV modules in the field. The result will be remarkably different than the naïve expectation that energy is simply the power times the duration of exposure to sunlight. Because of this, a 3 kW power plant cannot be expected to produce 3 kWh of energy for every hour of sunlight.

10.2.1 Effect of Module Orientation

The first and foremost effect that limits the energy generation of PV modules in the field is their orientation with respect to the position of the Sun. In fact, fixed installations receive light mostly from tilted angles that limit the total incident power per square meter. As a result, all over a single day the power input to the module grows from zero to the maximum irradiance at solar noon,[1] and then decreases again to zero.

In Chap. 2, the in-plane global irradiance $G(t)$, and its components (the in-plane direct irradiance $B(t)$, the in-plane diffuse irradiance $D(t)$, as well as the in-plane albedo $R(t)$) have already been defined. Let us ignore, for the moment, the diffuse component and the albedo and define the orientation of a PV module, whether in the field or on the top of a roof, as shown in Fig. 10.1a, where:

- the module *tilt* or *slope* β is the inclination[2] between the surface of the module and the horizontal plane;
- the module *azimuth* φ is the direction towards which the module is tilted: the usual convention in the PV community is to refer to as 0° azimuth the installation facing South, with positive azimuths referring to west-facing orientations.

From the definition of β it follows that a module with $\beta = 0°$ tilt lies horizontally on the ground (Fig. 10.1b) and another module with $\beta = 90°$ tilt is a vertically installed module, like on a building façade or on solar fences (Fig. 10.1c).

Let us now refer, for simplicity, to a module installed at the Equator at $\beta = 0°$ tilt, in *clear-sky* conditions[3] and at one of the *equinoxes* (i.e. when the Equator is aligned to the sun rays and the sun rays at solar noon are perfectly perpendicular to the 0° tilted module, see Fig. 10.2). The total flux of power absorbed by the PV module (the so-called *in-plane global irradiance* $G(t)$) in such conditions varies with time according to the relation

$$G(t) = S\cos\theta(t), \tag{10.1}$$

where S is the power of the sunrays per unit area, and $\theta(t)$ is the *angle of incidence* and is defined as the angle between the sunrays and a line perpendicular to the surface of the PV module (and, of course, this angle varies with the position of the Sun). The presence of the cosine of $\theta(t)$ in (10.1) (also referred to as the "cosine effect") accounts for the fact that the PV module absorbs no irradiance when $\theta = 90°$, i.e. at sunrise and sunset and $\cos\theta(t) = 0$. On the other hand, $G(t)$ is maximum at

[1] *Solar noon* denotes the time at which the sun reaches the highest point above the horizon in the precise geographical position where we are standing. It can be detected on a *sundial* and it is generally different from the *local noon*, which is the time we see on our smartphones and is set conventionally for each Time Zone, serving as a reference for the *local time*.

[2] β is the symbol for both the temperature coefficient for V_{oc} and the tilt of the module.

[3] Clear-sky condition is generally defined as the absence of visible clouds in the sky dome.

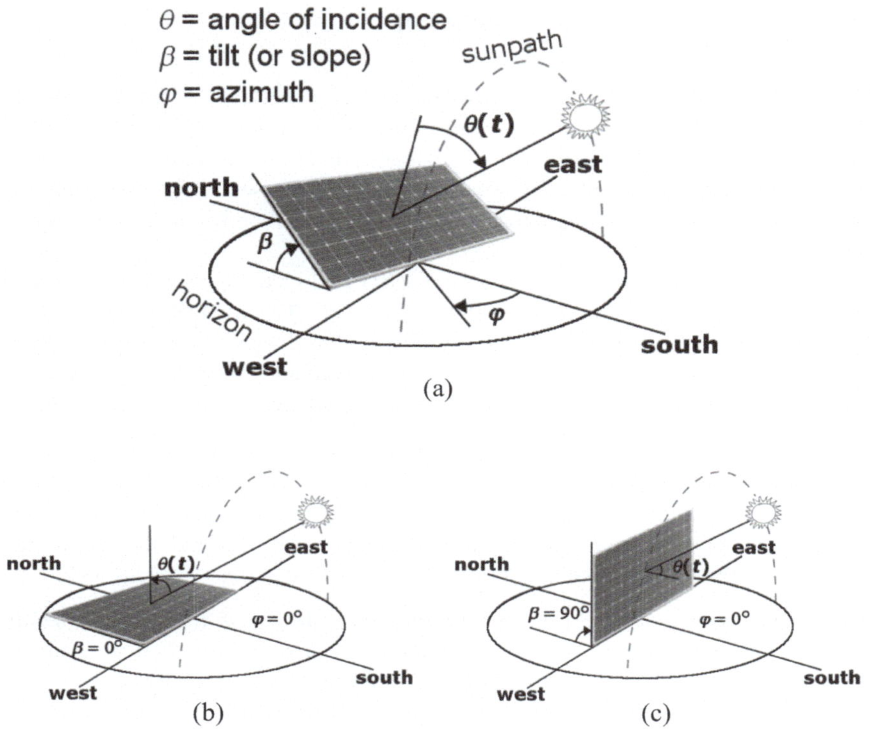

Fig. 10.1 **a** Orientation of a PV module, defined by the module *tilt* or *slope* β, the *azimuth* φ and the *angle of incidence* $\theta(t)$, a function of time, as it varies with the position of the Sun. Two special cases: **b** $\beta = 0°$, i.e. module lying horizontally on the ground; **c** $\beta = 90°$, i.e. module installed vertically

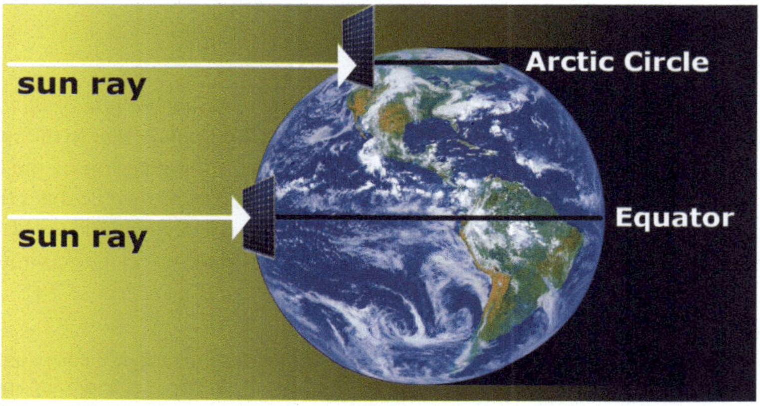

Fig. 10.2 Optimum tilt for PV modules during the equinoxes (i.e. when the axis of rotation of the Earth is perpendicular to the sun rays) at the Equator (latitude: 0°) and at the Arctic Circle (latitude: ~66° North)

solar noon, when (in our example, at the Equator during the Equinox) $\theta = 0°$ and $\cos\theta(t) = 1$.

More or less the same profile is expected at any given latitude when modules are installed at a tilt β, which is equal to the geographical latitude, as shown for example in Fig. 10.2 for the upper module installed with a $\beta = 66°$ tilt at the Arctic Circle (latitude: approximately 66°). In general the value of the maximum in-plane global irradiance varies with the seasons between a maximum and a minimum.[4]

So far we have implicitly considered PV modules installed at 0° azimuth (i.e. facing South in the Northern hemisphere or facing North in the Southern hemisphere). The correlation between the orientation and Sun height is more complicated if we consider installations with azimuth other than 0°: this is usually the case in residential installations, where modules are installed on the roof (typically laying parallel to the roof slope) and, thus, have both tilt and azimuth that depend on the roof orientation. But in general, the modules in real conditions see always varying global irradiance from 0 to about 1000 W/m^2. It is therefore important to know the performance of the module not only at 1000 W/m^2 (STC), but also at lower irradiances. In fact, no PV module is perfectly linear with irradiance: it has already been shown in previous chapters that PV modules of different technologies have different efficiencies at different irradiance levels.

Once the in-plane global irradiance as a function of time $G(t)$ (depending on the orientation as discussed above) and the module efficiency as a function of irradiance $\eta(G)$ are known, the power P of the module at time t could in principle be calculated as

$$P(t) = \eta(G)G(t)A \tag{10.2}$$

where A is the module area. Then the total energy generated would be

$$E = \sum_t P(t)\Delta t \tag{10.3}$$

where the sum is performed over time (typically over 1 year) and Δt is a reasonably short amount of time during which $P(t)$ can be assumed to be constant. But in the real practice (10.2) and (10.3) are correct only if the following assumptions hold:

- the efficiency of the module $\eta(G)$ does not vary with temperature;
- the efficiency of the module $\eta(G)$ does not vary with the angle of incidence $\theta(t)$;
- the diffuse component $D(t)$ and the albedo $R(t)$ are negligible;
- the efficiency of the module $\eta(G)$ does not vary with the spectrum of in-plane global irradiance (the so-called *spectral in-plane global irradiance*).

---

[4]Although the reader may have already argued that the *optimum tilt* is equal to the geographical latitude, this argument works only at the equinoxes: during the other seasons the optimal tilt is different and the general practice is to install modules at a tilt that is 70–90% of the latitude value.

Unfortunately, all these assumptions are not correct; therefore substantial corrections need to be applied to (10.2), to take into account the effects of temperature, of the angle of incidence (for both the direct and the diffuse components) and of the spectral mismatch between the spectral in-plane global irradiance and the STC spectrum. These effects will be discussed in the next sections.

10.2.2 *Effect of Temperature*

The effect of temperature on module efficiency is by far the most important effect. In clear-sky conditions, the temperature T_{m} of PV modules depends on both the irradiance G (the higher G, the higher T_{m}), the ambient temperature T_{a} (the higher T_{a}, the higher T_{m}) and the wind speed v (the higher v, the lower T_{m}): the relation that combines all these parameters to give T_{m} is

$$T_{\mathrm{m}} = T_{\mathrm{a}} + \frac{G}{u_0 + u_1 v}, \qquad (10.4)$$

where u_0 and u_1 are constants that can be determined experimentally and (10.4) is referred to as the *Faiman model* from the name of the researcher who first proposed it (see [1]).

With varying temperature, PV module efficiency decreases dramatically by several percent, with respect to the efficiency at STC, as explained in Chap. 9. This effect is a basic physical property of semiconductors. The theoretical details are far beyond the scope of this book, but consequently the voltage of PV modules drops significantly with temperature for practically all the PV technologies (thus the drop in efficiency), whereas the current is less affected by temperature variations.

Combining the irradiance and temperature dependence of the module efficiency η, the latter can be expressed as a function of two variables: the irradiance G and the temperature T. We obtain the graph $\eta(G, T)$ which is shown in Fig. 10.3. From this Figure the module efficiency η can be extracted for any value of irradiance G and temperature T.

10.2.3 *Effect of the Angle of Incidence*

The cosine relation of (10.1) is an idealized, simplified approximation, as at wide angles the incident light is, to a large part,[5] reflected by the surface of the modules. Optical transmittance properties vary from glass to glass, encapsulant to encapsulant and from technology to technology: in order to calculate the losses because of the

[5]The shape and dimensions of the metal grid on top the component cells can also affect light absorption at wide angles, where the metal grid can shade the active part of the cells.

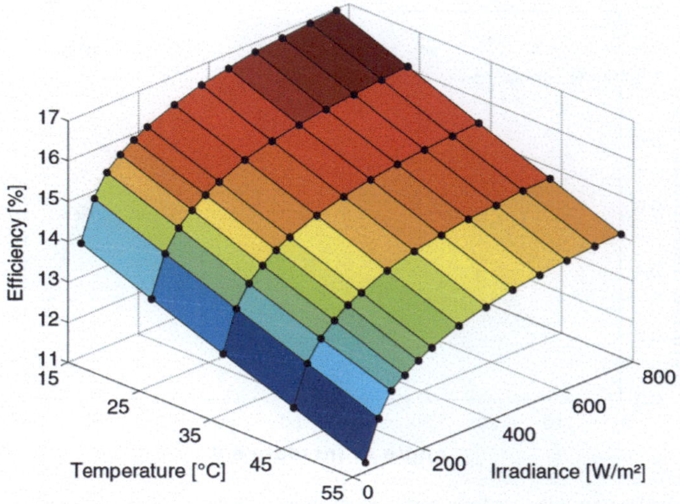

Fig. 10.3 An example of efficiency versus irradiance and temperature at the AM1.5 spectrum for a c-Si PV module; the dots indicate actual measurement points. *Source* [2]

angle of incidence, the following *angular transmittance* $\tau(\theta)$ can be introduced

$$\tau(\theta) = \frac{I_{sc}(\theta)}{I_{sc}(0)} \frac{1}{\cos\theta} \tag{10.5}$$

where $I_{sc}(\theta)$ and $I_{sc}(0)$ are the measured short-circuit currents, at the angle of incidence θ and at normal incidence, respectively. An example of $\tau(\theta)$ for c-Si is shown in Fig. 10.4.

The curve $\tau(\theta)$ is mathematically equivalent to the following Equation, proposed by Martin and Ruiz (see [3, 4]),

$$\tau(\theta) = \frac{1 - \exp\left(\frac{-\cos\theta}{a_r}\right)}{1 - \exp\left(\frac{-1}{a_r}\right)}. \tag{10.6}$$

where a_r is a dimensionless fitting parameter. The correction to the direct component of irradiance $B(t)$ is

$$B_{corr,AOI}(t) = B(t)\tau[\theta(t)]. \tag{10.7}$$

The contribution of $\tau(\theta)$ to the diffuse component is less straightforward; it obviously depends on the module orientation (via the tilt β) and has been calculated by Martin and Ruiz assuming an isotropic distribution of the diffuse irradiance. The related Equation can be found in the specialized literature (see [3, 4]).

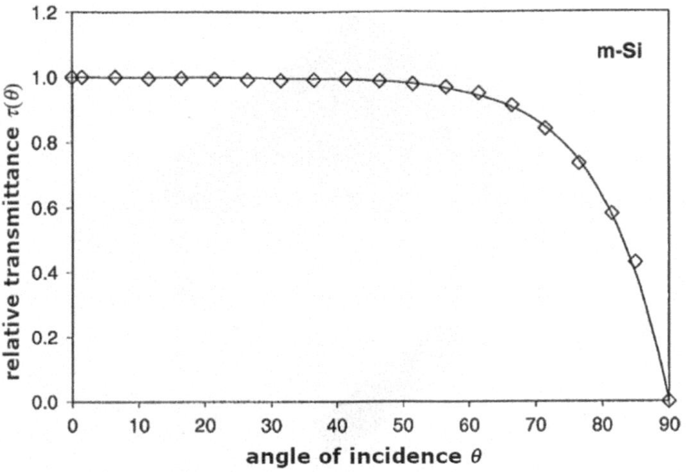

Fig. 10.4 Angular transmittance for a particular example of a c-Si module. *Source* [3]

10.2.4 Effect of Spectral Mismatch

No PV material absorbs the full spectrum of sunlight: light absorption of solar cells is limited by the characteristic bandgap of each PV technology—this determines the spectral response of the module. Spectral response can be expressed as current produced per incident power at a given wavelength (*spectral responsivity*, in A/W) or as number of photogenerated electrons that are collected at the terminals of the module per incident photon at a given wavelength (a dimensionless quantity—scientists refer to it as the *external quantum efficiency*, EQE).

When a PV module is tested at STC, the reference spectral irradiance is the AM1.5 spectrum; it is typically measured indoors on a solar simulator with a calibrated c-Si reference cell. In outdoor installations, on the other hand, the global irradiance G is usually measured with a *thermopile pyranometer* (a cavity radiometer, see Sect. 10.5.2), that is able to absorb and measure typically between 300 and 3000 nm, i.e. in the entire spectrum of sunlight and in a much wider range than that of common PV technologies.

Figure 10.5a shows the typical spectral response range (in terms of photogenerated electrons per incident photon) of a pyranometer compared to that of c-Si and CdTe. As we have already seen in Chap. 2, the global irradiance results from the combination of its direct component, its diffuse component and the albedo: but the combination of these three components on top of a PV module varies significantly all over the year and can be significantly different from the STC spectrum, even in clear-sky conditions, depending on the position of the sun, the orientation of the module and the altitude. As an example, Fig. 10.5b compares:

- the standard AM1.5 irradiance;
- the global spectral irradiance measured in Gonghe, China, on the 2nd May 2018 at 9.30 in the morning at zero tilt (also referred to as *Global Horizontal Irradiance*);

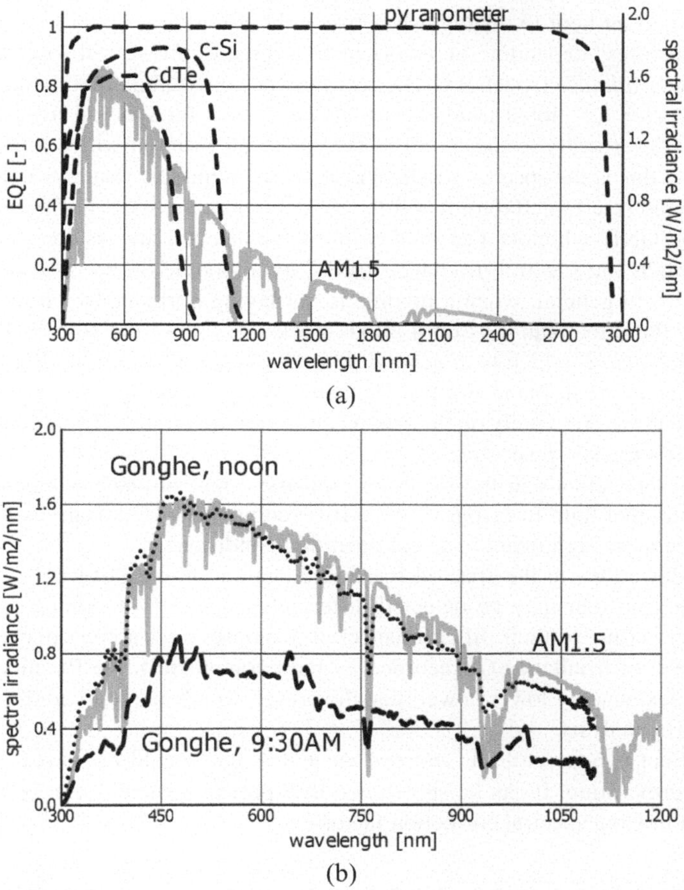

Fig. 10.5 **a** Spectral response (as external quantum efficiency, EQE, in electrons per incident photon) of c-Si and CdTe modules, and of a pyranometer, compared with the AM1.5 spectrum; **b** Comparison between AM1.5 spectral irradiance and real, field measurements. Measured spectra have been recorded at zero tilt in Gonghe, China (latitude $\phi = 36° \, 17'$N, horizontal irradiance, fixed installation), on the 2nd May 2018 at 9:30 in the morning and at solar noon

- and the spectral irradiance measured again in Gonghe at zero tilt, but at solar noon.

Both spectra were recorded in clear-sky conditions and at zero tilt, but not surprisingly, the measurement in the morning shows a larger difference from the AM1.5 spectrum, both in terms of absolute values and shape of the spectrum. The reason is the wider angle of incidence in the morning (hence, a lower irradiance due to the cosine effect) and a higher diffuse component than the direct one (hence a larger blue content in the spectral irradiance).

It is important here to distinguish between the different spectral irradiance that the module "sees" depending on its orientation. First of all we call spectral *Global Horizontal Irradiance* (spectral GHI), the irradiance measured pointing the detector (a *pyranometer* for global irradiance; a *spectrometer*, for spectral irradiance: see Sect. 10.5.2) vertically to the *zenith* of the particular location where measurement takes place: this is the spectral irradiance "seen" by a module that lays horizontally on the ground (see Fig. 10.6a). It is different from both the spectral *Global Normal Irradiance* (GNI), which is measured pointing the detector always directly towards the sun: this is the spectral irradiance "seen" by a module on a 2-axis sun tracker (Fig. 10.6b). In general, when a module is not laying horizontally on the ground, nor facing the direct sun rays on a 2-axis tracker, it "sees" the so-called *in-plane spectral irradiance*, which is in general different from both spectral GHI and GNI. (But of course, the in-plane spectral irradiance is equivalent to the spectral GHI if the module lays horizontally on the ground and to the spectral GNI when the module is on a 2-axis tracker, respectively).

Furthermore, clouds in the sky may limit the direct in-plane component B or scatter additional light in favour of the diffuse component D: deviations from STC can be even more significant in actual operating conditions.

These deviations in the spectral irradiance can cause relevant local differences in the spectrum that may be detected differently by the PV modules and by the pyranometers (due to their different spectral responses). As a consequence, under a given level of irradiance G (recorded by the pyranometer), the PV module may produce unexpectedly lower power than the power, which has been predicted based on the above analysis.

This effect is referred to as *spectral mismatch* and should be corrected for the correct energy rating. It can be shown that the spectral mismatch can be corrected with the following spectral correction factor

(a) (b)

Fig. 10.6 **a** Set of pyranometers measuring the Global Horizontal Irradiance (GHI); and **b** my friend Wim Zaaiman while aligning the 2-axis tracker for testing at Global Normal Irradiance (GNI)

$$C_s(t) = \frac{1000}{G(t)} \frac{\int_{300}^{\lambda_1} SR_{\mathrm{mod}}(\lambda) G(\lambda, t) \mathrm{d}\lambda}{\int_{300}^{\lambda_1} SR_{\mathrm{mod}}(\lambda) G_{\mathrm{AM1.5}}(\lambda) \mathrm{d}\lambda}, \tag{10.8}$$

where $SR_{\mathrm{mod}}(\lambda)$ is the spectral responsivity of the module, $G(\lambda, t)$ and $G_{\mathrm{AM1.5}}(\lambda)$ are the measured in-plane global spectral irradiance at time t and the standard AM1.5 spectral irradiance, respectively.

10.2.5 Calculation of the Annual Energy Output of a PV Module in Seven Steps

We now have all the inputs needed to perform the calculation of the annual energy output of a PV module in the field, knowing:

- the in-plane global irradiance as a function of time, $G(t)$;
- its direct component, $B(t)$;
- the in-plane global spectral irradiance as a function of time, $G(\lambda, t)$;
- the PV module temperature as a function of time, $T(t)$, which can be measured directly from the field or calculated via the Faiman formula (10.4);
- the PV module efficiency as a function of irradiance and temperature, $\eta(G, T)$;
- the fitting parameter a_r for the angular transmittance $\tau(\theta)$;
- the module tilt β;
- the spectral responsivity of the PV module, $SR_{\mathrm{mod}}(\lambda)$;
- the module area A.

The annual energy output can then be calculated with the following seven steps:

1. calculate the in-plane diffuse irradiance $D(t) = G(t) - B(t)$;
2. obtain the corrected in-plane direct and diffuse irradiances for the angle of incidence effect ($B_{\mathrm{corr,AOI}}(t)$ and $D_{\mathrm{corr,AOI}}(t)$);
3. calculate the corrected in-plane global irradiance for angle of incidence effect $G_{\mathrm{corr,AOI}}(t) = B_{\mathrm{corr,AOI}}(t) + D_{\mathrm{corr,AOI}}(t)$;
4. calculate the spectral mismatch correction factor $C_s(t)$, using (10.8);
5. calculate the spectrally corrected global in-plane irradiance $G_{\mathrm{corr,AOI+s}}(t) = C_s(t) \cdot G_{\mathrm{corr,AOI}}(t)$;
6. calculate the module power at time t, $P(t) = G_{\mathrm{corr,AOI+s}} \cdot \eta\big(G_{\mathrm{corr,AOI+s}}, T(t)\big)$;
7. calculate the annual energy output via (10.3), $E = \sum_i P(t)\Delta t$, where the sum is performed over 1 year of measurements and Δt is a reasonably short amount of time during which $P(t)$ can be considered constant ($\Delta t = 1$ hour is a good compromise for reliable results over 1 year of data).

In theory this approach works well, but requires as an input information that may be difficult to get, such as the in-plane[6] global irradiance $G(t)$, the in-plane direct irradiances $B(t)$ and the spectral irradiance $G(\lambda, t)$, all functions of time t. Section 10.5 below illustrates the needed experimental equipment to get this information, but also, in the absence of experimental data, where to find databases that contain modelled values for $G(t)$, $B(t)$ and $G(\lambda, t)$.

10.3 Three Relevant Exceptions

Before proceeding with the description of energy losses and failure mode mechanisms that limit the energy yield, as described in the previous section, three relevant cases require our attention, where the approach presented so far needs to be somewhat modified. These are the cases of module on trackers, of concentrating photovoltaic (CPV) modules and of bifacial modules: they are briefly discussed in this section. In all cases, the orientation of the modules has different characteristics from the orientation presented so far: therefore, the role played by the various in-plane irradiance components is also quite different.

10.3.1 Energy Rating for Modules on Trackers

Sometimes modules in utility-scale installations are put on trackers, to optimize the orientation and minimize the angle of incidence losses. Trackers move the modules facing the Sun around one or two axis during the day and turn the module back to their starting position at sunset: 1-axis tracking is simpler for many reasons, thus cheaper, and easier to maintain; on the other hand 2-axis tracking has a higher energy yield and is, in principle, easier to model. In this section we will refer for simplicity only to 2-axis tracking.

When PV modules are on a 2-axis tracker, the orientation is always optimal by definition and $\theta(t) = 0°$ at all times: we have already encountered this condition above and it has been defined with the acronym GNI, Global Normal Irradiance.

In this case, in clear-sky conditions the direct in-plane irradiance $B(t)$ varies less over the day than in fixed installations: this results in a much flatter $G(t)$ profile all over the day. Since it is now always $\theta(t) = 0°$, $B(t)$ does not need a correction for the angle of incidence, which is, thus, a less significant correction for modules on trackers.

[6]As introduced in the previous note, the *in-plane irradiance* is measured with the detector (a pyranometer for total irradiance, a spectrometer for spectral irradiance) placed in-plane with the PV modules. In most cases such equipment is not available on site, while meteorological data (or modelled ones) can be found on GHI and GNI.

Nevertheless, spectral mismatch correction can be relevant close to sunrise and sunset: furthermore, the temperature of the module is typically higher than in fixed installations, as $G(t)$ is higher at any time than in the case of fixed orientation.

10.3.2 Concentrating Photovoltaics (CPV) and Energy Rating

A second remarkable exception to the protocol for energy rating listed in Sect. 10.2.5 is concentrating photovoltaics (CPV), where high-efficiency PV cells are placed on the focus of an optical concentrating system, made either of Fresnel lenses or mirrors, in order to increase light intensity from 10 times the direct irradiance (low concentrations) up to 1000 times (high concentrations). The advantage is in principle twofold:

- it achieves PV material cost reduction—the relatively inexpensive optical concentrating system collects light on a 10–1000 times larger area than the area of the PV cells, which allows to use at a reasonable price the most expensive and highest efficiency PV cell technologies, such as the multi-junction alloys of III–V group cells (e.g. the triple-junction GaInP/GaAs/Ge, GaInP/GaInAs/Ge, or other alloys and higher number of junctions);
- it reaches higher efficiencies: thanks to more than 44%-efficiency cells, CPV modules can achieve efficiencies around 40% or above, thus, double those of commercial non-concentrating modules.

CPV was once referred to as the "third generation" of PV, aiming to reach high efficiency at low-cost. But the disruptive cost reduction of c-Si since the 2010s has put severe concerns on the real benefits of CPV, so that at the time of writing most market forecasts and stakeholders predict little or no future for it at all. Nevertheless, it is worthy here of mentioning where energy rating for CPV modules would differ from that of conventional flat-panel modules on fixed installations.

First of all, CPV modules need to be placed on a 2-axis tracker, to allow the direct irradiance to be collimated to the receiver: as a consequence, the diffuse component $D(t)$ is irrelevant for CPV, as it does not reach the cell, and energy rating should be calculated only based on the direct component $B(t)$.

Furthermore, since the module is on a 2-axis tracker, there is no effect due to the angle of incidence, which in principle should simplify the calculation. Another important simplification is temperature correction: since CPV cells are hit by extremely high irradiance, they would reach critically high temperatures, unless the cells were actively cooled during operation. In fact, the operating temperature of CPV cells is kept under control by active cooling and therefore the temperature correction for energy rating should be more straightforward.

But on the other hand, as CPV cells are almost certainly multi-junction cells, the spectral mismatch between the standard reference spectrum (either the standard global AM1.5 or the direct component of AM1.5 spectra) and the direct measured spectral irradiance has generally a stronger effect on energy rating than for flat-plane non-concentrating modules.

The last remarkable differences with respect to the list in Sect. 10.2.5 is that, due again to the sensitivity of CPV cells only to the direct component of irradiance, CPV systems strictly operate only in clear-sky conditions and are therefore subject to huge variations of operating irradiance depending on weather. As a consequence, CPV can be deployed only in regions where the DNI is high and weather conditions very stable all over the years, such as arid desert climates, to minimize the loss of energy production due to high diffuse component of irradiance.

10.3.3 Energy Rating for Bifacial PV Modules

So far we have always neglected the *albedo* (the component of irradiance reflected from the ground and surroundings). The reason was that at the optimal tilt, PV modules are always oriented towards the sky and the albedo is unlikely to make a significant contribution to the global in-plane irradiance on top of the module, which is during most of the sunny hours dominated by the other two components (the direct and the diffuse).

However, the albedo will make a relevant contribution to the global in-plane irradiance for bifacial modules. *Bifacial PV modules* (see [5]) are typically double-glass modules in which bifacial cells are able to absorb light also from their rear side: as a consequence, at optimal tilt the rear side of bifacial modules is largely illuminated by the in-plane albedo component $R(t)$, as illustrated in Fig. 10.7. This is even more relevant when bifacial modules are installed on trackers.

Energy rating for bifacial modules is still missing a protocol at the time of writing, but is receiving great interest from the PV community, as it is foreseen that bifacial modules will become more and more relevant in the PV market share, in the near future. With reference to Sect. 10.2.5, a future protocol for the energy rating of bifacial modules will need to take into account at least the following additional information:

- the global in-plane irradiance is now $G(t) = [B(t) + D(t)][1 + R(t)]$;
- the albedo needs also to be corrected for the angle of incidence effect, using an appropriate formula that can be found in the work by Martin and Ruiz [3];
- the effect of spectral mismatch described in Sect. 10.2.4 needs now to be quantified for both the front and the rear side. It will depend on the tilt of the installation and also now on the spectral albedo (for which a standard is undefined at the time of writing).

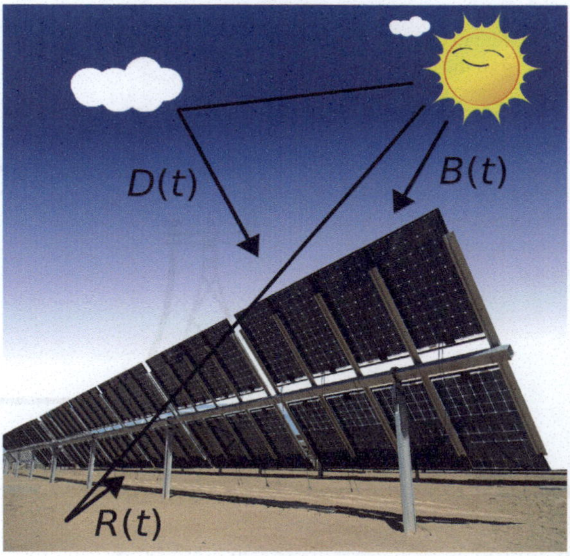

Fig. 10.7 Components of the global in-plane irradiance on bifacial modules: direct in-plane irradiance $B(t)$, diffuse in-plane irradiance $D(t)$ and in-plane albedo $R(t)$

10.4 Energy Losses and Failure Modes

Now you have all the theoretical tools to quantify how much energy your solar power plant generates in one year, provided you are in possession of all the parameters and functions listed in Sect. 10.2.5, right? Not really, not yet. In fact, there are remarkable losses to be taken into account, which can be grouped in two main categories:

- *power losses* from the module to the grid, which will be described in Sect. 10.4.1;
- and *failure modes*[7], to which is dedicated the second part of this section and that will be divided into "unrecoverable" (Sect. 10.4.4), "partially recoverable" (Sect. 10.4.5) and "recoverable" failure modes (Sect. 10.4.6) (Fig. 10.8).

[7]To avoid any misunderstanding, we need to clarify here that the term "failure mode" will be used in this chapter in a rather broader meaning than what is normally in use within the PV community. In fact the term "failure" will mean here more literally when a module "fails" to reach the expected power generation in the given conditions: thus, it will include certain mechanisms (such as soiling and partial shading) that do not necessarily represent intrinsic failures in the module components, but rather extrinsic causes of power degradation.

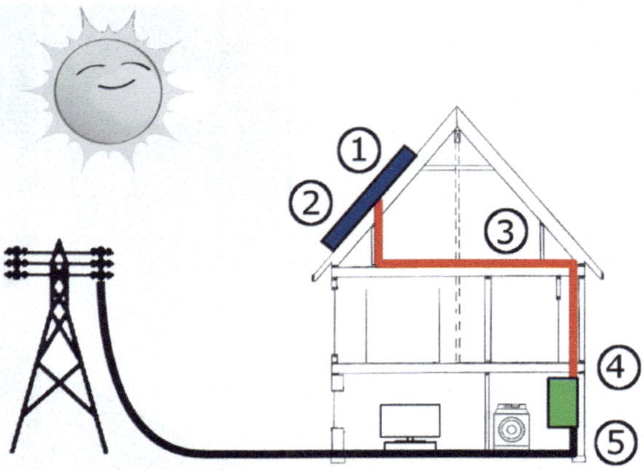

Fig. 10.8 Power losses from the module to the grid. These occur from the module down to the inverter and are due to: (1) quality of the module, (2) current mismatch and (3) Ohmic losses; and at the inverter level, due to: (4) inverter efficiency, (5) high or low power and voltage losses

10.4.1 Power Losses from the Module to the Grid

The first category represents all possible losses of energy from the total amount generated by the module to the "final destination", that may be the grid or, for stand-alone systems, the domestic appliances (both can be generally referred to as "the AC side" of the inverter).

In fact, the ***module quality*** itself represents the first of this kind of losses: not all the modules are the same and even from the same batch of production some modules may perform better or worse than others. In modern PV module manufacturing, modules are sold with stated positive power tolerance: for example, a 300 W module with –0/+5 W tolerance means that during the acceptance test in the production line all modules with power from 300 to 305 W have been labelled as "300 W modules". Hence, the customer is ensured that his module is "at least" a 300 W module. Each module in the production line nowadays is tested automatically and individually, but the simulator for the acceptance test is itself calibrated with a "golden module" that has once been calibrated with a given uncertainty (typically of a few percent). In the production practice, the golden module calibration uncertainty is usually taken into account in the power tolerance calculation. Nowadays PV module manufacturers put lots of effort in ensuring the quality of their products and therefore the power losses due to the quality of the module should be minimal.

But the fact that not all modules have exactly the same power generates a ***current mismatch*** between the highest quality module and the lower quality modules within a string of modules. This mismatch can determine a further reduction in the maximum power of the plant (as the power of all modules will be limited by the power of the one with the lowest quality). This loss mechanism is as unlikely as the previous (if all modules have positive tolerance, it is most likely that the system will perform better than expected), but, when it occurs, it amplifies the damage of the loss due to the quality of the modules. In fact, if one "bad" module generates less current than its production tolerance prescribes, the current generated by all the other "good" modules connected in series in the same string will be affected, as if all the modules were of "bad" quality.

If the previous two loss mechanisms depend only on the quality of the modules and should be minimal with good quality products, the ***Ohmic losses*** of the cables connecting the modules to the inverter is a physical phenomenon that always occurs. It is well known from basic physics that conductors dissipate heat when current flows through them, due to the so-called "Joule effect"

$$P = RI^2 \tag{10.9}$$

where P is the power loss, R is the resistance of the conductor (depending on the material of the conductor, its length, its section and its temperature) and I is the current flowing through the conductor. The power loss is larger at higher temperatures and for longer cables, but it is mitigated by a larger size of the cables. As an example, if the distance from the string and the inverter is 15 m (accounting for 30 m of total cable length), copper cables connecting commercial modules (current: ~8 A; cable section: 6 mm^2) show Ohmic losses of 5–6 W in the temperature range 30–60 °C.

So far, as to the power losses from the module to the inverter (i.e. on the "DC side"). But the inverter itself also exhibits power loss contributions. Figure 10.9 shows the efficiency characteristic curve of a commercial inverter (Kaco Powador 6600, a transformerless, single-phase inverter, on the market in 2012). As the chart shows, the ***inverter efficiency*** causes itself a minimum loss from 3 to more than 10%,

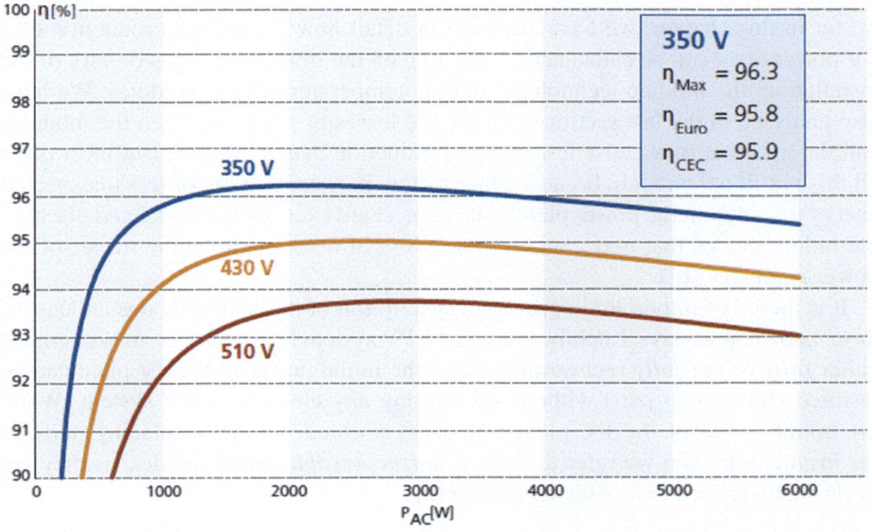

Fig. 10.9 Efficiency characteristic curves for a commercial inverter (Kaco Powador 6600: max input power: 6600 W; maximum output power 6000 W; maximum efficiency: 96.3% at V_{mp} = 350 V)

depending on the power generated from the PV modules. In order to minimize the efficiency losses in operating conditions, inverters are dimensioned to have maximum efficiency at the maximum power available from the modules, which is rather lower than the rated power at STC. (This rated power is rarely reached during operating conditions, as we have already noticed). As an example, the Kaco Powador 6600 of Fig. 10.9 is designed for 6.6 kW_p power plants, but has a maximum output of 6 kW, above which the power is cut off. It is also designed to have maximum efficiency of slightly more than 96% at 2 kW and 350 V (where the inverter efficiency loss is about 3%). But if the power plant of this example generates more than 6 kW, the inverter is capped at that value, and we can call the surplus power loss a ***high power*** loss. But for the same reason, on the opposite side there are also ***low power*** losses, when the power delivered to the inverter from the PV modules is too low to activate the inverter, and no power is generated at all.

Similar losses from the inverter occur also when the voltage from the modules to the inverter is above or below certain "cut-on" and "cut-off" thresholds, where the inverter works: we refer to these power losses as ***high input voltage*** and ***low input voltage*** losses, respectively.

While the inverter efficiency losses are unavoidable, the power and input voltage losses at the inverter are indeed avoidable, if the inverter is well dimensioned.

10.4.2 Overview of Module Failure Modes

So far in this chapter, we have analysed in detail how the energy production of a PV power plant can be calculated, depending on the irradiance, the geometry of the installation, the module technology and the temperature of the modules. We have also analysed in the last section what are the losses of power between the modules and the grid that may cause less energy production than predicted. But most often all this is still not enough, because other factors may cause more or less unexpected energy losses from the power plant: it is often a hard exercise to inspect and discover the failure modes that may cause a severe loss of investment for the owner of the power plant (see [6]).

It is therefore important to provide a description of the failure modes, at least of those known nowadays. Luckily enough, the PV system failure modes are sometimes either ***fully*** or ***partially recoverable***, when the initial power of the PV plant can be restored (fully or in part) without substituting any element of the system. When the initial power of the PV plant cannot be restored without re-placing parts of the installation, then we refer to them as ***unrecoverable*** failure modes, as they are intrinsically related to module manufacturing.[8]

---

[8] As certain partially recoverable failure modes are special cases of analogous unrecoverable failure modes, it will be more convenient to start with the latter (Sect. 10.4.4) and describe the partially and fully recoverable failure modes afterwards (Sects. 10.4.5 and 10.4.6).

10.4.3 The Timeline of Failure Modes

Before looking at some of the most relevant examples of the different failures modes, we will consider the time when failures occur. In fact, it is common practice to divide failure modes into three groups, depending on the time scale when they are likely to occur during the life of a PV plant, as shown in the "Bathtub curve" of Fig. 10.10.

- ***"infant" failures*** occur on PV modules and components before the installation, sometimes due to defects of production, to transport or installation;
- ***midlife failures*** occur during the expected lifetime of the plant;
- ***"wear-out" failures*** occur when the expected lifetime of the PV plant is over, and typically cause modules or components to be replaced.

Let us now look in some more detail at these three different time scales.

Infant failures are unfortunately very frequent, as the Bathtub chart shows. They may have originated in the production line (junction-box failures, defective interconnections, defective cells, early delamination). Other important cases of early, unrecoverable failures coming from transport and mounting are frame failures, glass-brakeage and cell cracks. Mechanical defects can also arise from natural or man-made hazards occurring during transport or installation (e.g. hail, lightning, module mishandling, transportation and crashes) (Table 10.1).

The most dreaded failures for plant owners are ***midlife failures***, as they always occur unexpectedly and can cause power and investment losses for the owner. In general, the failure rate (i.e. the number of failures detected per year) is smaller here than for infant failures. However, all components (junction-boxes, cables, frames, mounting structure, inverters, etc.) may deteriorate during the lifetime of the PV plant, most often due to poor module quality. This is less and less frequent in the last years, due to a generally broader recognition of the importance of quality, by both PV module manufacturers and by engineering, procurement, and construction companies. But the effect of poor-quality modules installed in the last two decades is now gaining importance in the installations that are still in their midlife. Most of these midlife failures are often difficult to detect, especially when they cannot be visually observed (as is the case with PID and with microcracks). A similar problem occurs in large power plants where visual inspection is often prohibitively expensive.

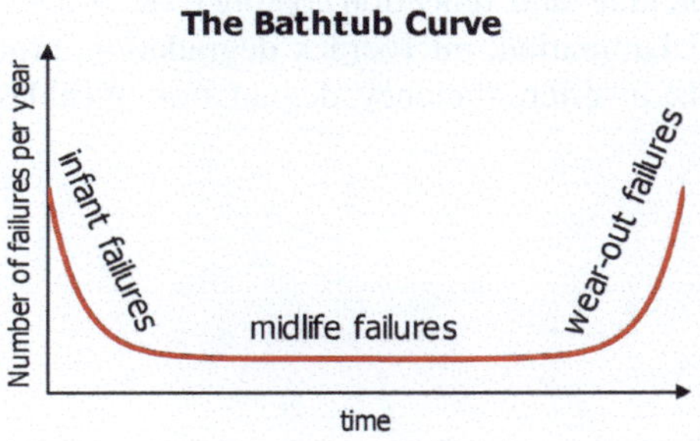

Fig. 10.10 The "Bathtub curve": infant failures, midlife failures and wear-out failures

Table 10.1 Overview of the best-known failure modes

	"Infant" failures		Midlife failures	"Wear-out" failures
	During production	Transport and installation		
Timeline	<Year 0	Year 0	Year 1 to ~25	>Year 25
Recoverable		Soiling	Shading Soiling Failures in the mounting structure Tracker misalignment	
Partially recoverable	LID		PID LETID	
Unrecoverable	Microcracks Junction box Interconnections Defective cells Early delamination	Microcracks Frame breakage Glass brakeage Cell micro- and macrocracks Natural hazards (snow, hail, lightnings, typhoons,...) Man-made hazards	Microcrack degradation UV degradation Inverter overheating and other failures Snail trail Hot spot Bypass diode failure Natural hazards	Microcrack degradation Anti-reflection degradation Discolouring Delamination Cracks Backsheet adhesion Cable degradation Junction-boxes

Midlife failures can, on the other hand, be detected by certain advanced optical non-destructive characterization tools, usable on the site, such as infrared thermography and electroluminescence imaging (see [8]).

"Wear-out" failures occur by definition when the expected lifetime of the modules in the field is expired, which conventionally corresponds to the expiration of the warranty. Of course, this does not mean that modules need always to be replaced after the warranty has expired, as in most of the cases, PV power plants perform very well even years after the end of their warranty [9]. In general, a power loss of $\leq 1\%$ per year is considered acceptable, due to unrecoverable failure modes. These include: anti-reflection coating (ARC) degradation, discolouring of the encapsulant, delamination, microcrack degradation, problems with backsheet adhesion, loss of the inverter efficiency, degradation of cables, of junction boxes, etc.

10.4.4 Unrecoverable Failure Modes

10.4.4.1 Microcracks

Among the most problematic unrecoverable failure modes are ***microcracks***, and the reason is that they cannot be detected by visual inspection: instead, only advanced imaging techniques (e.g. electroluminescence imaging, Fig. 10.11) can indicate their presence.

Modules with microcracks will typically show no initial degradation, but likely deteriorate in the medium and long term. Now, if microcrack degradation occurs typically in midlife, the generation of microcracks takes place most often during module production, transportation and installation.

During the lifetime of a PV plant, natural hazards may also generate microcracks.
Detection: *by electroluminescence imaging.*
Prevention: *quality control from production to installation; glass-glass modules are naturally less prone to microcracks; the effect of microcracks is also minimized with multi-wire modules.*

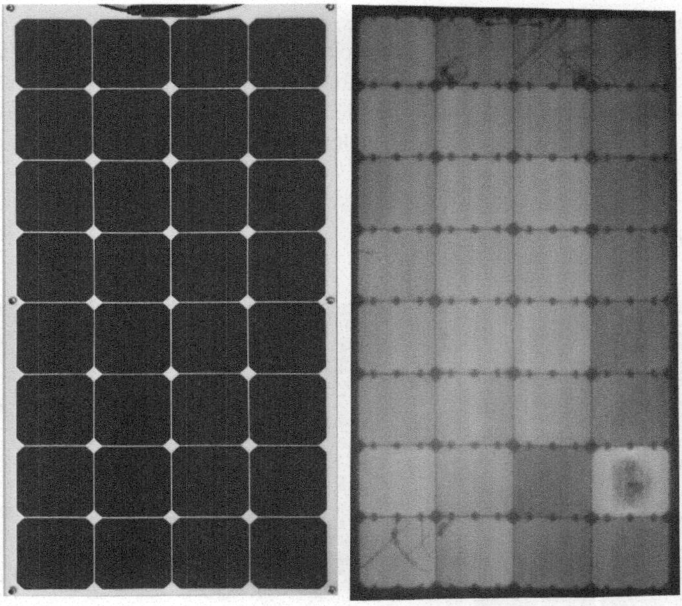

Fig. 10.11 Microcracks are an example of unrecoverable failure modes: they cannot be detected by the naked eye (*left*), but only by advanced imaging techniques, such as electroluminescence imaging (*right*, showing an electroluminescence image of the module pictured on the left)

10.4.4.2 High UV Degradation

The second example of unrecoverable failure is ***high UV degradation***, which has been observed to damage both the encapsulant, the backsheet and unprotected cables, causing: yellowing of the encapsulant; cracking and brittling of the backsheet; melting or brittling of the cable jacket (resulting in oxidation of the conductors, loss of insulation, higher resistance and power loss). Power plants exposed to high UV irradiance and high temperatures such as those in desert locations may require active cooling of the inverter: cases of ***inverter overheating*** have been observed to burn inverters, leading thus to significant investment losses.
Detection: *by visual inspection.*
Prevention: *by adequate system design.*

10.4.4.3 "Snail Trail" Degradation

An unrecoverable failure that is often an infant failure for PV plants is the so-called ***"snail trail" degradation***, where the name is due to the optical effect of this failure that recalls the humid track left by snails on surfaces (Fig. 10.12a). Certainly, this failure mode has nothing to do with snails, but it is caused by a discoloration of the surface of the cell near microcracks, due to a combination of heat, UV and humidity. It may occur in the first months after installation, but typically the degradation tends to stabilize: it is mainly an aesthetic problem. However, the microcrack from which snail trails originate can deteriorate in the long term and become the source of power loss.
Detection: *by visual inspection.*
Prevention: *almost impossible to predict.*

(a) (b)

Fig. 10.12 **a** Snail trails, also showing burns and oxidation of the busbar; **b** glass breakage, probably due to natural or man-made hazards

10.4.4.4 Failures Caused by Natural and Man-Made Hazards

Natural and man-made hazards (Fig. 10.12b) can again lead to infant, midlife or wear-out failures. These include: mechanical damages due to hail, heavy snow load, earthquakes, wind load, typhoons and any kind of natural disasters; thermal damages due to fire or lightning; corrosion of various components due to salt mist (particularly relevant in maritime installations), bird drops or atmospheric pollution.

If module qualification and type approval according to the international standard IEC 61215 has been carried out, we can, in principle, have confidence on the capability of the PV module to withstand at least some of the stresses described in this section. The standard itself is continuously improving to induce manufacturers to sell more and more reliable products, which can survive even the most severe environments.

Detection: *by visual inspection.*
Prevention: *impossible to predict.*

10.4.4.5 Failures Caused by Hot Spots

Finally, we should mention ***hot spots*** - a concept which has been already introduced in the previous chapter: it is the overheating of a cell due to partial shading[9] and reverse bias operation. It often occurs in combination with a ***bypass diode failure***. The bypass diode is responsible for limiting the effects of partial shading: but while we will see partial shading as an obvious example of recoverable failures, the effect of a hot spot is usually unrecoverable. Overheating of a part of the cell can cause burning and oxidation of busbar connectors, and consequently also of other components (encapsulant, glass and backsheet) with serious damage to the module. Furthermore, hot spots can be caused by defects within the module (cracks, shunts, poor solder joints leading to high series resistance, local delamination, etc.).

Detection: *by visual inspection and thermography.*
Prevention: *avoid partial shading, where possible; check the quality and robustness of the components.*

10.4.5 Partially Recoverable Failure Modes

We come now to three failure modes that we classify as "partially recoverable" because they lead to a degradation in the efficiency of the solar module, which can be reversed, although typically without being able to fully restore the original performance. These failure modes have the following characteristics in common:

[9]Hot spots can also be due to shunts inside the cell. A test on hot spot endurance is present in the international standard for module qualification and type approval, IEC 61215.

Fig. 10.13 Partially recoverable failure modes may need the intervention of a specialist (or, of a specialized device) to restore the original energy production capability. *Courtesy* Dji-Illustrations, Neuchâtel, Switzerland

- the efficiency loss of the module can be severe and depends mainly on the quality of the module;
- the quality of the module cannot be assessed by visual inspection and may require advanced measurement techniques (Fig. 10.13);
- the efficiency loss generally stabilizes after a certain lapse of time.

10.4.5.1 Light-Induced Degradation (LID)

The first failure mode that we will discuss is an infant-failure mode and gives rise to power degradation from 0.5% to 10%[10]: ***light-induced degradation***[11] ***(LID)***. The best-known form of LID takes place in *p*-type c-Si (which represents the majority of the c-Si modules on the market at the time of writing) and it is due to the combination

[10]Monocrystalline (mono-Si) *p*-type cells are more affected by LID (up to 10%) than policrystalline (poli-Si) cells (~3%). In comparison, effect on *n*-type c-Si cells is negligible (<0.5%).

[11]In amorphous silicon, an effect similar to LID, but not identical, is the *Staebler-Wronski effect* that we have already encountered in Chap. 6.

of the dopant boron with residual traces of oxygen in molten silicon during ingot formation (it is therefore also referred to as ***boron-oxygen LID***, or simply ***BO-LID***). The boron-oxygen molecules are able to capture the photogenerated electron-hole couples and, thus, reduce short-circuit current and maximum power. The activation energy needed for boron and oxygen to combine is provided by sun***light***, hence the name "***light***-induced degradation", or simply LID. The reaction is relatively fast and will start during the very first hours of light exposure at ambient temperature. Since the best-known form of LID is a boron-oxygen effect, *n*-type[12] c-Si was initially considered to be unaffected by LID. However, LID processes that are not related to the presence of boron and oxygen have recently been detected and therefore light soaking treatment of all c-Si PV modules has been inserted as a compulsory preliminary test for the standard qualification and type approval of commercial modules according to IEC 61215. It is also important to note that the nameplate power of all certified commercial modules nowadays on the market would have usually already taken into consideration LID and refers to stabilized conditions. Moreover, LID can be artificially reverted by annealing the module at relatively high temperatures, such as 150 or 200 °C: as these temperatures cannot be reached naturally in the field, LID is hardly considered a recoverable failure mode for modules within operation. But it is basically partially recoverable.
Detection: *power monitoring during the first weeks after installation.*
Prevention: *light-soaking (by the manufacturer).*

10.4.5.2 Light and Elevated Temperature Induced Degradation (LETID)

Another special form of light-provoked degradation has been observed at elevated module operating temperatures (well over 50 °C for long periods, therefore affecting modules mainly in hot climates); it takes place at a far slower degradation rate than the originally observed LID. It is referred to as ***Light and Elevated Temperature Induced Degradation (LETID)*** and represents the newcomer among the three recoverable failure modes described in this section. It has been reported that:

1. LETID occurs if the module production process had included high-temperature processing steps, followed by rapid cooling;
2. LETID does not manifest if the module operating temperature is kept below 50 °C or does not exceed that value for long periods;
3. the time span before LETID manifests itself is in the order of years of exposure to sunlight at elevated temperatures;
4. LETID naturally reverses (although not necessarily up to full recovery) after some time in the same high-temperature conditions that have generated it.

[12]*n*-type c-Si uses phosphorous instead of boron as dopant and leads in general to cells with higher efficiency; it is therefore expected to increase its market share in the future.

It can give rise to very slow degradation of maximum power of up to 5% or more (degradation up to 7% has been reported in hot Mediterranean climate, see [10]). The origin of LETID is suspected to be a high concentration of weakly bonded hydrogen near the rear side passivation layers (some authors therefore refer to it as ***Hydrogen-Induced Degradation***, or ***HID***). With high temperature and illumination, hydrogen is released and can bond to silicon independently of the dopant species, thus causing degradation that can occur in both *p*-type and *n*-type silicon[13] (see [11]). In reality a fully satisfying and scientifically robust understanding of LETID is still largely missing at the time of writing.
Detection/prevention: *no method fully agreed to date.*

10.4.5.3 Potential-Induced Degradation (PID)

Probably the most sadly famous example of midlife failure is the third one: ***Potential-Induced Degradation (PID)***. Several physical phenomena can cause PID, and unfortunately some are not yet fully understood (see [7]). What is commonly agreed is that PID is caused by leakage currents that flow due to the large potential difference in field installations (typically a few kilovolts), between the module frame[14] and the solar cells within the module (or vice versa). PID is more severe at high temperature and high humidity. It has been shown that PID can cause up to 30% loss of the initial power at the system level. It may affect both *p*-type and *n*-type modules: on the other hand, in HJT modules[15] the TCO layer is conductive and experimental results have shown that this prevents the PID effect.[16]

PID can be considered a partially recoverable failure mode. In fact, since PID originates from the potential difference between the cell and the module frame and mounting structure, a "PID box" (i.e. a voltage generator connected in series) can actively reverse this potential difference during the night. This will reverse the current flow that caused PID and restore the original PID-free state: the process can be repeated periodically every night to neutralize PID failures.
Detection: *electroluminescence in the field (but takes time and is expensive).*
Prevention: *by dedicated module (e.g. glass-glass, TPO or ionomers instead of EVA as encapsulant) and system design.*

---

[13]LETID effect in HJT modules is under investigation at the time of writing.

[14]Glass-glass frameless modules are in principle less prone to PID. Nevertheless, since glass-glass modules are usually mounted in the field with metallic clamps, a large potential difference between the cells and the clamp can still generate PID even in glass-glass modules.

[15]High-quality HJT modules are not laminated with EVA but with TPO, edge sealing is done with butyl and not with silicone and glass-glass modules are preferably used, instead of glass-backsheet modules. These very points lead more likely to PID-free behavior.

[16]The counter argument is that current migration into TCO layers (especially migration of Na^+ ions, the most popular candidate in theoretical explanations of PID), might lead—depending on the quality of the TCO—to TCO corrosion. This may be considered to be a side-effect of PID on those HJT modules, which use low-quality TCO. Good quality TCO layers are dense and so thick (100 nm, 1000 atoms), that Na^+ ions cannot easily migrate into the HJT cell.

Electroluminescence is probably the best-known method nowadays to detect both LETID and PID modules in the field at any phase of their lifetime. However, in the production phase neither LETID, nor PID are at the time of writing part of the international module qualification standard IEC 61215: it is therefore more difficult with PID and LETID, than with LID, to get assurance by the manufacturer that their modules will not be affected by such degradation mechanisms. While PID has already a dedicated technical specification (IEC TS 62804), which will most likely be included in the near future into IEC 61215 (with additional pass-fail criteria), a similar protocol for LETID test is still not available. Moreover, it is important for the plant owner to ask the module manufacturer for evidence that his products are LETID-free products.

10.4.6 "Recoverable" Failure Modes

Let us conclude this section on failure modes with two examples where the original functionality of the modules can be fully recovered. In what we have called "recoverable failure modes", detection can be done easily by visual inspection; prevention can be done by accurately eliminating the cause of failures when they occur (Fig. 10.14).

10.4.6.1 Partial Shading

Partial shading is the most trivial case of a fully recoverable failure mode that can be solved, by removing the shade, if more dangerous hot spot failures have not occurred in the meanwhile.

10.4.6.2 Soiling

The second and last example of failure modes that is naturally recoverable is ***soiling***, which can be due to a number of causes: deposition of sand (particularly severe in desert installations), snow and hail (in the continental climatic zones), leaves and bird drops (everywhere). In continental climates and at high latitudes where modules are oriented with an optimal tilt of 20 degrees or more, rain can be sufficient to remove most of the soiling and active cleaning is usually not needed. At lower tilts and in desert climates, where soiling is a major failure mode, it can be mitigated with periodic cleaning of the system and the full functionality of the power plant can be easily restored, provided the soiling has not yet generated severe hot spots or corrosion failures, as described in the previous section. Self-cleaning tools (Fig. 10.15) are under investigation for large installations in locations particularly prone to soiling.

Fig. 10.14 Soiling, a trivial example of a fully recoverable failure mode. *Courtesy* Dji-Illustrations, Neuchâtel, Switzerland

10.5 Simulation and Monitoring: Energy Yield Measurement

Now that both energy rating protocol and major failure modes have been presented, a new question arises on how the *energy yield*[17] of a power plant can be predicted in advance or measured *a posteriori*, in order to detect when a system is underperforming, due to possible failure mechanisms. The next two sections try to provide an answer to this question. In Sect. 10.5.1 the standard IEC simulation model for energy rating is briefly presented, together with a couple of tools available to the general public and to professionals. Furthermore, a working example of an energy yield measurement for a PV installation is given. Finally, Sect. 10.5.2 is dedicated to

[17]Energy yield and energy rating are often confused as they have similar meaning. In this book so far we used the term *energy rating* as a pure calculation of the energy produced in one year based on standard conditions, geometry, spectral mismatch, etc. *Energy yield measurement* is instead the measurement under real operating conditions over an unspecified time frame (which may be one year, for comparison with energy rating) and under any specific climatic condition.

Fig. 10.15 Soiling is the most obvious example of a recoverable failure mode: if the cause of soiling is removed in time, the full potentiality of the module can be easily restored and in some cases this occurs naturally with rain. For installations where rain is less frequent (e.g. desert fields), automatic cleaning tools, as the one pictured, are available on the market. When it is persistent or particularly aggressive and if cleaning is not efficient, soiling may generate hot spots or corrosion and may turn into an unrecoverable failure mode

professionals that may need dedicated measuring equipment for real-time monitoring of the PV plant performances, providing all the physical parameters described in Sect. 10.2.5 for an accurate energy yield measurement.

10.5.1 Standard Simulation, Open-Source and Commercial Tools

The international standard IEC 60853-4 for energy rating furnishes tabulated yearly irradiance data for six reference climatic profiles: *tropical humid* (total yearly global in-plane irradiance: 1677.7 kWh/m^2/y), *subtropical arid* (desert, 2295.5 kWh/m^2/y), *subtropical coastal* (1496.6 kWh/m^2/y), *temperate coastal* (972.9 kWh/m^2/y), *high elevation* (>3000 m, 2139.1 kWh/m^2/y), and *temperate continental* (1266.0 kWh/m^2/y). Data take into account the average weather variability in the given climatic profiles (that is why desert and high elevation have higher yearly global in-plane irradiance) and are based on historical observations, satellite measurements and statistical analysis. Data are tabulated as a function of time t (step 1 h) over one year, listed as day 1 to day 365, for a total of $365 \times 24 = 8760$ hourly data. The database includes also all the needed parameters for energy rating: global horizontal and in-plane irradiance at time *t*, direct horizontal and direct in-plane irradiance at time *t*, ambient temperature, wind speed, Sun elevation, angle of incidence, and global in-plane spectral irradiance. (All data refer to a fixed installation at $\beta = 20°$ tilt facing the equator).

For a more accurate analysis one may need data for a different geometry (i.e. different tilt and orientation) or for a particular geographical position that does not necessary match one of the six climatic profiles listed in IEC 60853-4. The open source tool ***PVGIS*** (see [12]) by the Joint Research Centre of the European Commission is a source of such information and provides additional features also for tracking systems and off-grid installations. The software requires some basic information on the precise geographical coordinates, geometry of installation, PV module technology, nominal power installed and system energy losses from module to the grid (discussed in Sect. 10.4.1: PVGIS default is 14%, but can be modified by the user). As a result, it gives the yearly PV energy production and its standard deviation (in kWh), and the yearly in-plane irradiation (in kWh/m^2). PVLIB (developed by Sandia National Laboratory, USA) and SAM (by NREL, USA) are other examples of free energy rating tools that are available online for download.

Other commercial tools are on the market, providing a variety of energy output. The best-known at time of writing is ***PVsyst*** (see [13]). PVsyst allows to input the 3D shading scene of the installation, thus highlighting the possible shading losses from nearby obstacles. The energy rating of PVsyst allows to calculate the distribution of energy throughout the year based on input data and Meteonorm[18], and gives as a result the total energy production (in kWh/y and in kWh produced per kW_p installed). It also shows the main energy losses involved in the simulation and other useful performance metrics.

Figure 10.16 shows a real case scenario, with simulation from PVsyst, for a 3.0 kW_p residential installation in the metropolitan area of Milan, Italy (tilt: 20°; azimuth: 35°; 9 Sunpower E20/327 modules; STC power: 327 W; module efficiency: 20.1%, connected to the grid in May 2012). Figure 10.16c illustrates the recorded monthly energy injected into the grid in 2013, compared with the PVsyst predicted monthly energy yield for the first year of installation. Figure 10.16d shows the yearly *AC Performance Ratio* (PR_{AC}), which is defined as the following dimensionless quantity

$$\mathrm{PR_{AC}} = \frac{\left(\text{yearly energy production}\left[\frac{\text{kWh}}{y}\right]\right) \times \left(\text{standard reference irradiance } 1\,\frac{\text{kW}}{\text{m}^2}\right)}{(\text{power rating at STC [kW]}) \times \left(\text{yearly in-plane global irradiance}\left[\frac{\text{kWh}}{\text{m}^2\text{y}}\right]\right)},$$

and indicates the fraction of energy that is actually injected to the grid after deduction of all the effects (orientation, temperature, angle of incidence, and spectral mismatch)

[18]Meteonorm is a commercially available online database with more than 8000 weather stations, five geostationary satellites and a globally calibrated aerosol climatology, providing results with high accuracy worldwide (see [14]).

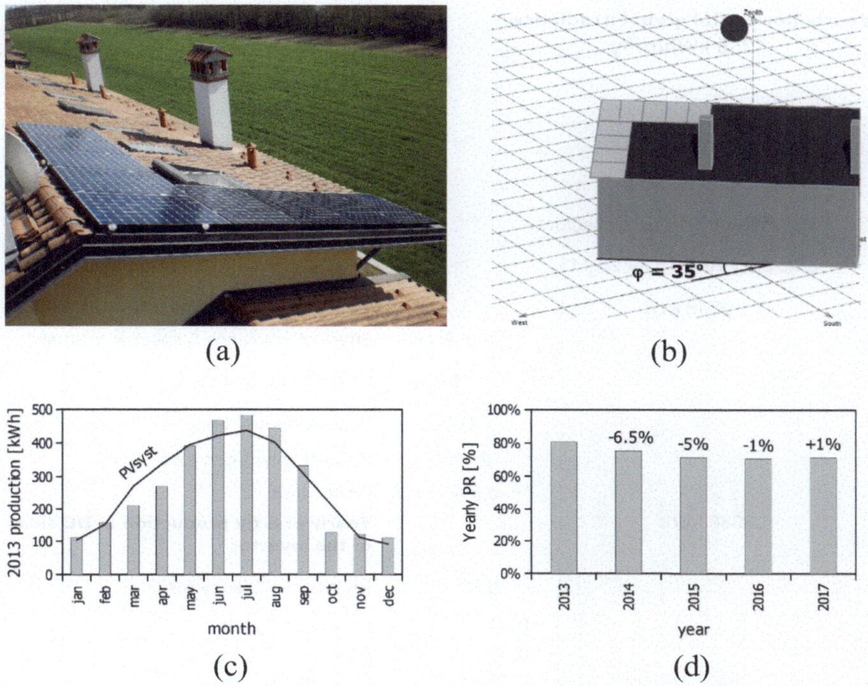

Fig. 10.16 Example of energy generation and energy yield measurement from a 3 kW_p residential plant (Milan, Italy). **a** A picture of the plant; **b** The 3D shading model by PVsyst; **c** The recorded monthly production in 2013 (total production for that year: 3237 kWh; PVsyst: 3192 kWh); **d** The yearly performance ratio PR_{AC}, showing percent variations from the previous year and possible evidence of infant degradation

and the energy losses, but without taking into account the failure modes. Values of PR_{AC} close to 80% are typical of a high-performance PV plant. The chart shows the higher degradation rate occurred in the first two years, due to infant failures.

Figure 10.17 illustrates the full energy rating analysis from PVsyst for the same installation of Fig. 10.16.

10.5.2 Measuring Equipment for Energy Yield Measurements

In real installations it may be difficult to discriminate whether the energy generation of a solar power plant is affected or not by midlife failures, based only on the history of the energy generation and models. A more accurate analysis can be performed if real irradiance and environmental data are available, measured on site. This section is, thus, dedicated to the professionals that are looking for an overview

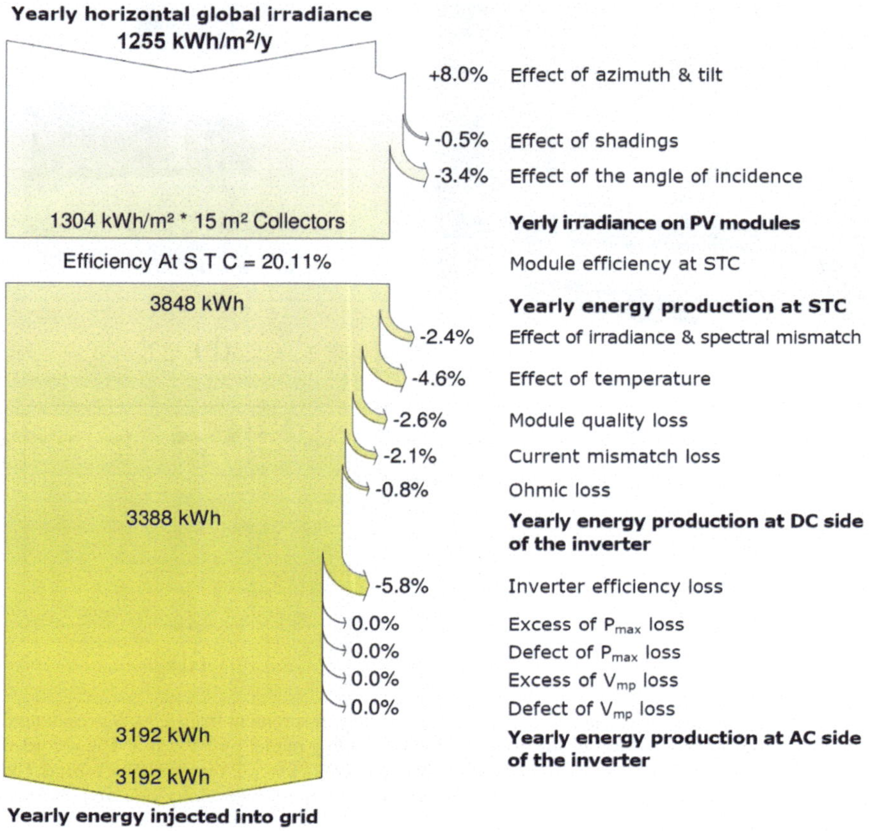

Fig. 10.17 Energy rating from PVsyst for the real case installation of Fig. 10.16. The starting +8% energy gain results from the orientation (20° tilt, 35° azimuth), that gives rise to a higher in-plane irradiance than the yearly horizontal global irradiance, which PVsyst estimated to be 1255 kWh/m^2. Energy rating model performed in 2012. *Source* [12]

of the measurement equipment needed for monitoring the energy production of a power plant: typically, this is more frequently the case for utility-scale solar plants, than for residential installations (Fig. 10.18).

Keeping in mind once again the protocol and list of physical parameters of Sect. 10.2.5, the first important instruments required are the tools for global and direct in-plane irradiance measurement. These are a pyranometer and a pyrheliometer, respectively. The ***pyranometer*** (Fig. 10.18a) is a thermopile[19] and is usually equipped with a double glass dome (to allow better thermal equilibrium and a 180°

[19] A *thermopile* converts thermal energy into electrical energy, generating an output voltage that is proportional to the temperature difference between its two terminals ("junctions"). In pyranometers, the temperature difference (and thus the potential difference) is measured between the black detector covered by the inner dome and the body of the instrument, which is protected by the sun shield.

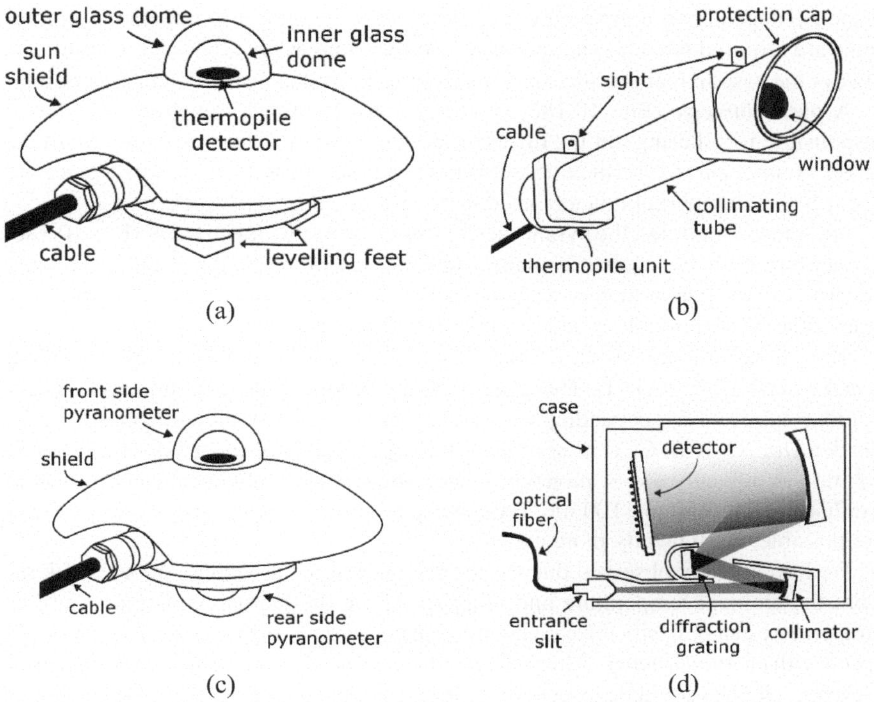

Fig. 10.18 Measuring equipment for monitoring the irradiance various components: **a** pyranometer, to detect the global irradiance; **b** pyrheliometer, to detect the direct irradiance; **c** albedometer, to detect the albedo; **d** spectrometer, to detect the spectral irradiance

field of view) and a sun shield, to protect from solar radiation the inner part of the thermopile, which remains in the dark and cold. With almost flat spectral response (Fig. 10.5a), a pyranometer is able to detect practically all the spectrum of the solar global irradiance, and its reading is substantially unaffected by temperature. If it is placed in-plane with the module (via levelling feet), the reading of a pyranometer gives directly the in-plane global irradiance $G(t)$ as a function of time, with no need of spectral corrections.

An alternative to the pyranometer is a ***photovoltaic reference detector***, with one advantage and some limitations. The advantage is a much faster response to changing irradiance than a pyranometer that typically require 5–20 s to measure irradiance: a PV detector instead has a response which is almost instantaneous. PV detectors (usually c-Si) have also typically the same spectral responsivity as the PV modules installed and, thus, spectral mismatch corrections can be negligible. But their

Since this temperature difference is proportional to the irradiance absorbed by the black detector, the pyranometer is calibrated to provide an accurate measurement of the global irradiance.

response depends on temperature and either active cooling or temperature corrections are required when the temperature variation ranges are above 10 °C (which is always the case in real installations), in order to limit the measurement uncertainty.

A ***pyrheliometer*** (Fig. 10.18b) typically consists of a collimating tube that is responsible for filtering out the diffuse irradiance and of a thermopile to record the direct component $B(t)$. Since the collimating tube needs to be pointed precisely towards the Sun, pyrheliometers need to be installed on 2-axis trackers.

For bifacial modules, the pyranometer is replaced by an ***albedometer*** (Fig. 10.18c) to measure both the global in-plane irradiance and the albedo $R(t)$: it typically consists of two pyranometers with opposite orientations, one oriented as the front side of the bifacial modules and one oriented as the rear side.

The global in-plane spectral irradiance $G(\lambda, t)$ can be measured by an ***optical spectrometer*** (Fig. 10.18d). There are a variety of spectrometers on the market for outdoor applications, depending on whether they are fast or slow-response and on the detector (CMOS, CCD or arrays of photodiodes; c-Si or InGaAs). The cheapest solution is typically a c-Si monochromator spectrometer, able to measure spectral irradiance from 350 to 1100 nm. Extensions to below 350 nm and above 1100 nm are possible, but typically more expensive.

We have already observed that the temperature of the modules can be calculated from the ambient temperature and wind speed via the Faiman equation. Ambient temperature can be easily measured with high accuracy with ***thermometers*** and wind speed with an ***anemometer***. Alternatively, ***thermocouples or resistance temperature detectors (RTDs)*** could be used to measure the temperature directly at the surface of the module, without the need of using the Faiman equation and ambient parameters.

Finally, while the geometry of installation and size of the modules are straightforward, both the spectral responsivity and the angular transmittance of the PV modules can only be measured by specialized testing laboratories. In most cases, this information can be provided directly by the module manufacturers.

References

1. D. Faiman, Prog. Photovolt: Res. Appl. **16**, 307–315 (2008)
2. M. Bliss, T.R. Betts, R. Gottschalg, Meas. Sci. Technol. **21**, 115701 (11 pp) (2010)
3. N. Martin, J.M. Ruiz, Sol. Energy Mater. Sol. Cells **70**, 25–38 (2001)
4. N. Martin, J.M. Ruiz, Sol. Energy Mater. Sol. Cells **110**, 154 (2013)
5. T.S. Liang, M. Pravettoni, C. Deline, J.S. Stein, R. Kopecek, J. Prakash Singh, W. Luo, Y. Wang, A.G. Aberle, Y.S. Khoo, Energy Environ. Sci. **12**, 116–148 (2019)
6. M. Köntges et al., in IEA PVPS, Report IEA-PVPS "Review of Failures of Photovoltaic Modules" T13-01 (2014)
7. W. Luo, Y.S. Khoo, P. Hacke, V. Naumann, D. Lausch, S.P. Harvey, J. Prakash Singh, J. Chai, Y. Wang, A. G. Aberle, S. Ramakrishnaab, Energy Environ. Sci. **10**, 43–68 (2017)
8. K.G. Bedrich, W. Luo, M. Pravettoni, D. Chen, Y. Chen, Z. Wang, P.J. Verlinden, P. Hacke, Z. Feng, J. Chai, Y. Wang, A.G. Aberle, Y.S. Khoo, IEEE J. Photovolt. **8**(5), 1281–1288 (2018)
9. A. Virtuani, M. Caccivio, E. Annigoni, G. Friesen, D. Chianese, C. Ballif, T. Sample, Prog. Photovolt: Res. Appl. **27**(4), 328–339 (2019)

10. F. Kersten, F. Fertig, K. Petter, B. Klöter, E. Herzog, M.B. Strobel, J. Heitmann, J.W. Müller, Energy Procedia **124**, 540–546 (2017)
11. D. Chen, P.G. Hamer, M. Kim, T.H. Fung, G. Bourret-Sicotte, S. Liu, C.E. Chan, A. Ciesla, R. Chen, M.D. Abbott, B.J. Hallam, S.R. Wenham, Sol. Energy Mater. Sol. Cells **185**, 174–182 (2018)
12. Photovoltaic Geographical Information System, http://re.jrc.ec.europa.eu/pvgis.html. Accessed 26 Dec 2019
13. PVSyst Photvoltaic Software, http://www.pvsyst.com. Accessed 26 Dec 2019
14. Metonorm, https://meteonorm.com/en/. Accessed 26 Dec 2019

Mauro Pravettoni is Senior Research Fellow at the National University of Singapore. Since 2017 he leads the Module Characterization & Reliability Laboratory of the Solar Energy Research Institute of Singapore, the leading testing laboratory of South East Asia. Mauro obtained his PhD at the Imperial College, London. He is a member of the International Electrotechnical Commission and of the British Society for the Philosophy of Science. Assessor of ISO/IEC 17025 for the competence of testing and calibration laboratories, Mauro has worked at ESTI, Joint Research Centre of European Commission, the European centre of reference for PV testing. In 2008, he characterized the world record efficiency luminescent solar concentrator. Mauro is a member of the international committee for the scientific conference at SNEC, the International Photovoltaic Power Generation Conference and the world's largest PV fair, in Shanghai.

Chapter 11
Solar Photovoltaics on Land, Water, and Buildings

Alessandro Virtuani

Abstract The aim of this chapter is to show examples of deployment of solar PV systems in different contexts. Focus will be given to the use of PV for power generation only, without considering the adoption of solar cells to power consumer electronic goods or other devices. The examples treated here include ground-mounted systems, floating PV systems;, integration of PV into the built environment, and micro PV systems for rural regions in Africa and other continents. We finally add a few examples to show how PV can easily be integrated everywhere, as in noise barriers on motorways, above water treatment plants, in car parks or on agricultural fields.

11.1 Solar Electricity for Powering the Word

In less than 20 years of effective solar PV deployment, starting in countries that had introduced the necessary regulations to allow the connection of solar systems to the grid, PV installations are growing massively and, with approximately 98 GW of new installations in 2019, have totalled ~ 585 GW (~0.6 TW) of cumulative capacity in the same year. Globally, the vast majority of the installations (more than 90%) are **grid-connected** systems, for which reliable statistics exist. On the other hand, **off-grid** installations are recently experiencing a considerable boom in several regions such as in Sub-Saharan Africa and India.

An additional segmentation of the market, distinguishes between **utility-scale PV** (ground-mounted) systems and **rooftop PV** (or **distributed PV**) systems. In 2018, globally 70% of the market is made of large utility-scale projects and 30% of smaller rooftop PV projects [1]).

Although no clear or sharp boundaries exist, generally plants larger than 2 MW are classified as utility-scale ones, and systems larger than 20 MW as large utility-scale plants. In very recent years, in order to benefit from economies of scale, plants much larger than 100 MW (even targeting GW scale) are being deployed in several

A. Virtuani (✉)
EPFL (PV-Lab), Neuchâtel, Switzerland
e-mail: alessandro.virtuani@epfl.ch

A. Shah (ed.), *Solar Cells and Modules*, Springer Series in Materials Science 301,
https://doi.org/10.1007/978-3-030-46487-5_11

countries, particularly where substantial solar resources are available (e.g. China, Mexico, Brazil, Gulf countries.)

Similarly, for rooftop PV, small-scale systems on residential units are called **residential PV** (typically with a size of 1–20 kW), larger rooftop PV systems (up to 2 MW) built on commercial/industrial warehouses or buildings are often referred to as **commercial and industrial PV**.

A recent flourishing market in developing countries is that of so-called **pico** or **micro PV systems**. These systems consist, generally, of a single solar PV panel with a power of about 10–300 Watts, connected to a battery and a charge controller. So-called pico or micro solar home-systems generate enough power to recharge mobile phones, illuminate households with LED lights, and can often also provide power to other appliances, such as refrigerators. Similarly, the integration of PV into infrastructures (e.g. motorway noise barriers, car parks, water treatment plants, etc.) unlocks the potential for the diffusion of solar PV virtually everywhere.

In this Chapter, organized in short informative boxes, we present a few examples of PV installations for the different categories.

11.2 Solar on Land

Ground-mounted PV installations are easy to realize and deploy and can be built relatively quickly. Often in 6–12 months, compared to hydro and nuclear projects that require over 4–5 years to be completed.

Examples of utility-scale plants of different dimensions are described in Boxes 11.1, 11.2 and 11.3. Additionally, a substantial decline in the cost of solar PV projects (approximately a 90% reduction over ten years) has improved solar PV's competitiveness, reducing the needs for subsidies and enabling solar to become competitive with other power generation options (e.g. fossil fuels and wind). This is particularly true for *large utility-scale* projects. For this kind of projects, the site selection is extremely relevant and is based on the following criteria: (a) the local availability of solar resources, (b) the proximity to the grid, and (c) the possibility to be easily connected to the grid.

For utility-scale plants, the plant owner (often an Independent Power Producer) sells power to the off-taker that can be a state-owned utility (or company) or the wholesale electricity market. The business models behind the development of this kind of projects are generally of two types:

(1) A model that has been fairly successful for large plants in several emerging markets (and that is nowadays adopted in some European countries too); here the terms (price and duration) of the Power Purchase Agreement is pre-determined through a tender process in which the developer submits a competitive bid, and competes against other bidders.

Fig. 11.1 Ground-mounted 60 MW utility-scale PV plant in **Agua Fria, Honduras.** Credits: Scatec Solar

(2) In a second model, solar power is sold to the *wholesale electricity market* (i.e., participating in *day-ahead* or *intra-day* auction markets), at fluctuating rates rather than at pre-determined fixed tariffs awarded for several years.

Box 1–Utility-scale solar plant 1

See (Fig. 11.1).

Plant description: Situated in the municipality of Nacaome in the Valle Department of Honduras, the Agua Fria plant is located in a very favorable region with high sun exposure. The plant was connected to the grid in 2015 and is made of poly-crystalline modules installed at a fixed-tilt. The average annual production is 103 GWh/y, covering the annual energy demand of approximately 80,000 households.

Business model: the electricity produced will be sold under a twenty-year Power Purchase Agreement (PPA) to the National Electricity Company (Empresa Nacional de Energia Electrica).

Box 2–Utility-scale solar plant 2

See (Fig. 11.2).

Plant description: The Oryx project is situated 200 km south of Amman in the municipality of Ma'an in Jordan, near the historical monument of Petra.

Fig. 11.2 Ground-mounted 10 MW utility-scale PV plant in Oryx, Jordan, 10 MW built under the Jordanian Renewable Energy Programme. Credits: Scatec Solar

The project was one of the first to be fully constructed in Jordan. The plant was connected to the grid in 2016 and is made of poly-crystalline modules mounted on a single-axis tracker. The average annual production is 25 GWh/y, covering the annual energy demand of approximately 5000 households.

Business model: The project has a 20-year power purchase agreement (PPA) with the state utility (National Electric Power Company).

Box 3–Giga-scale utility-scale solar plant 3

See (Fig. 11.3).

Plant description: located in a torrid semiarid region in Viesca (State of Coahuila), the plant extends over 2750 hectares and is operational since the end of 2018; it is composed of 2.4 million polycrystalline modules (of 320 W_p each). The average annual production is estimated at 1700 GWh/y. This project is representative of the trend of moving towards gigantic solar PV plants with capacities in the order of hundreds of megawatts or even gigawatts, particularly in the *sun-belt* countries. A high availability of solar resources, combined with economies of scale, allows Gigawatt-scale plants to considerably reduce the

Fig. 11.3 The Villanueva solar PV plant in Mexico has a capacity of 0.754 GW, approaching the Giga-Watt scale. It is the largest plant in the Americas. Credits: Enel Green Power

cost per kWh of solar electricity. On the other hand, to be profitable, these plants require excellent and very stable connections to the grid.

Business model: The overall 754 MW Villanueva facility is supported by a contract allowing the sale of specified volumes of energy for 15 years and clean energy certificates for 20 years to Mexico's Federal Electricity Commission (*Comisión Federal de Electricidad*).

11.3 Solar on Water

Floating solar systems can represent an alternative to ground mounted-solar systems. At present, the international market for floating PV is small compared to the total PV market and mostly limited to R&D projects [2]. A main driver for the growth of the market for floating PV is the reduced availability of land in regions with high population densities, in cities or on small islands (such as the Netherlands, Malta, Singapore, Hong Kong) and the cost of land that may become significant in other regions (e.g. Japan).

Most of the projects carried out so far on floating PV installations were done on closed fresh water bodies, such as water reservoirs or lakes. A positive side effect of deploying floating PV in these contexts is that, by casting shadows on the water, modules and floating platforms will decrease evaporation rates and reduce algae growth in the reservoirs.

Fig. 11.4 Floating PV plant with a capacity of 70 MW located in Bengbu, in the province of Anhui, China. Credits: Ciel & Terre International

Open-sea installations offer more challenges due to the generally harsher environmental conditions. In particular, water salinity may increase the probability of corrosion and salt deposits on the solar modules (generating soiling losses and potentially triggering PID[1] degradation); furthermore, waves and sea currents may cause considerable mechanical stress to the array structures and to the solar panels.

At present, Japan and China are the two driving markets, with Japan hosting a large number of projects and China being home to a number of very large systems, with the largest over 100 MW. In Europe, the UK, France and Netherlands host large floating demonstrative projects. The yearly installed capacity shows a quickly increasing trend from 3 MW in 2013 to more than 400 MW in 2017 [2–4]

Box 4–Floating PV (1)

See (Fig. 11.4).

Plant description: Realized on an artificial lake created to cover a pre-existing open-pit mine, the Anhui CECEP plant is divided into thirteen sub-arrays floating on a single reservoir and covering an extension of 63 hectares. The plant consists of a total of 194,731 PV modules, making this one the largest PV floating systems in the world, and has been operational and connected to the grid since March 2019.

[1]**PID**: potential-induced degradation. See Chaps. 9 and 10.

Fig. 11.5 Floating PV plant with a capacity of approximately 1.3 MW realized on an irrigation reservoir in Japan. Credits: Ciel & Terre International

Box 5–Floating PV

See (Fig. 11.5).

Plant description: This floating PV plant has been realized on an irrigation reservoir in Sasakuacho, Japan (Hyogo prefecture). The plant is divided into two sub-arrays (0.594 MW and 0.665 MW) floating on two neighbouring ponds. It consists of a total of 3,594 PV modules installed on a floating platform and has been operational and connected to the grid since December 2018.

11.4 Solar on Buildings

A substantial part of solar electricity generation could come from buildings, which are responsible on an average for about one third of a country's final energy consumption. *Rooftop solar PV* could provide electricity and, if combined with heat pumps, heating and cooling to buildings. Advantages are: (a) proximity to the end user; and (b) savings on the use of land resources.

As of today, the vast majority of these installations are so-called Building-Added PV (BAPV), in which the solar modules are "added" on top of existing building elements (such as a roof tiles, glazing, or façades) without being part of the building envelope. Conversely, in the case of Building-Integrated PV (BIPV) the solar modules and system are fully integrated into the "skin" of the building, becoming a structural element of the latter. In this case, besides generating electricity, BIPV modules need to provide to the building the same functionalities as provided by building elements, such as protection against noise and water ingress, and fire resistance. They are responsible as well for the building's performance in terms of heat retention and user comfort.

BIPV systems add complexity to the design and installation phase and are more expensive than conventional standardized PV products. On the other hand, in the case of the construction of new buildings and of the renovation of older buildings, BIPV installations could become competitive with more conventional BAPV systems [5], and possibly also with conventional building elements. BIPV installations should in principle be favoured over BAPV ones, as they are more easily harmonised into the built environment and urban landscapes.

Estimates show that well-oriented roofs and façades have the potential to provide up to 30% of the Swiss annual electricity demand (~20 TWh). In Europe, the potential of PV in buildings is estimated at more than 22% of the electricity demand expected in 2030, and similar studies exist for other countries including India [6], China [7], and Brazil [8]. In reality, the potential of solar PV on buildings is much higher. Therefore, solar PV on buildings could globally play a major role in decarbonizing the overall energy system, particularly in densely built and highly populated areas. Boxes 11.6, 11.7, 11.8 and 11.9 give examples of residential BAPV and BIPV systems and an industrial rooftop PV system.

Box 6–Residential rooftop PV: Building-Added PV (BAPV)

See (Fig. 11.6).

Plant description: The plant has a south-west exposure and a 14° tilt; it is installed on top of an existing roof made of traditional terracotta roof tiles. It generates an average of 5,000 kWh/y and was connected to the grid in 2008. It consists of 24 poly-crystalline modules.

Business model: The plant benefits from the 2nd "*Conto Energia*" feed-in tariff program. The power generated by the plant is awarded a premium price of 0.42 €/kWh for 20 years.

Box 7–Residential rooftop PV: Building-Added PV (BAPV)

See (Fig. 11.7).

Plant description: This 93.8 kW BAPV plant consists of 552 CIGS modules (from Solar Frontier) fixed on the rooftop with a tubular mounting system (from Schweizer). The plant has an estimated energy yield of 103,060 kW/y (per year). The solar power is used for domestic purposes by the 54 apartments located in the house. To increase self-consumption when the sun is not shining,

Fig. 11.6 Example of a 5.5 kW BAPV system installed on a residential building in the centre of Milan, Italy. Credits: Luca Morganti/Officina del Sole

the plant makes use of a battery pack with a capacity of 95 kWh of electricity storage.

Box 8–Residential rooftop PV: Building-Integrated PV (BIPV)

See (Fig. 11.8).

Plant description: The 27.7 kW (~171 m^2) BIPV plant is fully integrated in the roof of the historical building and has a multiple orientation (south, east,

Fig. 11.7 BAPV plant on top of the flat-roof of a residential building in Horw, near Luzern, Switzerland. Credits: Bernhard Etienne

west, north) and a 30°-tilt, which optimizes power generation in winter time. To provide a homogeneous appearance of the BIPV roof, several dummy modules were placed along the edges of the roof. The plant generates on average 27,600 kWh/y (covering 93% of the building's electricity consumption) and was connected to the grid in 2014. It is an excellent demonstration of how a PV system can be integrated into a protected building without affecting the historical and urban context.

Box 9–Commercial and industrial rooftop PV

See (Fig. 11.9).

Plant description: the plant is nearly horizontal (4° tilt) and has been realized on an industrial tin shed. It consists of 2,400 mono-crystalline PV modules and generates 1.11 GWh/y of electricity per year. It was connected to the grid in 2018.

Fig. 11.8 27.7 kW BIPV system fully-integrated in the roof of a historical building from 1880 in the old city of Neuchâtel, Switzerland. The building is home to the city associations (*Hôtel des Associations*), a foundation belonging to the municipality of Neuchatel, and it belongs to the Federal inventory of building sites of national relevance in Switzerland. Credits: SolarAgentur [9]

Business model: the electricity generated by PV plant is to a large extent (>75%) self-consumed. The surplus is exported to the grid in a *net metering*[2] regime.

Fig. 11.9 Example of a 721 kW rooftop PV plant (BAPV) built on top of an industrial warehouse in Vithlapuis, Gujarat, India. The company manufactures automobile parts and components Credits: B. D. Sharma

11.5 Solar in Developing Countries

In several parts of the word, people do not have access to electricity. Without being a primary need, access to electricity can dramatically increase the quality of life (lighting, food and medicine storage, communication, education, etc.) and promote economic development [10].

Currently, about 1,000 Million people lack access to electricity and approximately 50% of them are found in Sub-Saharan Africa alone, corresponding to 2/3 of the population of this part of the word [11, 12]. In these countries, the grid is often very unstable and poorly developed and power cuts are very frequent. Electricity production relies strongly on dirty and expensive diesel/petrol generators and the costs of electricity are extremely high. Fortunately, progress has been made in the past decade regarding the use of renewable electricity from water, solar and wind.

With solar system cost continuously declining, utility-scale PV projects are taking over and several plants have been realized in developing countries. Similarly, the

[2]**Net metering**: the PV system supplies the building load and exports any excess energy to the grid. At night (or with an overcast sky), building load needs are met by electricity imports from the grid. A bi-directional meter is usually installed to measure and record the net result. In net-metering, the electricity fed into the grid is awarded the same price as that of the electricity taken by the grid. At the end of the year, the owner of the PV plant pays only the net balance. This method is very advantageous to the owner of the PV installation. A tariff system that is much more diffused than net-metering, makes use of two separate meters. One for the electricity supplied to the grid and one for that taken from the grid. In this system the price of the electricity taken by the grid (P_G)—which includes system, distribution and other charges - is usually considerably larger than the price awarded to the electricity fed into the grid (P_F). To make an example, in some Swiss cantons, these rates correspond to $P_G = 0.25$ CHF/kWh and $P_F = 0.09$ CHF/kWh, respectively. This method is, understandably, less advantageous than net-metering, unless the system owner increases significantly the self-consumption of the solar electricity generated by the PV installation.

adoption of rooftop PV is spreading particularly in larger cities, as people start perceiving PV as a cost-effective and reliable technology [13]. On the other hand, also small-scale rooftop PV systems require a considerable up-front capital investment that in these countries may be out of reach for a vast majority of the population.

In very recent years, however, rural communities in Africa are witnessing the adoption of so-called *pico and micro PV* for low-income and rural households. These off-grid solar-home systems generally consists of a single solar panel (with a rating from less than 10–100 W), a small battery and a battery charger, and are used to power small appliances such as LED lamps or torches, a radio and a mobile phone charger. Larger solar home systems can power efficient stoves, smartphones and solar-powered TVs or refrigerators. These solar-home kits can have a large impact on the quality of life of people, allowing efficient and less polluting cooking, the use of smartphones to aid connectivity and business opportunities, and "clean" lighting powered by solar electricity (rather than costly and polluting kerosene lamps).

What makes today *pico and micro PV* accessible to a much larger customer basis, is the introduction of innovative business models combining mobile-phone and solar technologies. These business models, that avoid the need for costly one-off payments, are called *pay-as-you-go (PAYG) services.* Often customers make payments for a defined lease length until they fully own the system.

With PAYG, a basic solar-powered kit is purchased by the customer in installments, rather than paying a fixed sum in advance. An initial payment is followed by payments in daily rates that can vary between 0.20 US$ and 2 US$. The costs are calculated to be competitive with other forms of electricity generation, such as battery-powered torches and diesel generators. Additionally, many schemes allow adapting the daily rates (or even postponing them) making the scheme more accessible to rural communities (i.e. farmers, fishermen, etc.), who often have lower and irregular income sources.

Payments can be made in a variety of ways depending on the scheme, but mobile phone credit and scratch cards are normally adopted. In case of missing payments, the provider of the PAYG service and of the solar kit can deactivate remotely (using cell phone technology) the solar-home kit and reactivate it once the amount due has been paid. This possibility enforces dramatically the loyalty of the customers in submitting regular payments, and, as a consequence, the overall success of PAYG schemes.

Box 11.10 shows a couple of examples of *pico/micro* PV plants deployed in different countries of Austral Africa (South Africa and Zambia), whereas Box 11.11 shows an example of an off-grid installation on top of a nunnery in the Himalayan region of Himachal Pradesh in Northern India.

Box 10–*Pico* and *micro PV* systems in Austral Africa

See Fig. 11.10.

Fig. 11.10 Examples of a *micro (top)*, *pico* (centre) and residential rooftop (bottom) PV systems, respectively, in Stellenbosch (South Africa), in Northern Zambia and in Lusaka, Zambia. Credits: Riaan Meyer, GeoSun Africa

Fig. 11.11 Off-grid PV installation on top of the Pango nunnery in the Himalayan region of Himachal Pradesh in Northern India. The nunnery, situated near the Spiti River, hosts a community of Tibet Buddhist nuns. Credits: Ms Ishita Khanna, Ecosphere

Box 11–off-grid PV installation in the Himalaya

See Fig. 11.11.

11.6 Solar Everywhere

Besides the straightforward integration of solar PV into buildings and the deployment of floating PV installations, to save on the consumption of land and to avoid conflicts with other land uses, solar PV can be in principle integrated everywhere. We present below a couple of examples of solar PV integrated into existing infrastructures (Boxes 11.12, 11.13 and 11.14), into a solar car park (Box 11.15) and in a rural environment (Box 11.16).

Box 12–Solar motorway noise-barrier

See Fig. 11.12.

Fig. 11.12 PV noise barrier located on the A22 motorway and protecting the village of Marano (Trento, Italy) from noise pollution. The PV plant has a nominal capacity of 730 kW and was connected to the grid in 2009. Credits: Autostrada del Brennero Spa

Plant description: The plant is consists of 3,944 mono-crystalline modules covering a surface of ~5000 m^2. The PV barrier is displaced along the western carriageway of the motorway with an extension of 1.07 km and a height of 5.6 m. It has a variable orientation (along the carriageway) and two different tilts (35° and 60°, a compromise between acoustical and energy-yield performance). The modules are fully integrated into the noise barrier and constitute the sound absorption element. The plant generates on an average 690,000 kWh per year, enough to cover the annual electricity needs of approximately 600 households of the closest villages.

Box 13–Solar on a waste water treatment plant

See Fig. 11.13.

Plant description: This plant consists of 1,320 polycrystalline modules and is integrated into a retractable solar folding roof that can use the open basins of a waste water treatment plant to generate solar power. Key features are: extreme lightweight design, wide pole distance and retracting function to access the basins any time by crane. Waste water treatment plants have very

Fig. 11.13 Solar plant with a capacity of 634 kW covering a waste water treatment plant in Chur, Switzerland (Solarkraftwerk ARA). Credits: DHP

high power consumption, so that self-consumption of solar electricity is on average very high, allowing profitable business models. The energy yield of the plant is on an average 550,000 kWh/y.

Box 14–Cable-car PV station

Fig. 11.14 Both the top and base *Skway MonteBianco* cable-car stations in Courmayeur (Italy) integrate a solar PV system with a capacity of 13 kW in order to bring them close to the requirements of a zero-energy building. Pictured are some details of the base station in *Pontal d'Entreves*. The modules are bifacial high-efficient glass/glass modules. Credits: Giuseppe Virtuani

See Fig. 11.14.

Box 15–Solar car-park

See Fig. 11.15.

Fig. 11.15 This 18.9 kW PV plant is integrated into a projecting roof and serves as a solar car-park. The plant is located in Fiumicello close to Trieste (Italy) and was connected to the grid in 2010. Credits: Raffaella Lorenzi

Fig. 11.16 This 14 kW PV system is integrated in the canopy of a petrol station in the outskirts of Dakar, Senegal. The plant consists of bifacial modules. Credits: Alessandro Virtuani

Box 16–PV installations integrated in the canopy of a petrol station
See Fig. 11.16.

Box 17–Solar PV on agricultural fields: Agri-PV

Fig. 11.17 *Agricoltural-PV* (*or Agri-PV*) is a novel concept. In these systems, the PV panels, positioned above the crops, are automatically controlled by algorithms in order to continuously optimize plants' well-being (e.g. providing shade in hot summer months and protection against hail storms). The electricity produced by the panels is always a secondary objective, following the primary target of increasing the yield of the crops. From the perspective of solar system developers or owners, Agri-PV may allow accessing new lands, sometimes with the advantage of being close to the grid, and avoid conflicts with other land use (agriculture). For farmers, this may allow a more profitable and more resilient farming business, adding revenues and diversifying risk. Some studies [14, 15] indicate that, based on the type of crops, higher yields can be achieved, together with reduced crop loss and less water consumption, making Agri-PV particularly suitable for arid climates. The figure shows an Agri-PV system with a nominal power of 2.1 MW in Tresserre (domaine de Nidolères), near Perpignan, France. This large-scale demonstrator has an extension of 4.5 hectares and was connected to the grid in 2018. The full extension of the field will be covered by grapevines. Credits: Sun'R

See Fig. 11.17.

References

1. IEA (International Energy Agency) PVPS-2019 Snapshot of Global PV Markets, Report IEA PVPST1-35 2019: www.iea-pvps.org/?id=266
2. C. S. Won, W. Lawrence, D. Kim, B. Kang, K. Kim, G. Lee, *Floating PV Power System Evaluation Over Five Years* (2011~2016). In Proceedings of the 32nd EUPVSEC Conference (2016, Munich), pp. 1982–1984

3. M. M. De Jong, M. Dörenkämper, K. Sinapis, W. Folkerts, *An Overview of Floating PV Worldwide*. In Proceedings of the 35th EUPVSEC Conference (2018, Brussels), pp. 1437–1439
4. "Where Sun Meets Water: Floating Solar Market Report-Executive Summary", World Bank Group, ESMAP and SERIS 2018, Washington, DC: World Bank (2018)
5. C. Ballif, L.E. Perret-Aebi, S. Lufkin, E. Rey, Integrated thinking for photovoltaics in buildings. Nat. Energy **3**, 438–442 (2018)
6. A.K. Shukla, K. Sudhaka, P. Bareda, R. Mamat, Solar PV and BIPV system: Barrier, challenges and policy recommendation in India. Renew. Sustain. Energy Rev. **82**, 3314–3322 (2018)
7. P. Changhai, H. Ying, W. Zhishen, Building-integrated photovoltaics (BIPV) in architectural design in China. Energy Build. **43**, 3592–3598 (2012)
8. I. Santos, R. Rüther, The potential of building-integrated (BIPV) and building-applied photovoltaics (BAPV) in single-family, urban residences at low latitudes in Brazil. Energy Build. **50**, 290–297 (2012)
9. Swiss Solaragentur: www.solaragentur.ch/node/435; Energieanlagen: Hôtel des Associations, (2000) Neuenburg/NE
10. Sustainable Development Goals: United Nations Resolution A/RES/70/1 of 25 September 2015, par. 54
11. Africa Energy Outlook (WEO-2014), International Energy Agency (IEA) report (2014)
12. Solar PV In Africa: Costs And Markets, International Renewable Energy Agency (IRENA) report (2016)
13. A. Virtuani, G. Agostinelli, *Solar Electricity in Africa: Overcoming Barriers And Lessons Which May Be Learnt From Previous Experiences in Europe*, In Proceedings of the 35th Eu PVSEC Conference (2018, Brussels), pp. 2143–2148
14. www.pv-magazine.com/2019/08/12/agrivoltaics-good-for-agriculture-and-panel-efficiency/
15. www.pv-tech.org/news/fraunhofer-ise-proves-agrophotovoltaics-pilot-project-feasibility

Alessandro Virtuani is a senior researcher at the PV-Lab of EPFL (Neuchâtel) where he is responsible for the *PV module and reliability* group. He is co-founder of a start-up *Officina del Sole* (www.o-sole.eu), which offers testing/consulting services in the field of PV. His research interest covers performance/reliability testing, development of innovative module concepts, energy yield modelling of PV devices/systems, Building-Integrated PV (BIPV), and technology transfer to industry. In 2016, he has served as technical advisor to IFC (*World Bank group*) with a project aimed at promoting a sustainable diffusion of rooftop solar systems in Sub-Saharan Africa. Alessandro has worked at ESTI (Joint Research Centre of the European Commission), the European centre of reference for PV testing. He is a physicist with a PhD in Materials Sciences (University of Milano-Bicocca, 2004), obtained carrying out research activity on the optimization of CIGS thin film solar cells/modules at ZSW (Stuttgart). In 2014, he obtained a Master's Degree in Business Administration from SDA-Bocconi. Alessandro is the inventor of the *DragonBack* measurement method for the testing of high-efficient silicon devices, and is a member of the ETIP-PV (European Innovation & Technology PV Platform).

Chapter 12
Solar PV Systems

Urs Muntwyler

Abstract Solar PV systems are power supplies with PV modules as source. They can be "grid-connected" as an additional source of electricity for feeding the electrical grid.

12.1 Overview Solar PV Systems

Solar PV systems are power supplies with PV modules as source. They can be "grid-connected" as an additional source of electricity for feeding the electrical grid. On the other hand, they can be autonomous off-grid systems: Acting e.g. as uninterruptable power supplies (UPS) often with a battery as storage. Or, they can supply electricity to isolated houses. The power of a Solar PV system can extend over a large range from a few Watts (for small autonomous off-grid systems) up to several Gigawatts (GW_p) for large grid-connected systems (Fig. 12.1).

12.2 The Solar Generator

A solar generator consists of one or several PV modules. Each PV module consists in general of a large number of solar cells. Typically, there will be 60 or 72 solar cells per module (Fig. 12.2).

12.3 Off-Grid PV Systems

PV solar cell applications for energy production started with the development of the first crystalline Silicon solar cell by Bell laboratories in 1954. However, applications

U. Muntwyler (✉)
University of Applied Sciences, Burgdorf, Switzerland
e-mail: urs.muntwyler@bfh.ch

A. Shah (ed.), *Solar Cells and Modules*, Springer Series in Materials Science 301,
https://doi.org/10.1007/978-3-030-46487-5_12

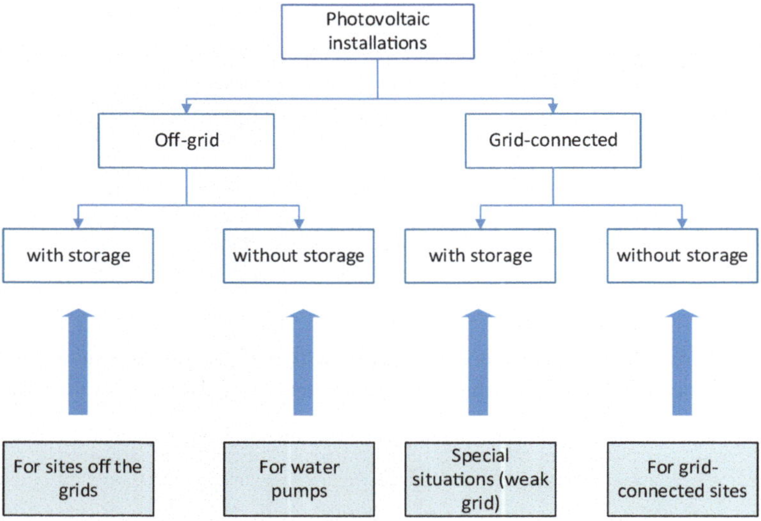

Fig. 12.1 "Off-grid and grid-connected" PV applications [1]

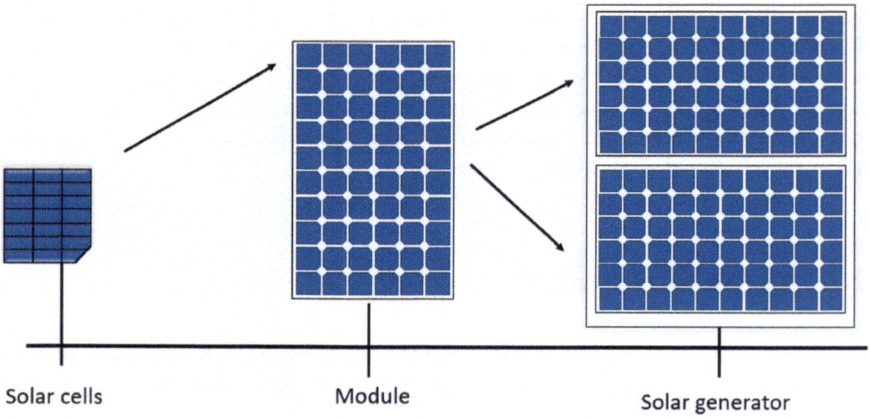

Fig. 12.2 Solar cell to solar generator [2]

were not around the corner as the new electricity source, "nuclear power" was pushed by the political campaign "Nuclear power for Freedom" by the US government. Therefore, solar cells were "trapped" in a few niche applications, such as power supplies for satellites, telecom repeaters and toys.

12.3.1 Off-Grid Stationary DC Applications on Earth

In the 70s, first applications with solar cells on earth were telecom power supplies and navigation buoys. After the so-called "First Energy Crisis" in 1973, some of the employees of satellite solar cell producers started specialized companies for terrestrial solar modules. The first terrestrial modules were produced by newly founded U.S. companies, such as Arco Solar/Solar Power/Solarex (USA). Sharp (Japan) and AEG-Telefunken (Germany) also entered the field.

12.3.2 Off-Grid Installations Without Storage Systems—Components/Sizes/Reliability/Yield/Economy

The standard size of a PV module was selected in such a way, so that the module could charge a 12 V lead-acid battery. This needs about 36 crystalline silicon (c-Si) solar cells in series. With such a PV module you can also build a simple but very interesting 12 V_{DC} uninterruptable power supply (Fig. 12.3).

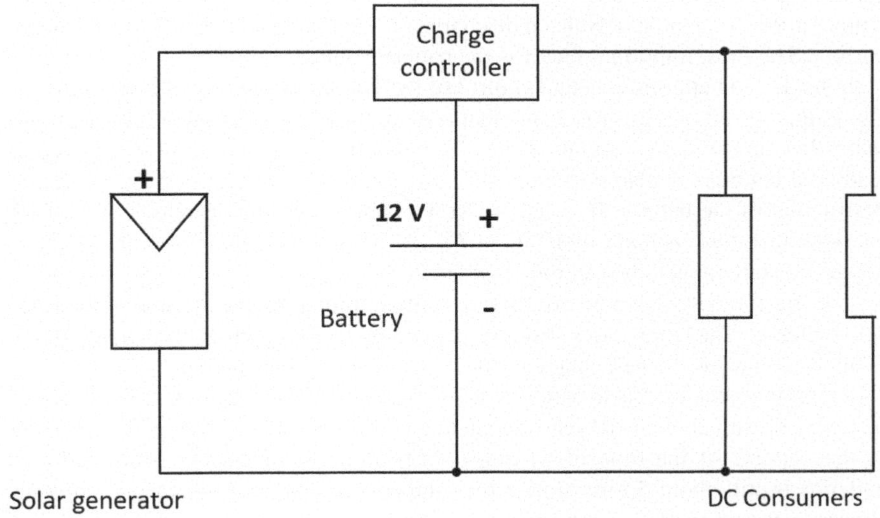

Fig. 12.3 Autonomous facilities with batteries consisting of at least four units: solar generator/charge controller, battery, DC consumer [3]

12.3.3 Components of Off-Grid DC PV Applications

The electric components of such a system are:

- PV generator
- Battery (so far mainly lead-acid)
- Voltage control of the battery
- DC consumers
- Mounting structures/lightning protection/cables/circuit breakers etc.

The critical technical component of such a system is the battery. The lead-acid battery should never be completely discharged. If this happens, the battery starts to loose active lead material and its plates become covered with sulfate. The battery capacity differs according to the temperature range of operation, and according to the value of the discharge current. For this reason, it is important to know the value of the battery discharge current, in order to be able to assess the capacity C_x of the battery. The term C_x denotes here the capacity of a battery, which is discharged in x hours. Thus, C_{20} is the capacity of a battery, which is discharged in 20 h. For photovoltaic off-grid applications, C_{20} or C_{100} are appropriate. Even regular strong discharges adversely affect the lifetime of the battery. The battery therefore needs an active protection against rapid discharge. The reference signals used are: battery voltage and battery temperature. Some more advanced discharge protections also consider the discharge current. The most important factor is the battery voltage.

In lead-acid batteries constant control of the voltage is important. If due to overcharging, the voltage rises to a value that is more than 15–20% higher than the nominal value, an explosion can take place. Note that the storage temperature T_{storage} of lead-acid batteries is also important. If $T_{\text{storage}} < 0$ °C, freezing occurs, leading to destruction of the battery. If $T_{\text{storage}} > 20$ °C, battery lifetime decreases by half for every temperature increase of 10 °C in T_{storage}. This is a challenge for photovoltaic off-grid installations in hot regions (Table 12.1).

The life-time of the whole installation is often limited by the lifetime of the lead-acid batteries. The latter varies strongly, depending on the depth of discharge (DOD). DOD is defined as the percentage of the capacity that is discharged.

At this moment, lead-acid batteries are well established in the market. They have to be replaced after about 6–10 years depending on the use and the average temperature of the battery. At the time of writing (2020), the cost of current from lead-acid batteries is still about 5 times lower than the cost of current from Li-ion batteries. However, the price of Li-ion batteries is steadily decreasing, so that we can expect more and more Li-ion batteries in the future. In the case of Li-ion batteries, we have to control each battery element separately, through a Battery Management System (BMS).

Further mobile DC—applications are: PV installations for gliders, glider trailers, boats. In future we will see electric cars with PV on their car body.

Table 12.1 Comparison of battery characteristics [4]

Type of batteries	Cycle life (h) 75% discharge rate	Cycle life (h) 30% discharge rate	Specific weight (Wh/kg)	Price/kWh *C*20 (€)	Cost (€/kWh) 30% discharge
Lead-acid car battery	Forbidden	800	40	300	1
Lead-acid heavy plate battery	200	800	40	200	0.5–2
Lead-acid OPzS battery[a]	1000	2800	36	470	0.15–0.4
Lead-acid bloc battery	1300	4500	16	570	0.18–0.6
Li-ion[b]	5 years	5 years	60	1000	1

[a]OPzS stands for stationary tubular plate battery (Types OPzS)
[b]Li-ion battery with a "battery management system (BMS)", an electronic system that controls the items of a battery pack)

Increasingly popular are off-grid installations with DC/AC inverters. For larger applications, e.g. farmhouses and micro-grids, hybrid installations combining PV with other sources of electricity (such as diesel-, wind-, and hydroelectric-generators) are being introduced (Figs. 12.4 and 12.5).

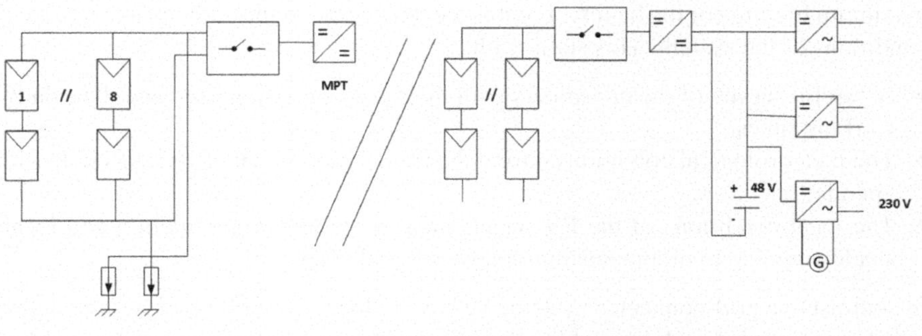

Fig. 12.4 Simplified electric block diagram of SAC (Swiss Alpine Club) Monte Rosa with DC/AC inverters [5]

Fig. 12.5 PV facade of the new Monte Rosa SAC refuge—a 48 V_{DC} off-grid hybrid PV system [6]

12.3.4 Design of an Off-Grid System

The design of an off-grid system aims for a high reliability of the system and a maximum life time of the battery. For this we have to know the ambient and working conditions of the system. This starts with:

- A careful survey of the consumers with their power consumption and their daily working times.
- The meteorological conditions during the time period in which the system has to operate.
- The local conditions of the PV panel, such as inclination/orientation and local shadowing due to mountains/buildings/trees and snow.

In contrast to a grid-connected system, we have to look, in an off–grid system, at the worst-case conditions during which the system has to operate. For a system, which works during the whole year, such as a telecommunication system, this is the Winter time (or the “rainy season” in tropical regions), i.e. the period during which solar radiation is at its lowest value.

Now we have to choose the system parameters: The first one to choose is the autonomy time t_{aut} of the system. t_{aut} is given by the worst-case conditions, when the PV panel does not produce any energy due to bad weather, snow etc. The following are typical values [in days] for t_{aut}:

- 3–5 days for a weekend house

- 7–14 days for lived-in houses
- About 3 days for a recreational vehicle (RV) and for a boat with an additional energy source, such as a wind generator
- 7–14 days for a micro-grid, or for hospitals (the value depends on the climatic zone)
- >14 days for a telecommunication installation.

12.3.5 Balance Off Systems (BOS)—Choice

To prevent the battery from discharge and to control the off-grid system we need further:

- A PV regulator which controls the energy flow from the PV module into the battery. The industry offers many different PV regulators which differ with respect to PV power, PV currents, and nominal battery voltage (typically either 12 V_{DC} or 24 V_{DC}). For small PV systems there are series-/parallel-regulators and for larger PV systems DC/DC-converters with Maximum Power Tracking (MPT).
- For larger systems one uses more and more DC/AC-converters.
- The mounting structure of the PV module is an important part of an autonomous PV system. Especially at high latitudes in the extreme northern (or extreme southern) hemisphere, the PV panel has often a very steep inclination, as one tries to maximize the energy yield for the "weakest" moment (Winter). Such steep inclinations can lead to high wind loads, which have to be carefully assessed. Snow loads also have to be carefully considered.
- Finally, we need electro-technical components, such as switches, cables etc. For DC grids special circuit breakers and fuses are needed.

12.3.6 Off-Grid Installations with Storage System and AC-Grids

As PV panels became cheaper and cheaper, one encounters larger "off-grid installations with AC-grids". They can power individual users, such as farm houses, alpine huts, etc.

Furthermore, as of today, we see more and more micro-grids for communities, airports in the desert, mines etc., which may have up to 1 MW_p PV panels. These systems replace the traditional decentralized production of electricity with diesel generators. The diesel generator will then only be used as an auxiliary power supply.

12.3.7 Off-Grid Installations Without Storage—Solar Pumps Etc

To eliminate the battery, there are some off grid—application without batteries. These are DC- or AC-motors for water pumping or for ventilation applications. If a ground water pumping system is exploited, a water reservoir can act as a cheap and reliable storage.

12.4 Grid-Connected PV Systems

A grid-connected PV plant consists of three parts:

1. The solar generator (totality of PV modules)
2. A one- or three-phase inverter
3. Some safety devices.

The inverter transforms direct current (DC) from the solar module into alternating current (AC), as circulating in the local grid (Fig. 12.6).

Grid-connected PV plants come in a whole range of different sizes:

(a) One PV module and one "module or micro-inverter": 10 W_p to 400 W_p
(b) String PV inverter systems: 600 W_p to 6 kW_p for lived-in homes
(c) Multi-String inverter systems: 6 kW_p to 120 kW_p—residential homes/multistory houses and commercial buildings
(d) Multiple multi-string PV inverter systems: MW_p size—commercial buildings and power plants.
(e) Central PV inverter systems: 10 kW_p to 2 MW_p size—commercial buildings and power plants
(f) Multiple central PV inverter systems: MW_p to GW_p size—commercial buildings and power plants

Their electrical diagrams are given in Fig. 12.7a–d [8]:

Due to their simple design, string inverters have obtained, especially in the last ten years, a higher market share. The central inverters which started 30 years ago, with 10 kW_p units and are now in the MW_p size for large utility-scale plants. The system voltage starts with one PV module with 30 V_{DC} and ranges up to many PV modules with up to 1500 V_{DC} (Fig. 12.8).

"PV inverters" are inverters, which are designed and built especially for use in PV systems. During the last 30 years their price has decreased by a factor 10, and their efficiency has improved from 95% to over 98%.

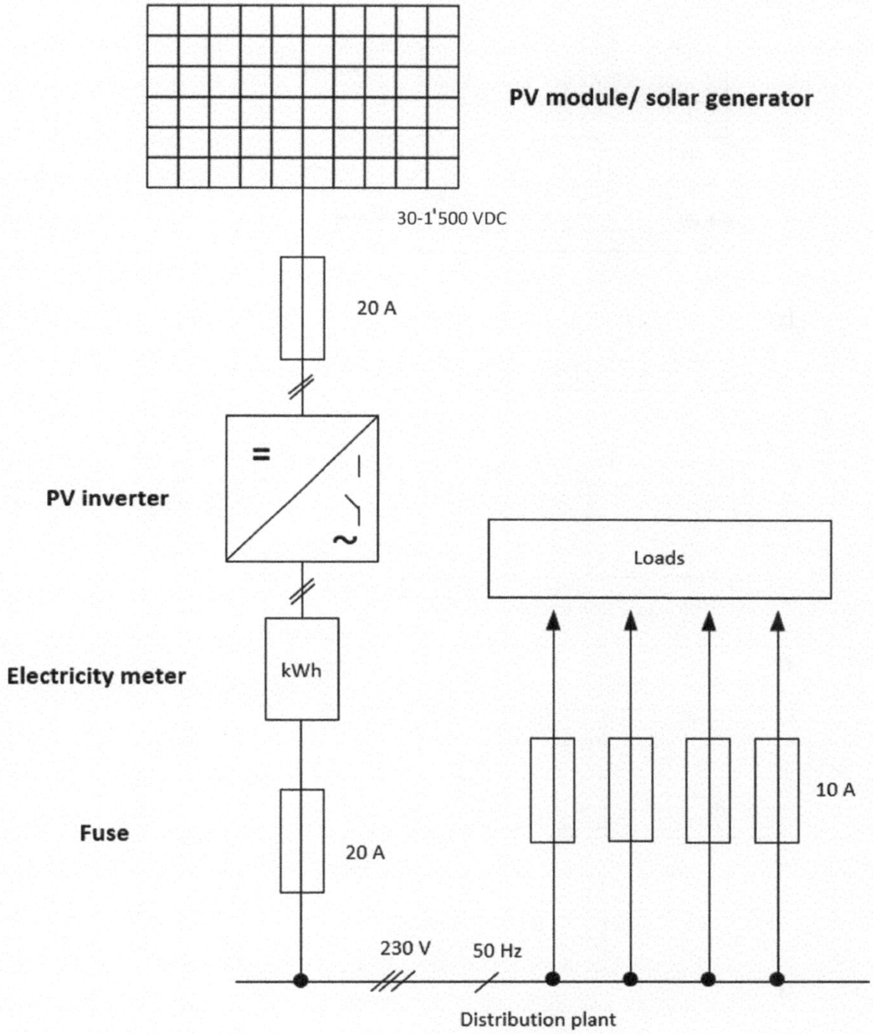

Fig. 12.6 Electrical diagram of a grid-connected PV Installation [7]

12.4.1 Energy Production of a Grid-Connected PV Plant

The calculation of the yield of a grid connected PV plant is less complicated than that of an off-grid installation. We have to know:

- Site/inclination and orientation of the PV generator → solar yield and temperatures
- The PV technology (c-Si/CIS/CdTe) → DC power and temperature behaviour

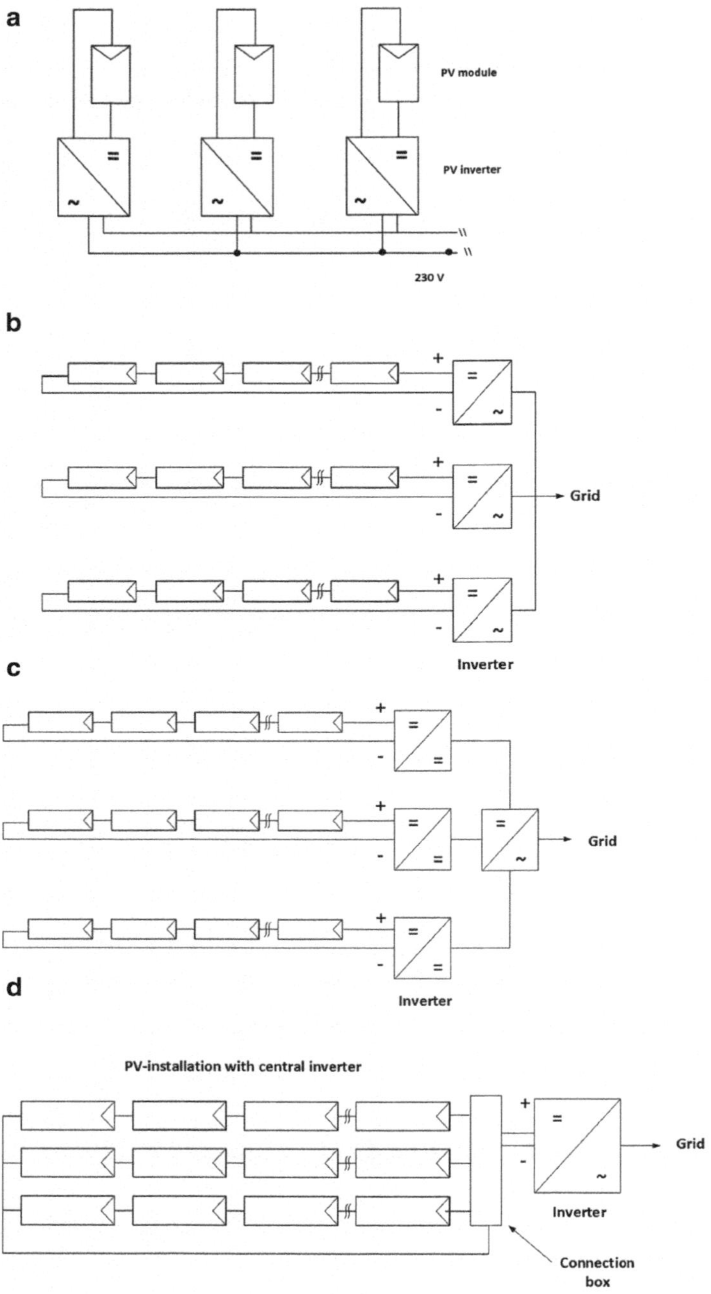

Fig. 12.7 **a** PV-installation with micro inverter. **b** PV-installation with string inverter. **c** PV-installation with multi string inverter. **d** PV-installation with central PV inverter

Fig. 12.8 PV installation in Kandersteg (CH) with multi-String PV inverters (*Photo* Muntwyler)

- The losses due to Inverter, DC- and AC cable losses, typically between 10 and 15%.

With this information we can calculate the AC-production of the PV plant. This is often expressed as the specific annual yield [kWh/kW$_p$]. Typical values are: "800–1200 kWh/kW$_p$" for Europe and places in the northern and southern hemispheres. Sunnier places get from "1200 kWh/kW$_p$" up to more than "2000 kWh/kW$_p$". For detailed calculations there is extensive calculation software, which offers additional options.

12.4.2 Planning Grid-Connected PV Plants

There are different ways to start the planning. We have to consider:

1. The amount of available space
2. The budget
3. The electricity consumption during the day
4. The maximum power of the cable of the grid, into which we want to inject electricity
5. The orientation and inclination of the PV field

6. The appearance of the PV field on a roof or on a façade.

In grid-connected PV systems, on buildings, one attempts to cover the whole surface of the roof or of the façade.

On the other hand, for economic reasons, there is a tendency to reduce the size of the PV plant, so as to obtain maximum "self-consumption". This leads to smaller PV installations; and per kW_p, to more expensive PV installations. Indeed, roofs should be covered fully on their surface with PV, giving, thus, maximum power and the lowest kWh/kW_p price.

Best orientations in the Northern hemisphere are south- and east-west—installations. In the Southern hemisphere it turns to north- and east-west installations. For flat inclined roofs in the Northern hemisphere north roofs are also an option. The same is for the Southern hemisphere, where south roofs for flat inclined roofs can be considered too.

12.5 Explanation of Symbols

Symbol	Name
	PV module
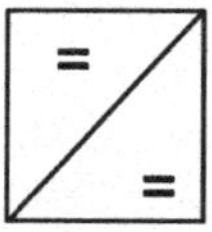	DC to DC converter
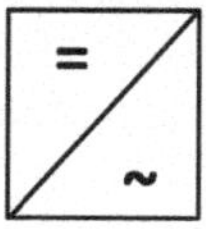	Inverter DC to AC
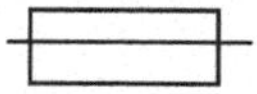	Fuse
	Generator

References

1. H. Häberlin, *Photovoltaik: Strom aus Sonnenlicht für Verbundnetz und Inselanlagen*, 2nd edn. (VDE/ Electrosuisse Verlag, Fehraltorf, 2010), p. 219
2. U. Muntwyler, *Praxis mit Solarzellen*, 1st edn. (Franzis Verlag, München, 1986), p. 43
3. U. Muntwyler, *Praxis mit Solarzellen*, 6th edn. (Franzis Verlag, München, 1993), p. 78
4. U. Muntwyler, *Muntwylers Solar Handbuch,* 12th edn. (Verlag Solarcenter Muntwyler Bern, 2010), p. 198
5. U. Muntwyler, *Muntwylers Solar Handbuch,* 12th edn. (Verlag Solarcenter Muntwyler Bern, 2010), p. 207
6. U. Muntwyler, *Muntwylers Solar Handbuch,* 12th edn. (Verlag Solarcenter Muntwyler Bern, 2010), p. 208
7. L. Herzog, U. Muntwyler, *Photovoltaik – Planungsunterlagen für autonome und netzgekoppelte Anlagen*, (PACER/ Bundesamt für Konjunkturfragen Nr. 724.243D, 1992), p. 22
8. H. Häberlin, *Photovoltaik: Strom aus Sonnenlicht für Verbundnetz und Inselanlagen,*2nd edn. (VDE/ Electrosuisse Verlag, Fehraltorf, 2010), p. 260ff

Urs Muntwyler is Professor for Photovoltaics and the leader of the PV Laboratory of the Bern University of Applied Sciences. His laboratory is active in PV system research, and in "long-term measurements" of PV installations. Urs runs the only accredited PV inverter test bench in Switzerland. PV and architecture is another main activity of his. He has a strong interest in the connection between Photovoltaics and Electric Vehicles. He was organizer of the „Tour de Sol, 1985–92" in Switzerland, the first solar car race in the world. Urs started three PV companies, the first in 1985. The main company Solarcenter Muntwyler AG was a market leader for off-grid and grid-connected PV installations in Switzerland. Urs published "Muntwyler's Solar Handbook" which had 12 editions from 1988 to 2011, with over 100,000 copies printed. From 1998 to 2018 Urs was chairman of the Technical Collaboration Programme "Hybrid- and electric vehicles" of the International Energy Agency (IEA). More details on his present R&D activities: www.pvtest.ch.

Chapter 13
Photovoltaics in the Future Energy System

Stefan Nowak

Abstract This chapter describes how the PV market is expected to grow significantly over time to become a key part of the future energy system. Expected developments in terms of volume, applications and costs are discussed, including how these relate to regulatory framework conditions. Moreover, relevant topics such as sustainability and energy system integration are addressed.

13.1 Market Development

13.1.1 Historical Development

After the discovery and the demonstration of possible applications of silicon solar cells in 1954 by Pearson et al. [1], the initial use of photovoltaics was to power the very first satellites in 1958, both by the United States (Vanguard 1) and by the former Soviet Union (Sputnik III).

Even at very high cost, the possibility to bring electric power to remote locations opened up the path to the terrestrial use of photovoltaics, starting with military and professional applications (navigational and telecommunication systems). In the 1970s, little by little, off-grid systems emerged as economically viable applications, opening up broader markets and leading to the industrial manufacturing of solar cells and modules. Photovoltaics also became an option for applications in remote places of developing countries where no electricity was available.

Deemed as hardly ever commercially competitive with costs of electricity from the grid, first grid-connected photovoltaic power systems were realized in the 1980s and formed the basis for a strongly increased research and development effort. In the 1990s, many public and private initiatives were taken worldwide, to demonstrate the large diversity of photovoltaic systems, from small-scale off-grid to grid-connected systems (integrated in buildings), and even up to "large-scale" (at that time MW_p)

S. Nowak (✉)
NET Nowak Energy & Technology, St. Ursen, Switzerland
e-mail: Stefan.nowak@netenergy.ch

A. Shah (ed.), *Solar Cells and Modules*, Springer Series in Materials Science 301,
https://doi.org/10.1007/978-3-030-46487-5_13

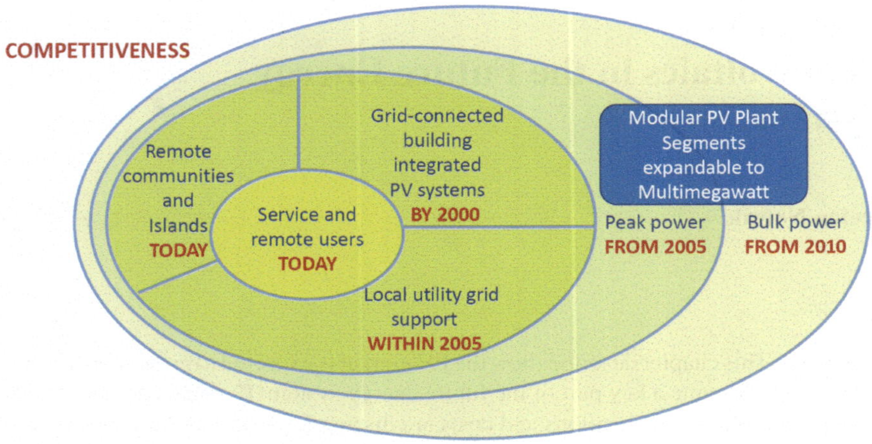

Fig. 13.1 PV market development and competitiveness anticipated by IEA PVPS in 1994 [2]

ground-mounted systems. In 1999, the global annual installed capacity had reached more than 100 MW_p. When the Photovoltaic Collaboration Programme of the International Energy Agency, IEA PVPS, was conceived in 1992, the "diffusion model" [2] shown in Fig. 13.1 was introduced; it schematically proposed how different photovoltaic applications could evolve over time. Retrospectively, this approach proved to be fairly accurate.

The turnaround to the twenty-first century marked the start of mass manufacturing and deployment of photovoltaics. Pioneered by Germany and followed by many other countries, the concept of the feed-in tariff (FIT) [3] was introduced in legislation (Erneuerbare Energie Gesetz EEG) [4], thereby providing a very effective market support scheme. Introducing a small surplus on the consumer electricity price, large amounts of funds become available which are used to pay for the actual solar electricity produced by PV systems over a defined, typically 20-year contract (power purchase agreement PPA) period. In 2002, the cumulative global installed capacity reached more than 1 GW_p, an important milestone in the photovoltaic market development. Although there was heavy political debate about the right design of a sustainable FIT support scheme, with some countries having only short peaks of market growth, partially followed by market collapse, the overall PV market continued to grow rapidly over the past 20 years: The worldwide photovoltaic market reached slightly above 500 GW_p of cumulative installed capacity in 2018. In 2016, photovoltaics already represented the largest annual installed electric capacity of all energy technologies (when decommissioning of fossil power systems are counted in) [5]. With an expected total installed capacity well above 600 GW_p by the end of 2019, 3% of the world's present electricity needs are now covered by photovoltaics.[1]

Table 13.1 provides some milestones in the development of the photovoltaic market over time and Fig. 13.2 illustrates the market development since the beginning

[1] As a global average over the whole year.

Table 13.1 Key figures of annual and cumulative PV market development over time

Year	Annual installed capacity	Year	Cumulative installed capacity
1999	>100 MW_p	1995	> 100 MW_p
2004	>1 GW_p	2002	> 1 GW_p
2010	>10 GW_p	2008	> 10 GW_p
…		2012	> 100 GW_p
…		…	
2018	>100 GW_p	2018	>500 GW_p

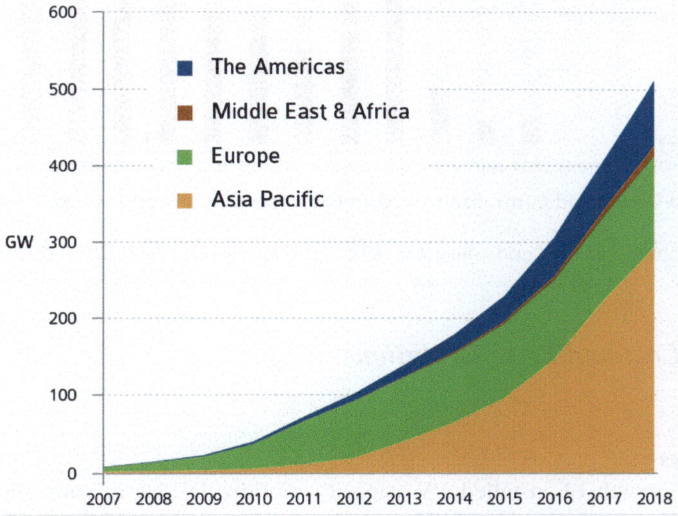

Fig. 13.2 Evolution of installed PV capacity in different world regions, IEA PVPS 2019 [6]

of the twenty-first century. Over time, driven by policy and economics, the market shifted gradually from mostly off-grid to mostly on-grid, as shown in Fig. 13.3.

In summary, it took more than 40 years of development to reach the first GW_p of installed PV capacity. The past 20 years have, however, brought photovoltaic technology from a negligible contribution to a relevant role in the energy supply.

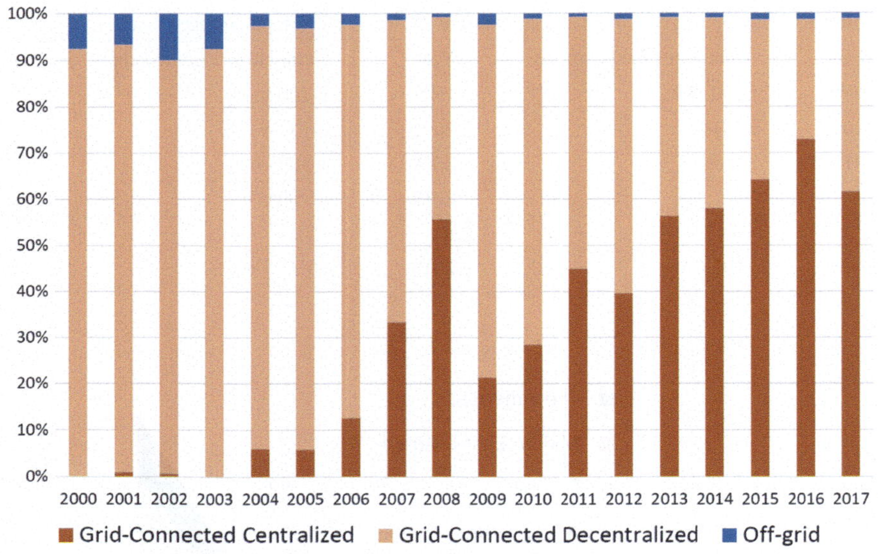

Fig. 13.3 Capacity share of grid-connected and off-grid installations 2000–2017, IEA PVPS 2018 [7]

13.1.2 Cost and Price Evolution

PV modules

The rapid growth of installed PV capacity worldwide was accompanied by an unprecedented cost reduction of PV modules, as shown in Chap. 1. Driven by important technology advancements as well as by rapid progress in mass manufacturing and scale effects, this cost reduction was by far greater than anticipated by many experts. The overall cost reduction over the past decades has been dramatic, resulting in present lowest PV module prices around 0.2 US$/W [8], down by a factor of 500, from 100 US$/W in the early 1970s. The historical learning rate for PV modules has for a long time been estimated to be around 20% (see Chap. 1). For other components of PV systems, the learning rates are typically lower. Learning curves are a powerful analytic tool to assess techno-economic progress, but should be used with care when forecasts are made. A more rapid learning rate is sometimes concluded for the past years, as illustrated in Fig. 13.4. However, in the authors' opinion, these very low price levels do not reflect a healthy and sustainable state of the PV industry.

Levelized cost of electricity (LCOE)

The levelized cost of electricity (LCOE) for a PV system is equal to the sum of all costs, which arise throughout the lifetime of the system for its construction and operation, divided by the total electricity generated throughout the life cycle. Key cost parameters are the initial investment costs, the operation and maintenance costs,

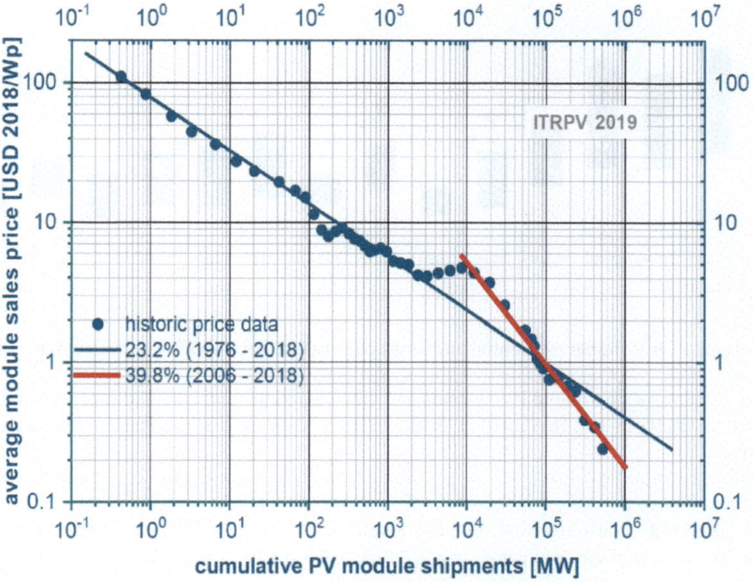

Fig. 13.4 PV learning curve: the two straight lines indicate two learning rates in % (cost reduction per doubled cumulative shipments), ITRPV 2019 [9]

the capital costs related to the interest rate and the lifetime (depreciation) of the system. On the generation side, key parameters, which affect the amount of electricity produced, are obviously the solar irradiation input and the conversion efficiency of the PV system. Moreover, the capacity factor describes the ratio of the average actual electrical energy output (over a given period of time) to the maximum possible electrical energy output (over that period).

Around 2005, the concept of "grid-parity" was introduced. "Grid parity" is attained when the LCOE of solar electricity reaches the level of electricity prices from the distribution grid: this marks a first tipping point towards competitivity for photovoltaics [10]. With the ongoing cost reduction of PV, grid-parity has been or is about to be reached in many countries, e.g. in Denmark, Germany, Italy, Spain, in more than 20 US states, in Japan, as well as increasingly in China.

In recent years, through tenders, power purchase agreements (PPAs) with very low PV prices and attractive contract durations have been announced. They have been realized in particularly favourable conditions of large utility-scale PV power stations, high solar irradiation, low capital costs and long contract durations. Thereby, PPAs at values below 20 US$/MWh were enabled in different countries, marking the increasing competitivity of PV generated electricity [11].

Higher yet still very attractive values of the LCOE have been reached for smaller, distributed systems and less favourable irradiation conditions. Figure 13.5 illustrates recent values of LCOE of PV systems in various countries [11]. It can be concluded that PV today often represents one of the lowest cost options regarding LCOE.

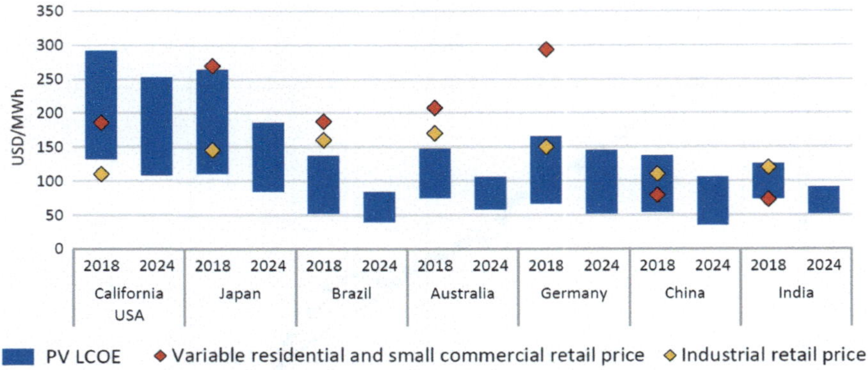

Fig. 13.5 LCOE for distributed PV systems and variable residential, small commercial and industrial retail electricity prices, IEA 2019 [11]. *Source* IEA based on IRENA, Renewable Energy Cost Alliance (2019)

Summarising, both investment costs and LCOE for PV systems have reached competitive levels. For the latter, the cost of capital becomes a key parameter which can affect the competitivity of the system [12].

13.1.3 Market Segments

The PV market segmentation has evolved over time. Early PV market assessments differentiated between off-grid (domestic and non-domestic) as well as grid-connected (distributed and centralized) applications. Over time, as shown in Fig. 13.3, grid-connected systems, when measured by installed PV capacity, by far outweighed off-grid systems. Off-grid systems are typically small sytems (below 100 W_p) or very small sytems (below 10 W_p), whereas the number of units installed is very high (several millions). Therefore, installed capacity is not the only parameter to adequately describe all markets.

Recently, utility-scale PV systems have become very large with PV power stations of several hundred MW_p up to more than 1 GW_p at one location and connection point. A more recent PV market segmentation, when measured by installed PV capacity, is therefore between grid-connected distributed and grid-connected utility-scale systems. The differentiation is now no more by system size, but takes the grid voltage level (high, medium or low) into account. Utility-scale PV systems have become a very important PV market segment. However, due to their increasing competitivity, the market for distributed grid-connected systems is expected to grow rapidly in the years to come [11]. Within the grid-connected distributed applications, distinction can be made between (small size) residential and (medium size) commercial or industrial applications.

Proposed since the early 1990s, Building Integrated PV (BIPV) has received a lot of attention as a desirable aesthetic application and future market segment (see also Chap. 1 of this Book). Limited product availability and high costs have, however, led to a slow growth of this market segment; the total installed capacity of BIPV systems was estimated to be 10 GW_p (2% of the global market) by the end of 2018. Though, more recently, many new products have been proposed, such as PV modules in colour, with different surface appearance, replacing traditional building elements (for roofs and façades) and making this market segment more attractive for the years to come. A recent example of a large BIPV system is shown in Fig. 13.6.

Fig. 13.6 Façade BIPV system on the Copenhagen International School, Nordhavn; building design and credit by C.F. Møller Architects, photo by Adam Mørk

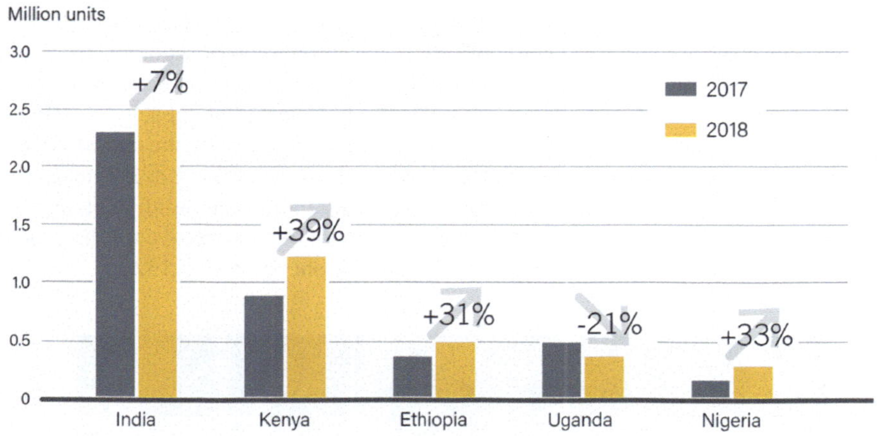

Fig. 13.7 Number of off-grid solar systems sold in top 5 countries 2017 and 2018, REN21 and GOGLA [13]

Off-grid systems in developing countries powering small applications for lighting, communications, refrigeration, healthcare or small businesses are gaining importance. Although small by total installed capacity, these markets are growing as costs are becoming affordable and, increasingly, financing schemes are available. Figure 13.7 illustrates the evolution of the number of solar off-grid systems in the top five markets [13]. The total off-grid market in 2018 (7.6 million systems) is estimated to about 60 MW_p.

Further promising market developments in this area concern "edge-of-grid" applications and microgrids. "Edge-of-grid" refers to areas where the main electrical grid is weak or unstable and the use of systems which include PV may serve as a solution. A microgrid is a group of interconnected loads and distributed energy resources, such as PV, within clearly defined boundaries that acts as a single entity with respect to the grid.

Emerging applications which are increasingly being realized are: Floating PV (larger than 1 GW_p of installed capacity) and Agri-PV, the combination of PV and agriculture (few hundred MW_p of installed capacity). Recently, vehicle integrated PV (VIPV) is discussed as a new application where PV meets mobility.

Concentrator photovoltaics (CPV) using solar cells with very high efficiencies, an optical system (lenses or mirrors) to concentrate the solar irradiation and a mechanical tracking system to follow the sun have been proposed for many years. Commercial development started in the mid-2000s with high expectations for applications in sun-rich countries with high direct sunlight. Due to the rapid cost reduction of flat plate (non-concentrated) PV, the CPV market development was slower than expected with a peak of more than 100 MW_p installed in 2012 [14], followed by a market slowdown in the years after. In total, the CPV market accounted for less than 500 MW_p of installed capacity by 2018.

Fig. 13.8 PV for space power applications: 32% Quadruple Junction GaAs Solar Cell by AZUR SPACE Solar (left), photo copyright by AZUR SPACE; north wing 11 kW solar array of EUROSTAR 3000 series communication satellite SES-14 (right), photo copyright by Airbus Defence and Space

PV for space power is a special application where availability, reliability and performance are more important than costs. Applications range from civil systems, e.g. for scientific, meteorological and telecommunication satellites, and space stations to military systems. Measured by capacity, below 1 MW_p is deployed in civil space systems annually, from very small to larger satellites, with a total installed capacity estimated to be between 5 and 10 MW_p. This market accounts for the highest solar cell efficiencies industrially produced, mostly using GaAs cell technology. Figure 13.8 shows examples of a GaAs space solar cell by AZUR SPACE and a solar array of the EUROSTAR 3000 series communication satellite SES-14 by Airbus Defence and Space. This PV market segment has the role of the "formula 1" automobiles in photovoltaics.

In summary, the PV market is highly diverse, serving multiple needs from very small to very large systems. As an energy technology, PV is, thus, quite unique by the fact that the same basic technology, from Watts to Gigawatts, covers a stunning, nine orders of magnitude.

13.1.4 Future Projections

Since photovoltaics entered the energy discussion, projections with large deviations about its future were made by numerous experts, research institutions, NGOs and international organisations, covering a wide span between visionary, very high expectations and very low, virtually irrelevant contributions to the future energy supply. More recently, as PV costs decrease, installed capacity grows and competitivity increases, there is a clear consensus about the important role that PV can and will play over the coming decades in contributing to a clean energy supply. With climate change becoming an increasing concern, PV is more and more discussed with respect to its prime role to decarbonize the energy system as a whole.

With the expected market development of the coming years [11], PV will have installed more than 1 TW_p (Terawatt) capacity in 2022, at the latest. Thus, PV is entering the TW-age very soon [15]. However, there is more discussion about how this development will continue: about whether the growth rate can be maintained and what the next milestones should be. There is in principle no doubt that from a technology and resource point of view, very high contributions to the future energy supply are possible. First, further cost reductions are expected to occur due to technology developments and industrial implementation. Second, very large contributions of PV electricity can be (technically and economically) integrated in the electricity and the energy system at large, if short and long-term storage options are implemented. This will also affect the cost of dispatchable energy in the future energy system. And finally, it is increasingly evident that PV electricity, when successfully transformed to other energy carriers (sector coupling, power-to-x technologies), will have large impacts well beyond the electricity supply with the potential to enter into areas such as mobility, heat, fuels or even chemicals [16]. Depending on whether the impact of a strong renewable energy deployment is anticipated for the electric power sector only or for a broad electrification of the entire energy system, a wide span for the possible future installed capacity of PV is opened, ranging from below 10 to more than 60 TW_p by 2050. Table 13.2 compares expected levels of PV deployment according to different sources.

Summarising present projections for the future deployment of PV, there is no doubt that PV, from today's perspective, will definitely play a crucial role and will thereby very likely be an important part of the backbone of the future energy system.

Table 13.2 Overview of different PV deployment scenarios, from very low to very high

Source/scenario	Cumulative installed capacity by 2030 (TW_p)	Cumulative installed capacity by 2050 (TW_p)
IEA stated policies scenario 2019 [18]	1.87	
IEA sustainable development scenario 2019 [18]	2.54	
IRENA 2019 [19]	2.84	8.5
Bloomberg New Energy Finance [20]	2.14	
ITRPV 2019 broad electrification, Breyer [9]		63.4
Shell sky scenario [21]	5.01	

Adapted from [17]

13.2 Regulatory Issues

Since a large majority of photovoltaic power systems are connected to the electric grid, regulatory issues play a dominant role in PV market development, both in technical and economic terms.

Feeding electricity at low power levels into the distribution grid is associated with the following aspects: safety issues in case of grid shut-down and subsequent potential islanding of the PV system, as well as voltage and frequency stability in the distribution grid. Technical grid integration aspects are therefore an important topic, in particular for high PV penetration rates (see Sect. 13.4 below).

Regulatory issues address the question how PV electricity is remunerated, and more broadly, how PV is integrated into the electricity market. In order to kick-start the PV market in the early 2000s, due to the initially very high costs of grid-connected PV systems, a regulatory framework was necessary which allowed for the market entry in economic terms. For this purpose, various support schemes were introduced by the policy makers in different countries over the past 20 years—they have since strongly evolved. Among these, one can distinguish in particular between pure investment subsidies (grants), remuneration schemes (e.g. FIT, premium tariffs, net metering, self-consumption), tax credits and measures (e.g. investment tax, VAT), green certificates and labels, renewable portfolio standards (RPS) and auctions (tenders, power purchase agreements (PPAs)).

The regulatory framework comprehensively describes how the electricity market is organized, the participating stakeholders and their responsibilities, the rules that apply, the prices, the time differentiation, etc. Figure 13.9 summarizes the share between different support schemes, as analysed by the IEA PVPS Technology Collaboration Programme [6].

Historically, besides investment subsidies, the most effective support scheme has been the concept of the feed-in tariff (FIT) described earlier in this chapter. Its main benefit is the investment security that it provides over a long contract period, typically 20 years. This also explains the success that this concept had in the early phase of PV market development. Some of the difficulties encountered are—in times of rapid cost

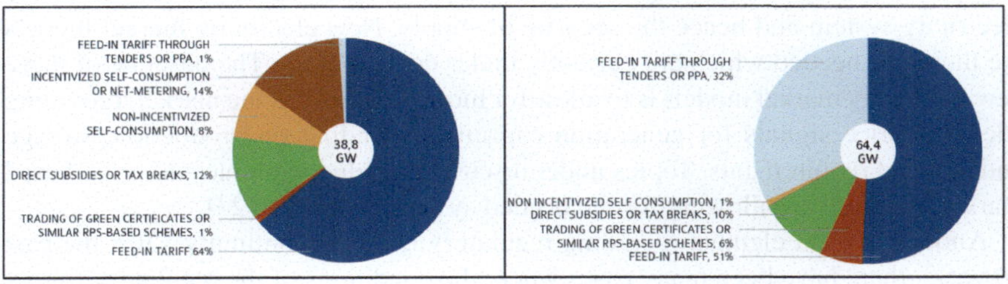

Fig. 13.9 PV support schemes in 2018, distributed PV market (left), centralized PV market (right), IEA PVPS 2019 [6]. Comment: "Non-incentivized self-consumption" is self-consumption without any financing incentive, possibly with additional burdens on the producing consumer (often called prosumer)

reduction—the advanced determination of the right level of the tariff scheme, possibly leading to overly generous tariffs, and—when uncapped—the growing burden on the electricity consumers. Together with the fact that LCOE costs of PV have drastically come down, often to below the consumer price of electricity, the time of the FIT-scheme, in particular in Europe, is coming to an end. The FIT scheme is increasingly replaced by other, more sophisticated schemes such as self-consumption.

The concept of self-consumption can be defined as the share of the total PV production directly consumed by the PV system owner him/herself. Benefits associated with self-consumption are a higher economic value of PV electricity and a lower stress on the electricity distribution grid. Over the past years, this concept has evolved and distinguishes between local self-consumption (a PV system owner consumes a part or all of his/her electricity production), collective self-consumption (a group of people consumes electricity from a shared PV system) and virtual self-consumption (generation and consumption take place at the same time but in different locations) [22]. These variations of self-consumption are expected to form an important driver of the PV market in the coming years. Self-consumption can be optimised by adding a local storage unit and by an effective management of the electricity demand (demand side management). This can, for example, be achieved by cooling the refrigerator or running the heat pump when PV electricity is abundantly available. Collective self-consumption offers the possibility to use PV electricity by neighbouring consumers which form a self-consumption community, formed by several buildings in a neighbourhood.

As the level of penetration of photovoltaic electricity increases in a given market area (country, state, region), new questions arise regarding the future development of the regulatory framework. On one hand, as PV becomes a relevant part of the electricity supply, its economic value on the market at a given time may change, depending on the price of electricity. For example, in situations of low electricity demand and high PV penetration, such as on sunny weekends, grid injection can become economically unattractive. On the other hand, if many consumers (often called prosumers) self-consume their PV electricity production, the costs of the grid may not be properly covered.

Moreover, increasing shares of electricity production from PV and wind power plants which vary with time pose new challenges regarding the flexibility of the electricity system and hence the security of supply. New electricity market models are therefore needed which are currently under development. The baseline of these new electricity market models is to allow for more flexibility in the market, providing adequate price signals for generation capacity depending on production, storage and demand requirements. Topics under discussion include volume and price-based market mechanisms allowing for the needed system flexibility [23].

Another critical element regarding regulatory issues is continuity. Over the past 10 years, there have been many cases where the conditions of the regulatory framework to support PV market development in a given country, particularly those with rapid market growth, were changed over time, sometimes with short lead times and in some cases even retroactively. This does not allow a secure and stable investment climate to be created; rather, it can be seen as an important policy risk for the further development of the PV market.

13.3 Sustainability

Environmental aspects of photovoltaics and, more generally, sustainability issues, form a very important subject for the further deployment of PV. These cover in particular materials and resource questions, industrial production processes, water usage, eco-friendly product design, land use, operation of PV systems, possible hazards, energy payback time, CO_2 emissions and recycling of PV components. PV is proposed as a clean energy technology with a high potential for reducing the emissions of greenhouse gases. It is therefore also being called upon to contribute to the decarbonization of the entire energy system. Great care needs to be taken to ensure that this is actually the case: an independent confirmation of the results obtained by the PV community is necessary.

Life cycle analysis (LCA) of PV systems is the science that provides quantitative results regarding key environmental indicators and possible critical issues. LCA enables one to take into account the entire life cycle stages, from cradle to grave, in measuring environmental and resource sustainability. In order to perform a meaningful LCA, Life cycle inventories (LCIs) are necessary and the availability of such data is often the greatest barrier for conducting LCA. Different solar cell technologies have different environmental characteristics, in particular regarding the chemical processes and materials used, as well as concerning the energy used to produce the PV modules. A complete description of PV LCA can be found in [24].

In industrial PV cell manufacturing, many processes are used that are common to the semiconductor industry, such as processes for cleaning, purifying and doping semiconductor materials. These processes employ hazardous substances such as chemicals (e.g. hydrochloric acids, sulfuric acids, nitric acids, hydrogen fluorides) as well as toxic elements and compounds (e.g. arsine, cadmium, gallium, lead, and phosphine). These materials require special handling and operating standards to prevent workplace hazards and exposure to toxics. Over the past decade, strong efforts were undertaken to reduce the amount of materials used for solar cells and modules and to reduce or avoid the use of critical processes. For silicon solar cells for example, silicon material usage was greatly reduced by thinner wafers (see also Fig. 1.5 in Chap. 1).

The dominant solar cell technology, crystalline silicon, requires an energy intense process to gain metallurgical-grade and subsequently solar-grade silicon from quartz sand (silicon dioxide). Depending on the energy mix (coal, nuclear, hydropower, or other renewables) at the specific location of the silicon manufacturing plant, large variations in the associated greenhouse gas emissions are "incorporated" into the solar cell material.

The life-cycle of photovoltaics starts from the extraction of raw materials (cradle) and ends with the disposal (grave) or recycling and recovery (cradle) of the PV

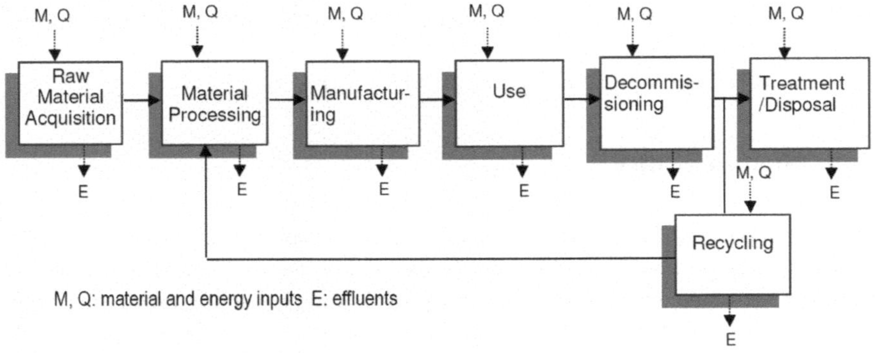

Fig. 13.10 Flow of the life-cycle stages, energy, materials, and effluents for PV systems, IEA PVPS 2015 [24]

components, as shown in Fig. 13.10. LCIs include detailed inputs and outputs during manufacturing of cell, wafer, module, and balance-of-system that are estimated from actual production and operation facilities. In addition to the LCI data that support the LCA results, data are gathered to enable analyses of various types of PV installations; these typically include operational data of rooftop and ground-mounted PV systems and country-specific PV-mixes. Some of the most important LCA indicators are (i) the primary energy demand needed for a specific technology, (ii) the energy payback time which is defined as the period required for a PV system to generate the same amount of energy, as the energy that was used to produce the system itself and (iii) the greenhouse gas emissions during the life cycle stages of a PV system, estimated as an equivalent of CO_2; the major emissions typically included are CO_2, CH_4, N_2O and chlorofluoro-carbons. There have historically been many discussions about the energy payback time (EPT) of photovoltaic systems. With the progress made over the past decades, the EPT of photovoltaic systems is typically of the order of 0.5–3 years, depending on the technology and the kind of system, in any case far below the typical life time of PV systems of at least 25 years.

Figure 13.11 provides an example of LCA results showing the greenhouse gas emissions of 1 kWh AC electricity produced with different PV technologies [25], based on industry data a few years back. As explained in Chap. 1, recent progress in industry, namely for crystalline silicon technologies, regarding efficiency, material usage and energy consumption for the different process steps leads to a substantial decrease in greenhouse gas emissions of the panel manufacturing processes, as indicated by the dashed blue arrows in Fig. 13.11. Similarly, both CdTe and CIGS thin film solar cells have also recently improved their industrial processes [26, 27], leading to lower greenhouse gas emissions, as indicated by the dashed red and green arrows in Fig. 13.11. For irradiation conditions in Central Europe, greenhouse gas emissions in the range of 20 g CO_2-eq/kWh and below thereby become feasible.

PV modules have a useful lifespan of approximately 30 years. With PV deployment increasing strongly, the number of PV modules that reach the end of their

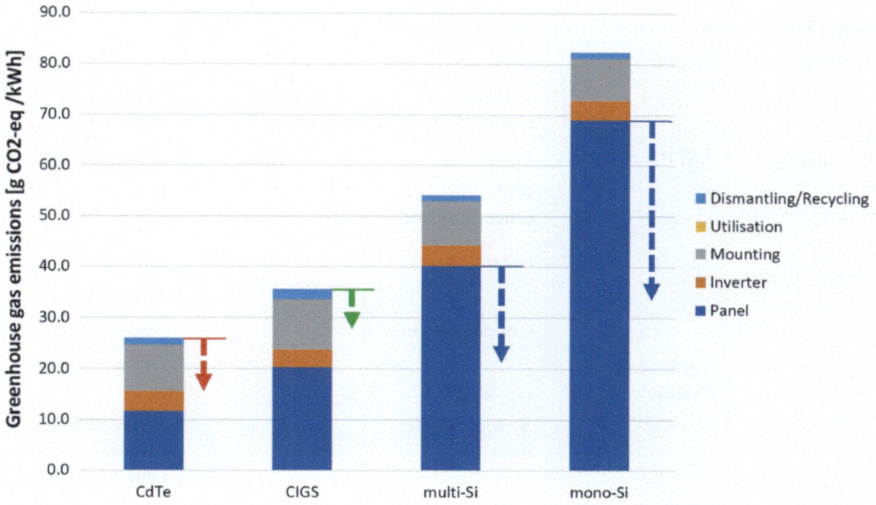

Fig. 13.11 PV LCA results: Greenhouse gas emissions by different PV technologies, after R. Frischknecht 2019 [25], Assumptions: 3 kW_p PV system, optimally oriented and mounted on a slanted roof in Central Europe; annual electricity generation 975 kWh/kW_p; lifetime 30 years (PV modules), 15 years (inverters); PV module efficiencies 16.0% (CdTe), 14.0% (CIGS), 16.5% (multi-Si), 17.5% (mono-Si) [25]. Comments: Dashed blue arrows introduced by the author of this Chapter indicate the estimated effects of recent industrial process improvements for crystalline silicon technologies regarding efficiency, silicon and glass material usage as well as energy requirement for the different process steps. Due to these improvements, the contribution of panel manufacture and its supply chain to the greenhouse gas emissions of 1 kWh AC electricity produced with crystalline silicon PV modules are likely to be reduced by at least 50%. The dashed red arrow introduced by the author of this chapter indicates the effects of recent industrial process and efficiency improvements for CdTe thin film technology according to [26]. Similarly, the dashed green arrow introduced by the author of this Chapter indicates the effects of recent industrial process and efficiency improvements for CIGS thin film technology according to [27]

useful life will also greatly increase—after the time lag of operation—accumulating as waste. PV module waste has been estimated globally to amount to 1.7–8.0 million metric tons cumulatively by 2030 and to 60–78 million metric tons cumulatively by 2050 [28]. In comparison, the world's total annual electrical and electronic waste (e-waste) reached 41.8 million metric tons in 2014.

Generally, sustainable waste management offers opportunities known as the 3Rs: reduce, reuse, and recycle. When a product cannot be repaired or reused, recycling is the next best option before disposing it as waste. In anticipation of the large volume of waste PV modules, PV module recycling has recently become an important topic. Although the amount of waste PV modules is still limited, recycling has already begun to be commercialized, and various technologies for PV module recycling are under development.

Recycling technologies for crystalline silicon PV modules and thin-film compound PV modules have different characteristics owing to the differences in the

Table 13.3 Classification of PV recycling technologies, IEA PVPS 2018 [28]

Target product	Module type	Analysis category		
		Targeted components	Processing method	Recovered materials
PV module	c-Si PV module	Frame	Mechanical Chemical Thermal Optical Electrochemical combination	Aluminum (Al)
		EVA (encapsulant)		Glass
				Cell
		Solar cell		Silicon (Si)
				Silver (Ag)
		Copper ribbon		Copper (Cu)
	Compound PV module	EVA (encapsulant)		Glass
		Semiconductor material		Semiconductor material (metal elements)
	Others (organic solar cell, dye-sensitized solar cell, thin-film silicon solar cell)	EVA (encapsulant)		Glass
		Semiconductor material		Semiconductor materia (metal elements)

module structures and in the metals contained in them. Table 13.3 gives an overview of the different PV module materials aimed for recycling. An overview of trends in recycling technology is provided in [29].

13.4 System Integration

When grid-connected PV systems started to become more common in the 1990s, this triggered a discussion how much PV capacity could be fed into the grid without causing problems for the electrical grid, e.g. regarding voltage and frequency stability. These problems could, in principle limit the amount of PV capacity that a specific grid can host. Here, a distinction needs to be made between the instantaneous PV capacity (power) in a distribution grid and the associated energy contribution to the electricity supply (energy = power × time). Contributions to the electricity supply of a few % typically correspond to instantaneous grid penetration levels above 50%, which have to be handled by the system. In the early days of PV grid-connection, more than 5–10% of PV capacity in a given distribution grid were seen as unrealistic due to the grid problems caused. Issues of concern were: the increasing bidirectional flow of electricity in the distribution grid, the necessity for grid reinforcements, the matching of supply and demand, and the variability of the PV supply pattern.

Since then, a lot of research has been undertaken in order to identify, demonstrate and implement much higher levels of PV penetration in the distribution grid. Over the past decade, given the strong market development of PV in many countries, situations of high PV penetration in electric grids have become a truly global issue in more and more regions around the world. As an example, the state of Bavaria in Germany has reached 15% average PV penetration.

Following the massive deployment of PV and other variable renewable energy sources (RES)—mainly wind—in recent years, high penetration is no longer a local distribution grid issue limited to a few regions or countries. Instead, today we are talking about "RES-dominated" grids, where variable, non-rotating power generation from renewable energy sources reaches levels of up to 100% of the load in the transmission system during longer periods. While the technical system integration of large amounts of PV capacity in a given electrical network is challenging and therefore requires adequate solutions, it is by no means a barrier towards the future large-scale deployment of PV as laid out in this chapter. New electrical network management techniques, digitalization and smart grids will be decisive for balancing load and demand in an increasingly decentralised electricity supply. Very high PV penetration rates where the PV capacity may well exceed the grid capacity, will require storage options and the coupling of different energy sectors (electricity, heat, gas, etc.), thereby increasingly connecting different parts of the energy system such as power generation, electrical grids, storage, including buildings and electric vehicles.

With the emergence of electric cars, a huge peak power and storage capacity will become available for the grid in many countries. Heat or cold can be stored in fridges, air-conditioning units, boilers, tanks, or in large underground caverns, even inter-seasonally (summer to winter). Hydropower, biomass, heat and cold generation are options already used worldwide for long-term storage. With the expected ultra-low cost of PV, the power-to-gas solution becomes an option, with the potential to compete with the price of hydrogen or ammonia made from fossil fuels. If CO_2 sources are available, synthetic methane can be produced, stored, and burned, ultimately in a cycle where CO_2 is recaptured.

These developments underline that PV has definitely arrived in the energy system. Today's discussion is therefore no longer limited to the integration of PV in the electricity grid but moves rapidly to the broader challenge of energy system integration.

References

1. D.M. Chapin, C.S. Fuller, G.L. Pearson, J. Appl. Phys. **25**, 676 (1954)
2. R. Vigotti, A. Claverie, in Proceedings of IEEE first world conference on photovoltaic energy conversion, 2247 (1994)
3. T.D. Couture, K. Cory, C. Kreycik, E. Williams, A policymaker's guide to feed-in tariff policy design. NREL (2010)
4. Renewable Energy Sources Act, Federal Ministry for the Environment, Nature Conservation and Nuclear Safety, Germany (2000)

5. Renewables 2017, Analysis and forecasts to 2022. International Energy Agency, IEA (2017)
6. Trends in Photovoltaic Applications, International Energy Agency Photovoltaic Power System Technology Collaboration Programme (IEA PVPS), T1-36 (2019)
7. Trends in Photovoltaic Applications, International Energy Agency Photovoltaic Power System Technology Collaboration Programme (IEA PVPS), T1-34 (2018)
8. https://www.pvxchange.com/de/price-index
9. International Technology Roadmap for Photovoltaic (ITRPV), 10th edn. (2019)
10. L.A. Hurtado Munoz, J.C.C.M. Huijben, B. Verhees, G.P.J. Verbong, Technol. Forecast. Soc. Chang. **87**, 179 (2014)
11. Renewables 2019, Analysis and Forecasts to 2024. International Energy Agency, IEA (2019)
12. E. Vartiainen, G. Masson, Ch. Breyer, D. Moser, Prog. Photovolt. Res. Appl. (2019) https://doi.org/10.1002/pip.3189
13. Renewables 2019, Global Status Report, REN21 (2019)
14. Current Status of Concentrator Photovoltaic (CPV) Technology, Fraunhofer ISE and NREL (2017)
15. N.M. Haegel et al., Science **364**, 836 (2019)
16. Global Energy System based on 100% Renewable Energy, Study by Lappeenranta University of Technology and Energy Watch Group, Lappeenranta, Berlin (2019)
17. L. Oberbeck, K. Alvino, B. Goraya, M. Jubault, D. Lincot, in Proceedings of 36th European photovoltaic solar energy conference and exhibition, 1954 (2019)
18. World Energy Outlook, International Energy Agency, IEA (2018)
19. Global Energy Transformation—A Roadmap to 2050, International Renewable Energy Agency, IRENA (2019)
20. Bloomberg New Energy Outlook, Bloomberg New Energy Finance (2019)
21. Shell Energy Transition Report (2018)
22. G. Lettner, H. Auer, A. Fleischhacker, D. Schwabender, B. Dallinger, F. Moisi, E. Roman, D. Velte, A. Huidobro, Existing and Future PV Prosumer Concepts, PVP4Grid Project (2018)
23. A. Bublitz, D. Keles, F. Zimmermann, Ch. Fraunholz, W. Fichtner, Energy Econ. **80**, 1059 (2019)
24. Life Cycle Inventories and Life Cycle Assessments of Photovoltaic Systems, International Energy Agency Photovoltaic Power System Technology Collaboration Programme (IEA PVPS), T12-04 (2015)
25. R. Frischknecht, PV update, Swissolar (2019)
26. P. Sinha, A. Wade, IEEE J. Photovolt. **8**, 793 (2018)
27. A. Briem, M. Held, B. Dimmler, in Proceedings of 35th European photovoltaic solar energy conference and exhibition, vol 1131 (2018)
28. End-of-Life Management: Solar Photovoltaic Panels, Joint Publication International Energy Agency Photovoltaic Power System Technology Collaboration Programme (IEA PVPS), T12-06 (2016) and International Renewable Energy Agency, IRENA (2016)
29. End-of-Life Management of Photovoltaic Panels: Trends in PV Module Recycling Technologies, International Energy Agency Photovoltaic Power System Technology Collaboration Programme (IEA PVPS), T12-10 (2018)

Stefan Nowak is Managing Director of NET Nowak Energy & Technology, a consultancy in renewable energy with special emphasis on photovoltaics, energy efficiency and sustainability which he founded in 1997. Stefan graduated in experimental physics in 1980 at the University of Fribourg (Switzerland), followed by a Ph.D. in plasma physics performed at the Swiss Federal Institute of Technology (EPFL) in Lausanne (Switzerland) and the University of Fribourg in 1986. His academic research activities covered thermonuclear fusion, plasma and surface physics as well as thin films. Stefan's biography is characterised by a continuous evolution from research to technology, innovation and business management. His broad technological expertise is enhanced by activities in communication, market research and policy support. Stefan received the Becquerel Prize in 2017. Stefan was programme manager for photovoltaic research for the Swiss Federal Office of Energy (SFOE) until 2018. He presently manages and advises various national and international energy related projects and programmes and serves as Chairman of the Executive Committee of the Photovoltaic Programme of the International Energy Agency (IEA PVPS, www.iea-pvps.org).

Index

A. Shah (ed.), *Solar Cells and Modules*, Springer Series in Materials Science 301,
https://doi.org/10.1007/978-3-030-46487-5

The manufacturer's authorised representative in the EU is Springer Nature Customer Service Centre GmbH, Europaplatz 3, 69115 Heidelberg, Germany. If you have any concerns regarding our products, please contact ProductSafety@springernature.com

Printed and bound by CPI Group (UK) Ltd, Croydon, CR0 4YY
30/09/2026
02225648-0014